Merchants of
Immortality

Merchants of Immortality

Chasing the Dream of Human Life Extension

STEPHEN S. HALL

Houghton Mifflin Company

Boston   New York   2003

FOR SANDRO AND MICAELA
May they live the experiment well

Visit our Web site: www.houghtonmifflinbooks.com.

Library of Congress Cataloging-in-Publication Data

Hall, Stephen S.
Merchants of immortality : chasing the dream of human life
extension / Stephen S. Hall.
p. cm.
Includes bibliographical references and index.
ISBN 0-618-09524-1
1. Longevity — Popular works. I. Title.

QP85.H255   2003
613'.0438 — dc21   2002192155

Printed in the United States of America

QUM 10 9 8 7 6 5 4 3 2 1

Portions of several chapters have appeared, in different form,
in the *New York Times Magazine* and in *Technology Review.*

Contents

Then the Lord God said, "See! The man has become like one of us, knowing what is good and what is bad! Therefore, he must not be allowed to put out his hand to take fruit from the tree of life also, and thus eat of it and live forever." The Lord God therefore banished him from the garden of Eden, to till the ground from which he had been taken. When he expelled the man, he settled him east of the garden of Eden; and he stationed the cherubim and the fiery revolving sword, to guard the way to the tree of life.

— GENESIS 3:22–24

Even though it is possible to design, manipulate, and orchestrate one's immortality in advance, it never comes to pass the way it has been intended.

— MILAN KUNDERA, *Immortality*

The Never-Ending Life

SEVERAL YEARS AGO, while spending a weekend in the country with my family, I stepped out onto the porch of the cabin where we were staying one night and looked up into the sky. It was unusually clear for a summer night in the Catskills, and every familiar jot and scrawl of the firmament was writ large — the Polestar, Little Bear, filaments of the Milky Way strewn like pulled cotton right down the middle of the dome, the entire landscape "apparelled in celestial light," as Wordsworth put it in his "Ode: Intimations of Immortality." A field of tall grass and wildflowers sloped down from the porch, and hundreds of fireflies blinked on and off in the middle distance, so that the line where our earthly, mortal light yielded to the celestial became beautifully blurred in the darkness. As I stood at the rail, I could hear the uniquely peaceful sound of untroubled sleep behind me in the cabin — my wife and two children. I scanned the dark sky for a shooting star. I even had a wish ready.

Since I didn't happen to see a shooting star that night, I don't think it will betray any cosmic confidences to reveal what my wish would have been, especially since it was so predictable. I wished for long, healthy, productive lives for my family, especially my children. In doing so, I know I was indulging a desire as ancient as our fascination with the heavens, a longing as timeless and fierce as the biological instinct to protect one's brood. A desire so old, in fact, that it is not only about nature, but a part *of* human nature, at least for the only species of life on earth known to be aware of its own mortality. And as an amateur student of aging, I also knew, as I stood at that porch rail, that my Darwinian warranty was about to run out. In strictly evolutionary terms, I'd just about outlived my biological usefulness to the species and would not much longer enjoy the built-in genetic protections crafted by eons of natural selection. Indeed, those two cherubs sleeping inside were the agents of my inevitable demise. Evolution protected me long enough for me (and my wife) to have children, but became biologically (and, in a sense, lethally) indifferent to us once we reached a certain age. From paramecia to primates, from the single-celled denizens of pond scum to poet laureates, natural selection stops caring about us once we have lived long enough to reproduce. Evolution in that sense is a strange ship: it moves ever forward through dark waters, keeping the species alive, even as it throws each and every member of the species overboard. I was nearing fifty years of age. My primary care physician had retired and I'd been forced to switch to a new doctor. His name — no joke — was Dr. Faust. And so this midsummer night's wish of mine, stripped of its conflicted humility and its faux altruism, revealed itself to be transparently self-referential. And here I'm tempted to add "like most of the wishes of my generation." Because what I was really saying was: Let us *all* live a long time, we're not quite ready to . . . to . . . I couldn't bring myself to utter the D word, even in a conversation with myself. I was content to reiterate the ancient ritual of submitting a time-honored petition to indifferent gods on dark, starry nights.

In the same way, I feel that an entire generation — a generation new to mortality, you might say — has been poised to file that same petition as a kind of generational class-action suit against the laws of nature. Many of us have been similarly poised at the railing of middle age, in the twilight of something more permanent than a summer night, launching that same fervent petition on behalf of our parents, our children, and, of course, ourselves. I am speaking in part of the baby boomers, 75 million strong in the United States alone, as well as our similarly entitled post–World War II siblings spread throughout the developed world. This is a generation, it goes without saying, that thinks of its petitions as somewhat special, a generation that is perhaps a little more insistent about answered prayers.

Or so it appears superficially. If you think beyond the demographic clichés, however, it's hard to believe with much conviction that the baby boomers are any more concerned about their mortality than previous generations and previous cultures. Can we possibly experience more feral emotions than the hunters and gatherers of 10,000 years ago, whose very mortality attached to the success of finding their next meal? Can we summon more urban angst than the average citizen of ancient Rome, who could expect to live only about twenty or twenty-five years? Can we honestly argue that we feel a more exalted fear of death than the soldiers of the greatest generation, teenage boys like my father, huddled in foxholes, dodging bullets? I have a hard time convincing myself that this is so. What makes this particular moment so unusual in the age-old posting of these timeless wishes is that they might actually be answered in an altogether different way, with altogether unexpected consequences, in the not-too-distant future. Perhaps I was looking in the wrong place for my shooting star, because in a sense the truly meteoric agency capable of delivering on these wishes may be found not in the world of cosmology but biology; the high priests of our secular age, the molecular biologists, have begun to address mortality in a way no group, no generation, and no society has ever dreamed of before.

They may not succeed, of course, and the purpose of this book is not to conflate promising science with the wishful thinking of an entire generation. It is enough to note that in the last decade the most skilled, ambitious, and indeed arrogant of our sciences has lined up to tackle the "problem" of aging (and its faithful sidekick, death) in a way fundamentally different from that of any previous era of medical intervention. This is happening at the very same time that an enormous demographic bulge in our population is burying parents and picking out gray hairs in the mirror. If nothing else, these trends make for a fascinating convergence of social desire and scientific ambition; of deeply personal psychological needs (and fears) and the shamelessly public promissory notes that issue from the lips of biologists, businesspeople, and other incurable optimists; of the inevitable decline of the human body (or soma) and the almost alchemical, regenerative capabilities of bland cells in plastic dishes; of the highest intellectual aspiration for basic knowledge that contemporary civilization can muster, alongside the most common and infinite capacity for greed and personal advantage that has ever sullied the name of human nature. Looking at this intersection from one perspective, nothing less is at stake than a partial or nearly total repeal of mortality; from another perspective, we might be witnessing a postmodern, molecular version of the Fountain of Youth tale, a spectacle of promise and hubris and failure that will make the Ponce de León story look like bad summer stock.

Medicine, especially in the last century, has consistently helped prolong life (or, if you prefer, forestall death), to the point where more people in developed societies are living to a greater age than ever before in human history. Because we've done such a spectacular job of minimizing the agents of premature death — diseases, accidents, poor hygiene, injuries, not to mention predation, starvation, and exposure — we are living so long that aging itself has only recently emerged as a subdiscipline of medicine. In a sense, we didn't even know aging existed as a biological phenomenon until we started living well beyond reproductive age, which is really all that evolution is interested in protecting. Now that we know aging exists as a separate, degradative phenomenon, and are beginning to understand it, we naturally want to see if we can tinker with the process. That is what we do, and that is what I have set out here to chronicle: an account of some of the people who have begun to revolutionize medicine's assault on aging, and the type of science they are doing. Inevitably, my encounters have also led to a cultural contemplation of what it might mean to us, as individuals and as a society, to repeal, even partially, the laws of mortality.

For most of the recorded past, humans could expect to live on average about twenty years (although that number is deceptively low because of the high incidence of infant and childhood mortality). A century ago, Americans born in 1900 could expect to live roughly forty-nine years. Some lived longer, of course, but many still perished at a very young age. Civilization — in the form of antiseptic medicine, sanitation and public hygiene, vaccination and other measures — has dramatically increased the amount of time we can expect to spend on earth. Indeed, as a prominent gerontologist, Leonard Hayflick, puts it, "Aging is an artifact of civilization."

☙

Some of these thoughts were on my mind on a sunny day in December 2000, when I headed north from San Francisco in a rented car. It was a professional pilgrimage, in that I was setting out to talk to Hayflick, a scientist well known within the biological community (indeed, almost infamous) and yet virtually unknown outside it. In 1961, Hayflick achieved a rarely attained degree of academic celebrity when he discovered that normal human cells grown in the laboratory have a finite lifetime — that is, they are programmed to divide a more-or-less fixed number of times (known now as the "Hayflick limit") and then simply stop replicating and senesce. *Senescence* is a word groaning with metaphoric throw weight in the context of human gerontology; cellular senescence begins a process of biological lassitude and decay that ultimately leads to cell death. Hayflick's discovery brought together a powerful mix of scientific interests: aging, life span, the biology of cells, immortality. It put the biology of aging — and therefore the biology of life and death — squarely in the crosshairs of the biologist's microscope.

Hayflick had sent me meticulous instructions on how to reach his home — a map marked with arrows, annotated directions of key crossroads, even aesthetic admonishments ("Go slow on Highway 1 for safety and to observe the beauty! Careful around blind curves . . ."). As I headed north on Highway 101 and cut across Mendocino County toward the Pacific Ocean, it was hard not to notice the everyday auguries of aging and mortality that color the way in which we view the world, even from a car window. Outside Guerneville, the road curved past — deferred to, actually — a number of towering redwoods crowding the asphalt. Some of those massive and long-lived creatures have lorded over this landscape for centuries (and yet they represent a lesson in complexity and paradox as well as longevity, for as Hayflick has pointed out in one of his books, only a tiny fraction of their cells are actually alive, the rest inanimate pulp). At another point, within spitting distance of the Russian River, several birds that I took to be buzzards — high-shouldered, glowering gatekeepers of the afterlife — perched on a wire, waiting, their patience seemingly informed by the knowledge that they never have to wait too long. Even when you weren't exactly looking for them, the signs and symbols of life and death were everywhere, just as they are every waking day, gentle but persistent reminders that mostly blend into the background of our busy days.

That's what made this a personal pilgrimage, too. At the time, I had just turned forty-nine and was about to trip an important threshold on my own actuarial odometer. My parents, both in their seventies, were alive and in reasonably good health. I had a daughter who had just turned five, a son soon to turn three. Those little details would normally be irrelevant intrusions in a scientific narrative; in this one, however, they form a kind of background matte to the portrait of science that occupies the foreground. It is our children, especially, whose lives may well be altered by this new science. Even without being crassly self-interested, it is impossible not to think about the science possibly to come in very personal terms.

In conversations with Hayflick and other scientists over the next few days and in subsequent months, I heard outlined, in sketchy but tantalizing detail, a medical future so bold in its ambitions, so profound in its potential impact, that if even a tenth of the promises pan out, it will fundamentally change how we think about life and what it means to be human. There was talk of genes that, when properly manipulated, might significantly extend life span. There was talk of stem cell therapy, a celebrated new technology that holds the hope of replacing aging or failing or diseased organs and other body parts. I even talked to several people whose cells were being used to clone them, in an attempt to create a short-lived, utilitarian embryo that could be harvested for stem cells and, perhaps, immunologically compatible cells and organs. In almost every instance, a biotechnology company had been formed, or was in the works, with dreams of commercializing a technology that would extend life or regenerate human tissues and cells. Indeed, the catchphrase of the day was "regenerative medicine," referring to a discipline that had its own meetings, its own funding and supportive foundations, its own ambitious agenda, and its own little swarm of bioethicists and journalists flitting around like gnats, trying to figure out what was going on and what it all meant. And it was happening very fast: on the ride to Hayflick's home, the news on the radio had been dominated by the still-unresolved Florida vote count in the 2000 presidential election. I think it is safe to say that no one, during those weeks of uncertainty, could have predicted that the new president's first major televised address to the nation would focus on, of all things, embryonic stem cells.

As I traveled around and heard these stories, it was impossible not to think back to that moment on the porch, to hear a little voice in my head say, with all the requisite self-interest of a baby boomer: What's in it for me? What will this mean in my lifetime? Will I live longer, or better? What's in it for my parents, who have both survived to about the predicted life expectancy of people born now (79.5 years for women, 74.1 for men in this country) but are not without medical problems that will need addressing sooner or later? And most of all, what will it mean for my children, for all children? When they reach middle age and beyond, will they indeed avail themselves of a vastly different pharmacopoeia, a spectrum of treatments that could well include cellular therapies, replacement organs grown from scratch, enzymes that immortalize cells? Just how satisfying will that longer life ultimately be, for myself and my children? And what will it mean if our society becomes disproportionately weighted on the elderly end?

It is too soon to provide any definitive answers to these questions, but it's a good time to begin asking them. And, as I quickly began to learn, there are plenty of strong and conflicting opinions about this future, beginning with the man who, in a sense, started it all.

֍

Early the next morning, I followed the final instructions — the last of three pages — to the Hayflick residence, a handsome two-story contemporary home on a little cul-de-sac overlooking the ocean. The natural wildness of the site was spectacular, but not nearly as spectacular as the scientific story Leonard Hayflick told inside. We spoke for about seven hours (the fruits of that conversation form the basis of chapter 1), but one moment particularly sticks in my mind.

It was late in the afternoon, after many hours of talk, and Hayflick was sitting on an ottoman in his living room. The silvery light off the Pacific, muted and dulled by high clouds on this December day, nonetheless seemed to ricochet off the white walls and high ceilings of Hayflick's home. As soft and cool as the afternoon light was, Leonard Hayflick was building up an indignant head of steam. To those who know him, including many who admire his remarkable career in science, the fact that he can still, at the age of seventy-two, climb up on his high horse is no surprise; he's never been one to hide his opinions, and for much of his life he's expressed those opinions without reservation and lived with the consequences. What provoked his ire on this day was a question I had asked. I admit to baiting him a little, because I suspected what his reaction might be, but I hadn't quite expected the magnitude of the reply. I asked about a single word that has increasingly crept into routine scientific discourse, into newspaper headlines, into New Age wish lists: *immortality.*

Hayflick has been a prominent cell biologist for four decades and is a former president of the Gerontological Society of America, so I naturally wanted to know what he thought about a stream of recent public statements by respectable scientists regarding the prospects of significantly extending the human life span through the related technologies loosely known as regenerative medicine, and the increasing use of the I word. (I can't claim to have come to this discussion with entirely clean hands; about a year earlier I had written an article for the *New York Times Magazine* about the discovery and commercialization of embryonic stem cells, and the illustrations — not my handiwork, I hasten to add — depicted octogenarians frolicking on scooters and in convertibles, accompanied by the words "Racing Toward Immortality.")

Several days before I spoke with Hayflick, for example, the first annual meeting of the Society for Regenerative Medicine convened in Washington, D.C. In his remarks to the group, William Haseltine, a cigar-smoking and ostentatiously optimistic bio-mogul who serves as chairman of the company Human Genome Sciences, predicted that several emerging technologies — stem cell therapy, tissue engineering, and the use of gene-related proteins — would change the way medicine is practiced, and would forever change our expectations of how long we might live. More to the point, Haseltine had been quoted several times as predicting that twenty-first-century medicine would achieve a kind of "practical immortality."

Perhaps inevitably, the West Coast version of this genre of meeting took the form of the annual gathering of the Extropy Institute in Berkeley, a meeting attended by several excellent hard-core molecular biologists and later amusingly chronicled by Brian Alexander in *Wired* magazine. Michael Rose, an evolutionary biologist at the University of California at Irvine, was quoted as saying, "I am now working on immortality . . . Who gives a fuck what people consider flaky! If it's the truth, it's the truth." Cynthia Kenyon, a well-respected molecular biologist at the University of California at San Francisco, spoke of her work identifying a "grim-reaper gene" and a "fountain of youth gene" in nematodes, and was quoted as predicting that dramatic life-span extension would become a reality in the twenty-first century. Michael West, the head of a company called Advanced Cell Technology, did not attend the meeting but was definitely there in spirit. "We are close to transferring the immortal characteristics of germ cells to our bodies and essentially eliminating aging," he told *Wired.* "That sounds spectacular, but I believe those are the facts." In what passed for scientific caution and restraint, Calvin Harley, head scientist at the biotech company Geron, said he believed it was not inevitable that our "somas" — our bodies — are dead-end carriers. "We are all born young," he said. "There is a capacity to have an immortal propagation of cells. The way we have evolved is to go from germ line to germ line, with our somas the dead-end carriers." "But," he added, "that is not inevitable."

As I recited each remark to Hayflick, I could see him alternately stiffen and squirm. "How shall I put it?" he began after a long pause, clearly offended by the hubris of his colleagues, several of whom he considers close friends. "I've been in this field longer than any of the people that you've mentioned, which," he conceded with a laugh, "probably doesn't mean a helluva lot. But I'll say it anyhow. Every five years, for the past forty years, there have been pronouncements made by people with names other than those that you mentioned, and with expectations identical to the ones that those people made. I'm still waiting. And I'm afraid I'm going to wait not only through my lifetime, but probably forever.

"The problem," he continued, shifting into second gear of his dudgeon, "is that there is a failure to understand the universality of a phenomenon. If they can show me the simplest way to prevent aging in their own automobiles, to have them live for a hundred years, then there will be some reason to buy into the biology argument. But they cannot do the simple thing, like keep their cars from aging for a twenty-year period, to say nothing about biology. And furthermore, what makes them think that the molecules that compose living things are any different from the molecules that compose inanimate objects, in respect to deterioration over time? And finally, and probably the most telling argument, which will never ever surface in articles like that, is the stupid question, Why do you want to do it in the first place? What is the benefit? People have this underlying, tacit belief that increasing human longevity, or curing aging, or however you want to characterize it, is a good. They've never asked themselves or never described what that good is. And I challenge *all of them* to provide a single scenario that makes sense. Any scenario that they're liable to describe will come closer to science fiction than probable scientific reality."

"But," I replied, "it is in the air now."

"It's been in the air since human history has been written in caves!" Hayflick almost shouted. "It's *always* been in the air. It's no different between now and any other period of time. There just happen to be more people involved. More people who haven't taken the time to understand this field, unfortunately."

Hayflick had especially unkind words for the genetics of aging, a field that has recently exploded with discoveries both in model organisms like fruit flies and in human centenarians. "There are no genes for aging," he insisted. "I'll say that categorically, and I'll defend it despite what you have heard and will hear from Cynthia Kenyon and others. People like Kenyon and Leonard Guarente and others are not working in the field of aging at all. They're working in the field of, to be liberal, longevity determination —

to be more specific, developmental biology. Aging is a deteriorative process, as most people should know, and those folks are not working with that aspect of the animals they're working with that involves the deteriorative changes that occur during aging. They're manipulating biological development with the beautiful experiments that they're doing, and there's no denying that and I'm not speaking to their experimental design. I'm speaking to their understanding of what aging is and what aging isn't. The fact is that everything *will*, whatever the hell you do. Everything in the universe ages."

Hayflick has earned the right to express these opinions, because he arguably laid the groundwork for the entire field of molecular gerontology — the notion that aging, its causes as well as potential remedies, might fruitfully be attacked at the level of cell biology and molecular intervention. And he is well versed in the demographics and statistics of longevity determination; in fact, not long after my visit he became so infuriated by the reductionist hubris of some of his fellow biologists that he teamed up with gerontological demographer S. Jay Olshansky and dozens of other prominent aging experts to prepare a manifesto decrying the misguided messages imparted to the public about antiaging research.

The would-be practitioners of "practical immortality" were spinning out a far more optimistic, revolutionary view of the future. A couple of days after visiting Hayflick, I paid a visit to the laboratory of Cynthia Kenyon at UCSF. Kenyon looks younger than her forty-six years and, despite locutions that sometimes flirt with Valley Girl diction, possesses a breadth and depth of knowledge that is immediately apparent and instantly intimidating. She cut her teeth working under several of the most celebrated molecular biologists of the last half-century, including Sydney Brenner at Cambridge and Mark Ptashne at Harvard, and has narrowed her focus to several intriguing genes in a small worm known as *Caenorhabditis elegans*. These tiny nematodes, when viewed through a microscope, appear to have no other purpose in life but to endlessly carve sinuous arabesques in their growth media, their movements mesmerizing and beautiful. Whatever their purpose in life, Kenyon and her colleagues have found a way to extend that life — quadruple it, in some cases — by altering a single gene. She is unapologetically exuberant about the possibilities this might hold for human biology and human medicine.

"You know, if you look at an old worm under a microscope, it has all these tissues, and the tissues have all our genes in them — you know, myosin or transmitters, whatever — and the worm looks awful. And then you change one gene and the *whole* worm, all the tissues, looks good. So you'd never think you could do that with one gene. And once you see it happening in a worm — the impossible has already happened. What you would think would be absolutely impossible is *not* impossible. You can do that. Now, whether you can do it in a human and blah-blah-blah? Well, the big jump has been taken. You can do it in an animal. That's the main thing. Whether or not you can do it in a human? Maybe, sure, I could see maybe you couldn't for some reason," she said, pausing to give this possibility its due. "But I doubt it." Although she hedged her words scientifically during our conversation, she has not hedged her bets entrepreneurially. In the fall of 2000, she formed a company with MIT scientist Leonard Guarente that has as its ultimate goal the creation of medicines that would extend the human life span. One venture capitalist with whom I spoke called it "the hottest technology around right now."

In all the years she spent doing elegant experiments on the genetics of nematodes, Kenyon told me, hardly anyone outside the scientific community paid any attention to what she had accomplished. But as soon as she began to tackle the molecular biology of aging, she was inundated with requests for interviews. "Night and day," she said, "night and day. The public is absolutely fascinated by aging. They don't want to get old. And you can see — read Shakespeare. Read the sonnets. They're all about aging. A lot of people have an interest in biology that really doesn't extend much further than their desire to cure a disease, I think. But no one likes to get old, and no one likes to see their parents get old, or their grandparents . . . You know, it's just . . ." — and she reached for the right sentiment — "it's a very, very powerful human desire, I think, not to get old. And you really feel that in a *big* way when you study aging." That emotion has become tethered to the most sophisticated science of our time.

Longevity genes, replacement body parts, stem cells, immortalizing enzymes — you won't find reference to any of them in the sonnets of Shakespeare. But Kenyon's remark inspired me to go back and read the sonnets; I found that she was right. There in abundance you will find the timeless, anticipatory human sadness about aging, about "winter's ragged hand" and "that churl death," that, four hundred years later, fires our social fascination with the topic. We prick our ears at any breakthrough, whether marketed by clairvoyants or molecular biologists, that purports to arrest or reverse the inevitable process of aging, or even to extend the human life span in such a way that it no longer seems preposterous to speak of a certain, practical immortality.

But as Hayflick's exasperation suggests, there is an abiding division and tension, even among biologists, on whether the human life span can be extended through better biology. It is a debate that is going to be played out before an extremely attentive audience over the next decade or so.

✑

In my journey up to Sea Ranch, I later realized, I had unintentionally followed an earlier pilgrim, someone who has perhaps understood the link between the emotional, cultural longings for an extended life span and the science that might deliver it better than anyone else in this story.

In the summer of 1992, a young man named Michael West drove those same roads, took those same cliff-hugging turns past Monterey pine and Douglas fir, passed that same wild and ravishing seascape on the way to visit Leonard Hayflick. For West, this truly was a pilgrimage, a journey to pay homage to a master, for Hayflick was the scientist whose work had prepared the bed in which all of West's dreams had begun to take root. A few months earlier, in the fall of 1991, West had blown away a roomful of jaded West Coast venture capitalists with his vision of creating a business to develop medicines that would treat the process of aging, based on several cutting-edge molecular technologies just then coming out of academic labs. The money people had watered the seeds of West's ambitious ideas with millions of dollars, and by March 1992, he had a company on paper. It was called Geron, and it was the first biotechnology company explicitly devoted to the molecular biology of aging.

West had much on his mind in those days — hiring scientists, finding lab space, riding herd on research, scouting out new technologies. But one of the first things he did was drive up to see Hayflick. And Hayflick was thrilled to have him. "He spent the weekend up here," Hayflick recalled, "and stayed in the guest room. He was so riveted by this concept. I took him out to dinner and he hardly touched his food, he was so busy talking about what he wanted to do. It was very refreshing to see somebody who — I've known a few scientists who burn with a white-hot flame, and Mike is one of them."

They talked all weekend long — both are excellent talkers and storytellers. They talked about aging research. They talked about personalities in the field. They even talked about some of the classic medical textbooks in the field of aging, and exchanged copies of first editions of these seminal books. Hayflick showed West the famous letter he had received from a Nobel laureate, rejecting for publication a 1961 paper that subsequently became one of the most widely cited in twentieth-century science. "He was really one of the first young people to enter the field who had a sincere interest in the history of aging research, and that was extremely impressive," Hayflick told me. "I thought to myself, 'Here's a fellow to be cultivated.'

"It's immodest of me to say it," Hayflick continued, "but he knew the history of the field, and he was fascinated by my discovery of the limits on cell replication in culture" — the discovery, that is, that cells grown in a lab dish don't, and can't, live forever. "And he just wanted to talk to me about that. How did I discover it? What went through my mind? What were my views on the company, on its direction, on people who might be hired? I think that weekend stimulated him, because he knew he could rely on me to provide help, suggest people to contact, and so on." Hayflick paused here, then added, in a speculation rife with implications, that he might also have served as "kind of a father figure, maybe" for West.

That speculation may actually get closer to the reality of things, not least because it hints at the way in which that relationship may have influenced a pitched public-policy debate nearly a decade later. While there is no question about the crucial role Leonard Hayflick played in the early days of molecular biology's attack on aging, and how his early experiments have inspired a fabulously productive area of contemporary science, what's far less appreciated is how he also served as a role model and inspiration for Michael West — not simply for his science, although that was important, but for his attitude, his temperamental readiness to defy authority, and his willingness to pay a price, an enormous and almost unconscionable price, to do something he believed was right, even when everyone else in the world believed he was wrong. And it's quite possible that after his weekend at Sea Ranch, West appreciated — as does almost anyone who speaks at length with Hayflick — a deeper moral to Hayflick's story. Although he can be stubborn and antagonistic and even bombastic, perhaps Leonard Hayflick's greatest sin and scientific transgression was that he was way ahead of his time. It was a lesson that West, who shares many of the same qualities, took to heart.

1

THE HAYFLICK LIMIT

THE CANISTERS STOOD in the corner of the garage, amid garbage cans and gardening tools, right next to Hayflick's champagne-colored Lexus. Battleship gray, 30 liters in capacity, the two Union Carbide liquid nitrogen containers looked so utilitarian that they might easily have been mistaken for a pair of wet-dry vacuum cleaners, but they have a slightly more exalted purpose than cleaning up messes: they are designed to keep biological materials frozen at a constant temperature of −192 degrees centigrade. These particular containers were full of inch-long glass ampules, hand-filled and sealed with the flame of a Bunsen burner nearly forty years earlier. The ampules contained an unusual item for garage storage — human cells, and in particular a human cell line called WI-38.

We had been talking for the better part of the day, but inevitably the conversation kept circling back, as does Hayflick's hectic and remarkably eventful scientific career, to those cells. And as he stood there in the garage in the late afternoon December light, in his white shirt, tan sweater, and gray slacks, the sun easing into the Pacific Ocean just beyond the front of his Northern California home, there was a look of contentment, of sheer triumph, on Hayflick's face that went far beyond pride of ownership. Beyond smug, even. It was as if, not inappropriately, his face had been rejuvenated by the sheer adolescent thrill of what he had pulled off. *After all these years, he still had the cells.*

These are the cells from a famous colony he created from human fetal tissue in 1962, the very cultures that had overturned half a century of flawed dogma about the secret life of cells. These are the prototypes of the cells used to make vaccines that have educated the blood of virtually every American citizen under thirty years of age. These are the cells that traveled in the backseat of Hayflick's family sedan when he "absconded" with them and drove across the country from Philadelphia to California in 1968 to join the faculty at Stanford University. These are the cells that United States government investigators seized when they raided Hayflick's lab at Stanford in 1975, the cells that formed the crux of his career-crushing six-year lawsuit against the government (a lawsuit settled out of court in 1981, to Hayflick's evident satisfaction). Most important, these are the cells that changed modern biology's view of the aging process, the cells that, directly and indirectly, opened up several rich avenues of research and inadvertently led to the biotechnology of aging. "Those are the cells that I stole," Hayflick said, lifting a rod containing ampules out of the cylinder through a dramatic billow of supercooled condensation, like some Oz of the microverse. "And I'm proud of it."

For four decades, Hayflick has been schlepping the cells around wherever he goes. In fact, he's kept more than his famous WI-38 cell line on ice. He also has a cell strain in there called WISH, for Wistar Institute Susan Hayflick, because they were derived from the amniotic fluid surrounding his gestating daughter. He has the prostate cancer cells of his old boss, Charles Pomerat, a famous cell biologist who died of the disease. He has the Chinese hamster ovary cells in which the clot-busting drug "tissue plasminogen activator" (or TPA), a staple now in virtually every emergency room in the developed world, was first produced by genetic engineering. This may seem like a daft or even macabre variation on stamp collecting, but even in his seventies, Hayflick has never stopped being what he has always been: a cell biologist, a scientist fascinated by the life and care and well-being of interesting cell lines, each as different as children, each with different personalities and habits, each with different potentials. And after topping off the liquid nitrogen every couple of months for all these years, maintaining what he calls the longest continuous culture of human cells in the history of biology, he confessed to me that he's been trying to find someone else to look after them. "As you know," he said with a sly half-smile, "I'm not going to live forever."

Precisely the point of those cells, you might say. For all the fame and grief they brought Len Hayflick, they told the world of biology that life — at least the life of the normal cells that make up the tissues and organs of our mortal bodies — may have built-in limitations on longevity; specifically, they told biologists that normal human cells growing in a dish can replicate only a finite number of times and no more. They hit a wall. They simply lose the biological vigor for life and stop dividing. And that simple fact, rippling through molecular biology and boardrooms (and, perhaps, through every cell in each of our bodies), began to change the way scientists think about the process of aging.

⌒

Sitting in the living room of his home at Sea Ranch, speaking in the gruffly authoritative but precise rhythms with which he had recounted this same story so many times before, Leonard Hayflick still conveyed timeless excitement and ageless pain in his voice. He is a courteous, even courtly, septuagenarian, a little heavyset these days, hair salt-and-pepper and close-cropped; his dark eyes, guarded by large heavy-rimmed glasses, have the wet attentiveness of an animal that has been mistreated. And yet there is also something physically pugnacious about him, an almost ballistic geometry to his head. By turns he expresses humility (and, occasionally, insecurity) about his modest, blue-collar upbringing and then the deep knowledge and genial arrogance of someone who has not just stood on the shoulders of twentieth-century giants of science but has rubbed shoulders with many of them and butted heads with a few. His most prominent physical feature, however, may be the permanent chip on his shoulder, almost a bone spur of bitterness and resentment, about the way he has been treated during his scientific career. Much of that woe can be directly traced to those cells in the garage. He has had an amazingly productive career — he discovered the cause of a prominent human disease, pioneered the isolation of cell strains that have been used to create an estimated one billion doses of vaccine, accidentally founded an industry based on the creation of the powdered substances used by laboratories throughout the world to nourish growing cells, served as president of the Gerontological Society of America, and fought a celebrated but personally devastating lawsuit against the federal government in the 1970s that, Hayflick claims, made entrepreneurial biology safe for academic researchers. If he doesn't exactly go looking for a fight, he seems to have bumped into them all along the way. "Len has a real bullheaded streak," said one old friend. "Always has."

To understand how Hayflick reached this point of advanced dissatisfaction, and at the same time to understand why he persists as an important figure even in the age of stem cells and cloning and life-extending genes, it helps to consider how he came to science in the first place. Born in May 1928, he grew up in West Philadelphia; his mother was a typist and bookkeeper, his father a "dental mechanic," constructing prosthetic devices for the mouth. As an eight-year-old, he became fascinated with chemistry ("I still have the instruction booklet that came with my Gilbert chemistry set," he volunteered at one point in our conversation), and he ultimately built what he describes as a college-grade laboratory in his basement. With a friend from the neighborhood, he would ride his bike on the hour-long trip to the campus of the University of Pennsylvania, where a venerable chemical supply house called Dolby's had been providing research materials to the university for more than a century. Hayflick and his friend routinely purchased dangerous and volatile chemicals, like metallic sodium, and then, of course, played with them ("We did a lot of work with explosives" is the way Hayflick put it).

Childhood science is a rarefied kind of troublemaking, however, because mixed in with the mischief is a precocious, self-preserving reverence for materials, for ingredients, for *stuff*, and it's a lesson Hayflick has never forgotten. The boys were granted access to the basement of Dolby's, for example, where they retrieved one-of-a-kind nineteenth-century handblown glassware. "In fact, I still have some pieces in the house here," Hayflick said. And as we continued talking, he walked over to a cabinet in the dining room and returned with a strikingly beautiful four-chambered handblown retort from the 1800s. "This is the kind of stuff that turned us on, okay?" So it wasn't just the cells in the garage; he has carted around with him a museum of all his formative interests, a portable archive of his own intellectual passions.

By the time Hayflick went to high school, his reputation as a chemistry whiz had preceded him. He began to correct his teachers during class, and after one such remedial episode, his chemistry instructor shrewdly arranged for him to work in the chemistry department stockroom instead. Hayflick graduated from John Bartram High School with honors. Although he received a full four-year scholarship to Temple University, he longed to go to Penn — even though he would have to pay (and he still remembers the tuition to the dollar, a sure fingerprint of lower-class strivers: $250 a semester). So he came up with a clever strategy — he registered at Penn in January of 1946, then immediately took a leave of absence and enlisted in the army, thus qualifying for the GI Bill. In 1947, when his eighteen-month tour was over and he returned to Penn, his entire tuition was paid by the government. These were more than simple economic decisions; they reflected a kind of lifelong tension between the financial and intellectual modesty of Hayflick's upbringing and the high-flying careerism of the scientific circles in which he began to travel, between feelings of personal insecurity and the confidence that one needed to flourish in Ivy League classrooms and, later, on the gridiron of science.

And then Hayflick discovered biology. Although he was still interested in chemistry, a chance occurrence in a bacteriology class made a profound, oddly aesthetic impact around the time he had to declare his major. A laboratory assistant walked into the classroom one day with a tray of test tubes known as "slants" — so-called because agar, the yellow jelly-like biological growth medium, had been allowed to solidify at a slant. "She gave me this tray of slants, and I looked at it, and of course I'd never seen agar slants before," Hayflick said, as if describing a Matisse he had stumbled upon in a museum. "And on them were these wiggly lines. Each had a wiggly line. And they were different colors. Brilliant colors. White, yellow, purple, green — all the colors of the rainbow." The colorful scribbles had been produced by nature's microscopic palette; they were simply different strains of bacteria, growing on the nutrient-rich agar. "That hit me like a bomb. And turned me on to bacteriology." Microbiology became his major.

Following his graduation from Penn, Hayflick did drug company research at Sharp and Dohme (a division of Merck), returned to Penn to get a master's and a Ph.D., worked briefly at the Wistar Institute in Philadelphia, got married, and went to do a postdoctoral fellowship at the University of Texas Medical Branch at Galveston, in the lab of the prominent cell culturist Charles Pomerat (it is Pomerat's cancer cells that Hayflick still carts around). But during these peregrinations, Hayflick battled the feeling that he didn't quite belong in the company he was keeping. "I had an inferiority complex, I suppose, and never thought I had the brains to get a Ph.D." was the way he put it. It was a revealing remark: the need to prove himself, combined with a stubborn, tenaciously independent cast of mind, set the stage for all the scientific triumphs and personal disasters that were destined to unfold.

While he was still in Texas, Hayflick learned that a forceful, charismatic figure named Hilary Koprowski had been hired to reorganize and reinvigorate the venerable Wistar Institute of Anatomy and Biology, and that there was a job for him if he wanted it. So in 1958, Hayflick, with his wife and two children, moved back to Philadelphia.

〜

In August 2001, exactly one week after George W. Bush announced his policy on embryonic stem cells, I met Leonard Hayflick at his old stomping grounds. The Wistar Institute occupies several buildings in Philadelphia, most notably its original, three-story structure, built of a muddy-mustard brick in 1894. Located on Spruce Street, the complex is tucked into a corner of the Penn campus but is independent of the university. The central atrium of the original building, where Hayflick met me, features an ornate open-lattice iron staircase, painted white, beneath a large and airy skylight, but what Hayflick vividly recalled from the 1950s — and what made this visit not unrelated to the political passions animating the stem cell debate half a century later — was the parade of beaming mothers carrying newborn babies up those stairs to thank the institute's then director, Edmond J. Farris, for helping them get pregnant. In addition to being the oldest private biological research institution in the United States, the Wistar was also one of the earliest (albeit unofficial and indeed somewhat surreptitious) centers for assisted reproduction. As a Penn student working at the institute in the early 1950s, Hayflick had on occasion donated sperm for the artificial inseminations that Farris performed with a turkey baster. "We had to make money to get through school," Hayflick said with a shrug, "and that's how we did it."

The reproductive assistance was particularly successful at the Wistar because Farris had developed an extremely accurate test, using the Wistar's albino rat colony, to determine when his patients were ovulating. Unfortunately, the women technically were not his patients, because Farris was not a doctor, and that was his undoing. An enterprising reporter got wind of the business, Hayflick told me, and wrote an exposé in one of the local newspapers; the archdiocese of Philadelphia became involved and "just shut them down." It may sound like an amusing anecdote, but the incident in fact represents one of earliest clashes between organized religion and reproductive technology in this country — a conflict that since Hayflick's first sojourn at the Wistar has influenced, and in many ways shaped, the direction of scientific research, from the early days of artificial insemination through research on fetal tissue (of which Hayflick became an early and infamous practitioner) to in vitro fertilization and, most recently, stem cell research.

When Hayflick returned to Philadelphia in the late 1950s, he found the Wistar to be a "ghostly" place. It retained the nineteenth-century, naturalist feel of its founder, Isaac Wistar, whose ashes, along with those of Philadelphia's scientific elite, filled brass urns lining the entryway. Generations of Philadelphia teenagers had sneaked onto the premises to ogle the collection of Siamese twins, a cyclops, and other grotesques of human development, sealed in jars of formaldehyde and housed in the first-floor museum. Hayflick at first occupied an elegant, bay-windowed lab on the second floor, which he later shared briefly with Stanley Plotkin, to whom he provided a critically important cell line for the development of a vaccine for rubella. Soon he ran a large cell culture facility in the middle of the second floor and became an expert in that arcane craft — the art of cultivating and maintaining living cultures of cells in glassware (literally, in vitro) so that the cells can be studied. At the time, cell culturists often seemed to provide little more than a supplementary technical support service, like sous chefs to the executive chefs of science. The ascendant science of the day was virology, and virologists required vast amounts of living cells grown in vitro as fodder for their rapacious viruses. That Hayflick also was in charge of cleaning glassware and preparing growth media suggests that he was a kind of supertechnician. But cell biology also represented the wave of the future, and any time Hayflick needed to console himself, he could recall the view from the railing outside his lab, which in the early 1950s took in the enormous, sixty-foot skeleton of a finback whale suspended from the ceiling and, down below in the museum, those large jars containing a cyclops and Siamese twins and other human abnormalities. That was the old biology, nature viewed with the naked eye. But things were changing.

It is difficult to convey the heady atmosphere that prevailed at the Wistar in the late 1950s and early 1960s, under Hilary Koprowski's autocratic, energetic, and ambitious leadership. Koprowski, a flamboyant Polish-born biologist, was one of those larger-than-life figures who divided the world into friends and enemies and never forgot the list. He was said to be a concert-level pianist (he kept a grand piano in the seminar room at the Wistar), drove a sporty MG, maintained an apartment in the room adjoining his office (with, everyone recalls, a separate entrance and exit), and kept a well-stocked bar in his office ("His secretary made the best martinis and

Bloody Marys in town," Clayton Buck, the Wistar's current director, told us). When Koprowski arrived in May of 1957, he said in an interview, the Wistar "was a graveyard. It had the largest number of human and animal skeletons in the world." He promptly shipped the whale off to the Field Museum in Chicago and two Indian mummies to the Smithsonian, and began to get rid of all the bones. "He was an old-country, Herr Doktor type," recalled David Kritchevsky, one of seven "refugees" who came with Koprowski to the Wistar from Lederle Labs and watched this "benevolent despot" go to work. "There were no labs," Kritchevsky recalled. "There was no cohesion. No one knew what anybody else was doing, nor did they care. There was no 'before' there." Under Koprowski, the Wistar reinvented itself as one of the preeminent centers for the study of viruses, at a time when the study of viruses was the most exciting (and, appropriately, the most swash-buckling) area of biological inquiry. In going from whales to viruses, the Wistar exemplified biology's midcentury transition to reductionism, with Koprowski its single-minded Ahab chasing this smaller prey.

Koprowski insinuated himself into the thick of the race to create a po-lio vaccine (indeed, his efforts to test early polio vaccines in Africa led to re-cent accusations, subsequently disproved in the scientific literature, that these trials may inadvertently have introduced the human immunode-ficiency virus to Central Africa and sparked the start of the worldwide AIDS epidemic). Vaccine makers from Sharp and Dohme consulted closely with Wistar researchers during the golden age of vaccine discovery. Every-one who was anyone in cutting-edge biology passed through. "When Hil-ary came to the Wistar," Hayflick recalled, "it was a magnet; it was a cross-roads for the world's cell culturists, cell biologists, and virologists, so I got to meet everybody in sight there." And by dint of an experiment that ap-peared at first to be a biological dead end, Leonard Hayflick soon discov-ered that everyone was coming to see him, too.

The discovery of what became known as the "Hayflick limit" grew out of a series of experiments that began for all the wrong reasons. To study vi-ruses, you have to be able to grow them under controlled conditions; and since viruses can only replicate in living things, these conditions demand one fundamental ingredient: living cells. At first, virologists grew their viral cultures (including the viruses used to make human vaccines) in untidy tis-sues like monkey kidney cells, but this was less than optimal because the animal tissues often harbored, unbeknownst to researchers, their own contaminating tribes of viruses. It fell to Hayflick, as the Wistar's chief cell culturist, to find better cells in which to grow viruses. In the late 1950s, this was especially crucial to the development of safe vaccines.

But Hayflick wanted to do something more significant than simply provide the raw material for other people's experiments, so he took up one of the most fashionable, and ill-fated, ideas of the 1950s: the notion that cancer was caused by viruses. In 1958, he planned a series of experiments in which he would grow normal human cells and then expose them to extracts from human cancer tissue, to see if material from malignant tissue could convert, or "transform," normal cells into cancer cells. First, Hayflick had to figure out how to cultivate normal human cells, which in itself was tremendously difficult. Deranged cells, such as the famous HeLa line of cancer cells, sometimes grow like gangbusters in the lab; normal cells, by comparison, are exceedingly tough to grow. So he began to search for human tissues that might be used to create a permanent, self-perpetuating colony, or cell line.

"Of course, getting human tumor tissue from operating theaters was no problem," Hayflick told me. "You could get that stuff easily, every day. But getting normal human tissue *was* a big problem. First of all, I wanted human fetal tissue, and this is crucial." At that time, scientists were beginning to realize that even normal adult human tissue — excised tonsils, for example — was an invisible bestiary, teeming with human viruses. But fetal tissue, unexposed to the outside world, was pristine. "It was known that human fetuses were likely to be free of these garden-variety viruses, and that's where I wanted to get my tissue," Hayflick said. "Well, as you can imagine, that was a bit more difficult, although surgical abortions were certainly conducted in this country at that time, and at Penn. I did get tissues from Penn frequently." It was an international connection, however, that proved most useful. Sven Gard, a well-known personality in the polio field, was based at the Karolinska Institute in Stockholm but had spent a sabbatical at the Wistar. In the watercooler logistics of exchanging biological materials, proximity became destiny. "This fellow had worked in a lab across the hall from me," Hayflick said, "and through Hilary he said, 'Well, we can get you all the fetal tissue you want from Stockholm. You know, we do this every day there, and it's all legal. Let me talk to my colleagues, and we'll set up a system.' So we set up a system. And we got lots of stuff from Stockholm."

"They didn't send the whole fetus," Hayflick continued. "By that time, I knew what tissues I wanted. I wanted a lung or a kidney. And so they removed these organs from the fetuses, chopped them up in relatively big chunks, put the material in cell culture fluid, in test tubes or small flasks, packed it in wet ice, and sent it by air. Just ordinary airmail." Once the fetal tissue arrived at the Wistar, Hayflick would chop it into minuscule, confetti-size pieces of 1 to 4 millimeters, or tease it apart with forceps. Then he used a standard enzyme to digest away the connective tissues, until all that was left were the cells. The next step shows how the most lumpen of equipment can make or break an experiment: Hayflick's lab placed these fetal cells in special glassware, first in flat vessels called milk dilution bottles, and then into flat-sided flasks known as Blake bottles. These broad, flat bottles, which normally lie on their sides, were provisioned with basic nutrients and fetal calf serum. The composition of these potions, known generically as growth media, was in itself a black art. For some reason, the combination of glassware and nutrients worked.

A successful cell culture is a bit like a time-lapse nightmare of urban sprawl. Within an hour, the fetal cells from Stockholm would, like tentative newcomers, explore their vitreous environment and, if they found it to their liking, begin to attach to the floor of the glass bottle. The cells had the telltale spindle shape of what are known as fibroblasts. And once settled in, they would do what cells are naturally inclined to do: grow, spread, indulge their inherent biological version of manifest destiny. After about three days of incubation at 37 degrees centigrade (roughly body temperature), a sheet of human cells would cover the entire flat surface of the bottle, all human, all alike, all normal; as this living carpet of cells expanded to cover the entire surface of the bottle, technicians would split the population into smaller portions and introduce them into new bottles. The carpet of cells had a special sheen. As Hayflick and a colleague later wrote in a landmark 1961 paper, they possessed a "quilt-like appearance." It is no exaggeration to say that this turned out to be a rather magical quilt.

ڡ

The best cell culturists are like improvisational sauciers, constantly tinkering with their arcane recipes, changing the amounts of ingredients, their quality and purity, when and how they are added to the brew in their odd little bottles, all to create a reasonable facsimile of the kind of biochemical environment in which cells take on a life of their own. If allowed to grow unchecked, cells in culture continue to multiply, or double, every week or so. But if you remove the cells from the bottle, place them in sealed ampules, and then put the ampules in a freezer, as Hayflick began to do, they remain frozen in time, the odometer of "cell doubling" (or replication) stuck in place, until they are thawed and cultured again. In this way, the cells can be distributed throughout the world to researchers. And that is exactly what began to happen. With four or five technicians working for him full-time, Hayflick's lab at the Wistar became a semi-industrial operation. The cells from Sweden and from the hospital at Penn grew beautifully, and within a few months Hayflick had twenty-five separate strains of normal human cells growing, meticulously numbered from WI-1 to WI-25.

So what? What was so special about having normal human cells growing in a dish? The answer is coursing through the veins of virtually every person reading these lines. The fetal cells that Hayflick tamed and trained proved to be spectacular for viral research and, more to the point, vaccine production. They were essentially pure, with no viral contamination. Roughly a year after he started, Hayflick enlisted the help of a colleague at the Wistar, Paul Moorhead, to scrutinize each cell strain to make sure it possessed the normal complement of human chromosomes — and no more. Because the cells indeed possessed two copies of all 23 human chromosomes, or 46 in all, they were said to be both normal and "diploid."

As word began to spread, Hayflick laid down hundreds of ampules in the freezer; he was inundated with requests and distributed the cells to all comers. So it became an unofficial business of sorts, keeping the virologists happy with a steady supply of clean, normal human cells. But then came the shocking and confounding observation: Hayflick began to notice that certain populations of cells suddenly stopped growing. "After several months, I began to realize that some of these cultures were . . . their former flourishing condition had changed to one of stasis," he said. "They stopped dividing. They didn't die immediately. In fact, they will stay that way for a year or more, without dividing. They'll metabolize, but they won't divide. And frankly, I didn't think very much of it initially. I just thought everybody knew this. I thought everybody knew that normal cells could not divide indefinitely, and that only cancer cells were immortal." He paused, sitting on the ottoman in his living room, and settled into a posture of self-recrimination. "Now, *why* I thought I knew everybody knew that — I'm still struggling with an answer to that, to this day."

It was the staggered and unpredictable arrival of the fetal material from Sweden that helped the lightbulb go on. Cell cultures that had been growing for roughly ten months suddenly stopped dividing; cultures begun from fetal material received more recently continued to flourish, even though the same recipe, the same ingredients, and the same type of glassware had been used by the same technicians. The pattern of these staggered, aberrant interruptions of growth suggested something intrinsic to the biology of cell growth, not a failure of technique. "Either at Penn or at the Karolinska Institute, they did these abortions at odd times," Hayflick explained. "And so all of a sudden — and this is actually a key part of the story — the randomness of the delivery was an absolutely key part of my discovery. Had it been uniform, I don't know that I would have made the observation I ultimately made."

At the moment he noticed this curiosity, Hayflick's puzzlement intersected with two forms of psychological vulnerability. One involved his own lack of self-confidence in the face of this unexpected observation. The other involved the failure of the scientific method on a mass scale, fueled by a collective lack of self-confidence in the biological community stretching back to the end of the previous century. Cells in culture had stopped growing, just like the Wistar cells, for sixty years, but the cell culturists had blamed themselves for the shortcoming. "For the previous sixty years, from the time when cell culture technology began, in the late 1890s or 1900s," Hayflick explained, "it was believed virtually from day one that all cells put into culture were inherently immortal. That's a key statement in this whole story."

If Hayflick was slow to see the full implications of this thoroughly accidental discovery, it was because a very big shadow fell over the experiments. In a classic triumph of charismatic authority over rigorous experimental inquiry, Alexis Carrel, a prominent doctor at the Rockefeller Institute, had mesmerized all of biology early in the century with the claim that cells grown in a culture dish would live forever. The French-born Carrel, who received a Nobel Prize in 1912 for experiments related to transplant surgery, had in that same year placed a smidgen of heart muscle from a chick in a dish, replenished it almost daily with nutrients, and purportedly kept the cells growing and dividing right up to his death in 1944. It wasn't just Carrel's results but the theatrics surrounding his technique that created almost a cult of invincibility about the experiments. He insisted that the walls of a cell culture lab had to be painted black, and while performing their manipulations, his technicians donned long black gowns and hoods, so they looked like a photographic negative of a Ku Klux Klan gathering. As a result, everyone assumed that normal cells, whether isolated from chickens or humans, were immortal. Carrel's mistake, science historian Jan Witkowski suggests, was that in replenishing the cell cultures with nutrients collected from freshly killed chicken embryos, he and his colleagues may inadvertently have supplied fresh new cells to the culture each time they fed them. In other words, they had merely created the *illusion* of immortality through sloppy laboratory technique. The gerontologist Steven Austad puts it more strongly: "It is now clear that the errors and incompetence were Carrel's own, although a more charitable interpretation is that in his case it was the laboratory assistants who were incompetent or that they spiked the dishes occasionally with fresh cells because they were just too afraid to tell him that the cultures had died." By the time Hayflick made his observation in the 1960s, nearly a quarter-century after Carrel's death, there were hundreds, if not thousands, of naked emperors parading through the world of cell biology, all convinced that the immortality of cells was an unassailable scientific fact. "If it were true that all cells are immortal, as the previous sixty years of research allegedly demonstrated, then aging has nothing to do with intracellular events," Hayflick said. Put another way, if Carrel and the others were right, biology had nothing to say about aging.

The sociology of this mass self-delusion held biologists in such a powerful grip that if cells died in culture — and indeed, they often died — it was routinely attributed to mistakes in scientific technique, primary among them the use of inadequate or "toxic" growth media. This culture of blame became so entrenched that even Hayflick succumbed to it at first. "That was dogma for the first sixty years," he told me with a shrug. "And as a consequence of that belief, the first sixty years were substantially spent in perfecting media that would prove this belief that cells were immortal. It wasn't done in a conscious way. It was just . . . I was by no means the first person to discover that cultured normal cells don't live forever. That had been discovered before I was even born! But nobody ever realized that it was probably a normal state of affairs, and *not* attributable to failure to understand how to culture."

Hayflick noticed that as a general rule his cells stopped growing after about fifty divisions, or "population doublings." They continued to eat, excrete wastes, and perform all the metabolic housekeeping necessary to stay alive. They just didn't replicate any more. Eventually, debris attached to them, and they ultimately suffered "degeneration." As they prepared to describe these sensational observations to other scientists, Hayflick and Moorhead had a more than modest disagreement in interpreting this phenomenon. Hayflick insisted on speculating that this peculiar behavior by normal cells in culture represented a "finite limit" to the number of times a cell could divide, an observation that "may bear directly upon problems of aging, or more precisely, 'senescence.'"

Moorhead disagreed. He told me later that he felt Hayflick over-reached in his speculation. "He tends to exaggerate, and in writing the paper, we frequently went back and forth, but that's Len. He's the kind of person who will fight tooth and nail." Four decades later, the scientific community is still going back and forth over the same issue — whether cell senescence has anything to do with aging. But Hayflick prevailed in that first early disagreement, and the connection to "replicative senescence" appeared in the published paper. With that phrase, Hayflick had begun to re-invent himself as a gerontologist.

⌒

It's never easy to challenge established dogma, especially when you're thirty years old, now with three kids, living on soft money and prone to alienating colleagues with frequent outbursts of candor. After pondering this situation, Hayflick made things even a little harder on himself: he decided to challenge the leading authority on cell culture at the time. He traveled to a scientific meeting in Atlantic City with no other purpose in mind but to pose a single question at the end of perhaps the most important lecture of the meeting.

The occasion was the annual gathering of the Federation of American Societies for Experimental Biology in, as best he can remember, April of 1959. It was a short car ride from Philadelphia, but in many ways it was the longest trip of Hayflick's life. "I was still this young kid," Hayflick told me, "and, you know, I didn't know anybody there. They were all strangers to me, because I was just new to the field. And the reason I went there is that Ted Puck was giving a plenary lecture." Theodore Puck was a prominent, Polish-born biologist who was widely considered the presiding guru of growing cells in the lab; the invitation to give a plenary lecture at the annual meeting of FASEB testified to his stature. On the day of the talk, hundreds of biologists poured into Haddon Hall, the old beachfront resort, and filled the auditorium. "My memory is that there must have been a thousand people there," Hayflick continued. "I'm sure there weren't that many, but that's my vision: just a huge hall, a huge number of people."

Puck spoke for about an hour. He had created a cell line from Chinese hamster ovaries in 1958 (a line still widely used today), studied the chromosomal complement of these cells, and even developed some cell cloning techniques. He had published much of this research in the *Journal of Experimental Medicine*, which, as Hayflick correctly states, was "the Cadillac of journals" for cell biologists. Hayflick stood nervously at the back of the room. "When he was finished," Hayflick said, "I had the temerity to raise my hand and ask a question. I said, 'Dr. Puck, in your experience with normal human cells, have you ever found that they stop dividing and then you're unable to work with them any more?'"

Puck considered the question and then, according to Hayflick, replied as follows: "Oh yes, that happens often, but it's not a problem. We simply go back to the freezer and reconstitute what we want."

"And I *knew* that he had missed it!" Hayflick said, leaning forward, speaking with as much giddy, adrenalized triumph as he must have felt on that spring day forty years earlier. "My major concern was having screwed up somehow. I wouldn't publish without knowing an answer to that question," Hayflick continued, "because he was the dominant person working with the cells that I was working with. So when he gave me that answer, I knew he had seen what I had seen, just like everybody the previous sixty years had. But" — and here Hayflick luxuriated in a thrilled, conspiratorial whisper — "*but he didn't know what was going on!*"

Hayflick didn't bother to stick around for the rest of the meeting. "I went there *just* to find out whether Ted Puck had seen the same thing I had," he said. "Now I *knew* that he did, and that gave me confidence."

Oh, there were still a few minor wrinkles to overcome. In a bid to convince the more influential scientists in cell culture of this improbable result, Hayflick sent some of his human cells to a number of elder statesmen in the field and predicted, exactly and correctly, in which month the cells would stop growing. This stunt initially met with great skepticism, but nine months later the phone began to ring, with scientists reporting that Hayflick's cells had indeed hit the wall and stopped growing. That convinced Hayflick and Moorhead; they wrote up their paper about the discovery and sent it to the *Journal of Experimental Medicine.* After a long delay, the journal famously rejected it.

"The largest fact to have come out from tissue culture in the last fifty years," wrote Peyton Rous, the journal's editor, in a scolding letter of rejection that Hayflick still possesses, "is that cells inherently capable of multiplying will do so indefinitely if supplied with the right milieu in vitro." This parroted the conclusions of Alexis Carrel. Rous added, "The inference that death of the cells is due to 'senescence at the cellular level' seems notably rash." On May 15, 1961, the paper received a somewhat less hostile reaction at a somewhat less prestigious journal called *Experimental Cell Research,* where it was promptly accepted for publication. Soon the world would learn of twenty-five separate cell strains of normal human cells, all of which reached a point of no return in their life under glass: they simply stopped dividing, having bumped into a wall of replication that would ultimately become known as the Hayflick limit. Hayflick made sure to supply me with subsequent scientific discussions of this paper, which has been cited by hundreds of scientists over the years and is considered a "citation classic" by the Institute of Scientific Information, which keeps track of these things.

There was, however, one more minor problem. After the paper had been accepted, but two months before it appeared, Hayflick's freezer in the basement of the Wistar broke down, and all twenty-five precious cell strains perished. So the day the paper describing these marvelous human cell populations came out, none of them existed. "So here I am," Hayflick recalled, "with a paper being published, and none of the stuff that I discovered is available. Paul and I were devastated. But not terribly so. By this time, we knew that the general character, and the characteristics, of these cell populations was pretty much the same . . . And so although we had described twenty-five, it was not important now to do twenty-five. It was now important to do one, and do it *well.*"

They had to hurry. "People were clamoring for the cells all over the world," Hayflick said. And he was eager, too. He was intoxicated with the possibility of using fetal cells to create vaccines. With the first batch of cells, he and Moorhead realized that human fetal tissue was "absolutely clean." The cells contained no natural viruses and yet could easily be infected by more than thirty human viruses tested in the lab: measles, polio, herpes, Coxsackie . . . the list went on and on. The implications for vaccine production were immediately apparent to Hayflick; polio vaccines were currently being produced in monkey kidney cells (and injected into American schoolchildren), even though the monkey tissue was "horribly contaminated" with simian viruses that might get into the vaccines and infect human recipients, recalled Maurice Hilleman of Merck Laboratories. "There were about forty indigenous viruses in the monkey cells," he said, "and some of those cells were used for the Salk vaccine. They were in the primary cell cultures that Sabin used also." The terrifying possibility of viral contamination became starker and grimmer in 1960, when Hilleman reported the discovery of yet another monkey virus, known as SV40, which could potentially infect human vaccine recipients. Hayflick maintains, correctly, that for this reason alone, conducting research with human fetal tissue, especially with the aim of creating human vaccines, is not only morally sound but avoids alternatives that, if pursued with today's knowledge of viral contamination, would be tantamount to medical malpractice. Koprowski made much the same point in an article in the *Journal of the American Medical Association* in 1961, in which he argued for the use of Hayflick's human cell lines in vaccine production. There was, surprisingly, a great deal of resistance to using normal human cells, not because of ethical concerns but because of fears that the "normal" cells might actually be cancerous. With this concern in mind, Koprowski concluded his article with typical flair: "For those who fear — a Thurber moral: 'You might as well fall flat on your face as lean over too far backward.'"

So Hayflick's team went back to create a new cell strain from scratch. They created WI-26, but the supplies were exhausted almost immediately because of the demand from scientists. They created WI-27 but had to dump hundreds of already-filled ampules when Moorhead discovered that one of the chromosomes had an anomaly. After these two false starts, they finally created a cell line for the ages. "By this time," Hayflick said, "we knew what we wanted." They wanted a female embryo, to avoid any confusion with the previous, male strain. They wanted to know a good deal about the donors, because of the potential use of the cells in vaccine preparation. Although the pedigree was not sterling, it certainly offered more medical history than that of a monkey captured in the jungles of Southeast Asia. "The tissue came from a woman whose husband was a mariner,"

Hayflick said, "and he was apparently a drunkard and she didn't want more children, and so this was a surgical abortion. And that's how WI-38 was born."

As before, Hayflick cultured the cells, using primarily fetal lung tissue, and then his technicians — working in tiny isolation rooms barely larger than phone booths, using mouth pipettes and Bunsen burners — began laying down the ampules in June 1962. This may sound like humdrum piecework, but Hayflick showed me a room in the basement of the Wistar whose ceiling had been pocked with shards of glass. The Wistar workers discovered, to their considerable chagrin, that if the ampules were not properly sealed, liquid nitrogen would leak into the small glass containers. That was no problem as long as they remained frozen, but as soon as these ill-sealed ampules were removed, Hayflick explained, the liquid nitrogen would instantly expand as a gas and "the ampule would explode within five to eight seconds." He added, "This ceiling has glass embedded in it from exploding ampules. I used to get blown up once a month." As a result, Hayflick insisted that he be the only staff member to remove ampules from the freezer. Those improperly sealed ampules had another consequence, one that would later haunt Hayflick: some of the cells became contaminated.

Outside science, the WI-38 human cell line is virtually unknown, its name as poetic as a brand of motor oil. Within the rarefied confines of cell biology, however, these cells — which could be frozen, shipped, shared, swapped, and distributed throughout the world, and used for endless experiments in both basic and applied research — revolutionized the creation of human vaccines and gave birth to the field that Hayflick likes to call biogerontology. "WI-38 is," Hayflick claimed, "the most highly characterized normal human cell in the world, and its biology is better known than any other normal human cell population in the world." The cells allowed the biology of aging to come in from the cold and take its place alongside other investigations of the internal workings of cell biology. And Hayflick won't let you forget that these cells practically ruined his career, his life, his family. But, as he likes to say, "We'll get to that later."

The availability of WI-38 triggered an incredible period of practical advances and unending controversies. When people began to complain in the early 1960s that they couldn't get Hayflick's cells to grow, he suspected that local variations in the mix of nutrients, or growth media, were causing the problem, and he ended up inventing a powdered medium that could be shipped anywhere in the world. He and several colleagues published the formula for the medium in *Nature* in 1964, and although he pointedly told me that he had received "not a single penny" for this discovery, it has evolved, with the biotechnology industry, into a business worth hundreds of millions of dollars a year. The Wistar Institute, meanwhile, tried to patent Hayflick's cells as a "self-replicating system" in the early 1960s, but there was no legal precedent for such a claim, and the application was turned down (in 1980 the United States Supreme Court ruled, in a landmark case, that living organisms could indeed be patented, but that decision came nearly two decades after WI-38). As early as 1960, Plotkin, Koprowski, and Hayflick had created the first polio vaccine using human fetal cells, and Plotkin showed that it was safe and immunogenic. While all this was unfolding, Hayflick also found time, in 1961, to discover the organism (*Mycoplasma pneumoniae*) that causes primary atypical pneumonia, the human disease commonly known as walking pneumonia. And as WI-38 became the object of almost cultish craving among biologists worldwide, the National Institutes of Health gave him a grant to produce, store, and distribute the cells to other scientists.

For the next decade or so, Hayflick was consumed with two tasks. One was supplying the cells to researchers. The other, more controversial, was his attempt to convince regulatory authorities at the NIH and the Food and Drug Administration that WI-38 would be a much safer starting material for the preparation of human vaccines than contaminated monkey cells. This is a particularly sensitive and complicated part of the story, rife with allegations of self-interest and conflict of interest, because several key officials at the federal Division of Biological Standards, which licensed vaccine preparations for general use, were at the same time busily creating their own rival vaccines as NIH researchers. In 1965, for example, the DBS rejected a rubella vaccine created by Stanley Plotkin of the Wistar using WI-38; instead, it approved a competing vaccine, made with monkey cells, by some of the very researchers who served in the DBS. That vaccine, introduced by Merck in 1969, caused persistent side effects, including fever and irritation at the site of vaccination. European regulators, meanwhile, agreed to license and use the Wistar vaccine because it was both safer and more effective, and American officials only later made the switch. "The side effects became so unpleasant," Hayflick said, "that Merck was pressured to replace its vaccine with Stanley's vaccine," which occurred in the 1970s. The DBS and its parent organization, the NIH, resisted the use of WI-38 for vaccine preparation for nearly ten years.

There has been no systematic effort to go back and see if these early vaccines prepared in animal tissue may have seeded the later development of disease, but recent research has suggested a link between polio vaccines given to schoolchildren in the 1950s and a form of cancer called mesothelioma, which might be triggered by SV40. "There was opposition within the FDA," Plotkin said in an interview, "and there was a prestigious figure in opposition, namely Albert Sabin, who was against the use of diploid cells, because he felt they must be cancerous, and there must be something wrong with them." Hayflick and Plotkin battled the NIH for ten years, which was good for vaccine recipients but not for Hayflick. He told a Senate committee in the early 1970s, for example, that the Division of Biological Standards should be moved from the NIH to the FDA because of inherent conflicts of interest (it was later moved). The battles gave him a profound sense of being underappreciated for his signal achievement, and they created almost an institutional enmity at the NIH for someone who complicated and denigrated the scientific lives of its researchers. The animosity, some contend, would later come back to haunt him.

But the Wistar researchers had the last laugh. "Up to this date," Hayflick told me in December 2000, "three quarters of a billion — with a *b* — people on this planet have received vaccines produced in my cell strain, WI-38, or its imitators . . . All of the rubella vaccine, the German measles vaccine, produced in the Western Hemisphere is produced in WI-38, and rubella vaccination is a prerequisite for school admission in every state in this country, so you can go figure out what the numbers are here. And I can tell you that I received phone calls from the White House during the Reagan administration and during the first Bush administration from staffers saying, 'We finally tracked you down. We understand that you are the one who developed the cell strain derived from an aborted human fetus, and we want to confirm something that we simply cannot believe is going on. And that is that vaccines have been produced in those cells and are available for sale in the United States.' And my reply was, 'Where the hell have you been for the past twenty years?'"

In what little free time remained amid all these battles, Hayflick became increasingly intrigued by the implications of these cells in aging.

Later, with a graduate student named Woodring Wright, he performed a clever experiment showing that what they called the replicometer — the mechanism that counted the number of cell doublings and triggered cell senescence after fifty or so rounds of replication — was almost certainly located in the cell nucleus. By the late 1970s, the phenomenon of replicative senescence even had a name; Macfarlane Burnet, the legendary Australian immunologist, was the first to refer to it as the Hayflick limit.

ᥜ

For all the elegance and influence of the Hayflick limit, Leonard Hayflick's cells carried him to the cutting edge of several less pleasant aspects of modern biology. Ever since the Wistar tried unsuccessfully to patent the cells in the mid-1960s, the issue of intellectual property dogged the cultures, as well as the related issue of who "owned" them. Biologists rarely considered these questions then, and in Hayflick's case the questions took on even sharper relief when he moved out to California and found himself living among the electrical engineers and computer pioneers in Silicon Valley, who translated lab research into profit-making private ventures. Hayflick was among the first biologists to ask "Why not us?" and was certainly the first to be severely punished for violating the norms of the tribe.

No account of the Hayflick limit is complete without at least a brief recitation of the legal wrangling and personal travail that consumed Hayflick's life in the wake of his groundbreaking scientific work in the early 1960s. It began around 1968, when he learned that the Wistar Institute was offering the WI-38 cell strain to a major European pharmaceutical company, along with instruction in their growth and consultation services, in exchange for "a large financial benefit" for the Wistar — a benefit, according to Hayflick, "that would accrue to the nucleus of people who came to the Wistar with Hilary [Koprowski]." This development — set against an almost cultural discrimination in the biology of that time, with virologists as the stars and cell biologists as the glorified technicians — fueled Hayflick's long-standing feeling that he hadn't received sufficient recognition; it left him feeling, he said, "like the concessionaire at the ballpark." He was never consulted about the business arrangement, although he claimed that a colleague at the Wistar showed him a memo describing the deal.

"I almost fell through the chair when I saw that," Hayflick told me, "because I had brought more attention to the Wistar Institute than anybody except Hilary up to that point, but the people he had brought from Lederle had permanent positions, and I did not. He planned to use the money generated by my efforts to benefit the people who had permanent positions, and it didn't make me very happy. I was becoming more and more bitter," he admitted. He began to investigate other job possibilities, and soon he accepted a tenured position at Stanford. His imminent move further soured relations at the Wistar; when his pending departure became known, Hayflick said, Wistar officials asked him to sign a document stating that he had to leave the WI-38 ampules behind because they belonged to the Wistar. This might well be considered one of biology's first Material Transfer Disagreements: now the very same cells that revolutionized the molecular study of aging had also become among the first human biological materials over which a serious and complicated tussle about intellectual property and title of ownership arose.

In the year 2000, the conditions set by the Wistar would be impossible; indeed, the entire biotech industry would collapse overnight if enterprising biologists were prohibited from transferring the fruits of their academic labors to the private sector for commercial development. But in 1968 the Wistar's demands played right into the complex mosaic of Leonard Hayflick's personality: fiercely proud of his research, stubborn by nature, he insisted that he had as much right to the cells as anyone else. And then he had what he calls his epiphany. "Why in the hell are we giving these cells to all these pharmaceutical companies, who are making billions of dollars?" Hayflick asked himself. "I can tell you that Merck was making three quarters of a billion dollars a year using human diploid cells as a substrate for vaccines," he said. "So if you do the math, and multiply that by the past forty years, and the number of other companies worldwide using those cells, you get some sense of the magnitude of the profit being made. And I began to realize that the spectacular dinner I got once a year from a pharmaceutical company was inadequate payment."

But in a perverse way, it was always about more than money. Hayflick began, literally, to think of the cells as his children. "They were my babies," he said at one point, recounting this episode. "I had my whole reputation invested in them. And I also reached the conclusion that I had as much right to their title as anyone else. I won't say that at that time — nor will I say now, actually — that I had any greater title to them than anybody else who might lay claim to them. But I certainly had a reasonable argument for *equal* claim. That was my position, and I think it was a rather modest position. It was, however, a unique position, because no one else in the entire field of biology that I know of had reached the conclusion at that time, that what they had discovered using federal funds could be remotely regarded as belonging to them." Before leaving for Stanford, Hayflick quietly made arrangements with the American Type Culture Collection to transfer the WI-38 cell lines; according to the agreement, Hayflick would retain ten ampules, Merck would receive ten ampules, and the remainder, hundreds of frozen vials of WI-38, would go to the ATCC, the national clearinghouse for the dissemination of cell lines. But at the last minute, Hayflick had a change of heart, deciding he wanted to hang on to the cells until the ownership issue was resolved. And so, when he and the three oldest of his now five children piled into the family's black Buick Century for the long cross-country drive to California in 1968, there in the backseat between two kids, secured with rope, was a large container of liquid nitrogen filled with a distinctly modern form of contraband: hundreds of ampules of WI-38. The *Philadelphia Evening-Bulletin,* a now defunct paper that Hayflick once delivered as a kid, later chronicled the transfer in a headline Hayflick recalls as, "Philadelphia Scientist Drove West with More Than His Luggage."

And ultimately, there were indeed headlines. Hayflick's improvisational cross-country riff on technology transfer did not cause conspicuous problems at first. He and his family moved into faculty housing on the Stanford campus; from his tenured position in the Department of Microbiology, Hayflick built a large lab that, to hear him tell it, brought in a cataract of federal research grants that all but drowned out the rest of the departmental budget. With twenty graduate students and postdoctoral researchers, his lab gravitated strongly toward the study of aging in cells. Hayflick was invited to biogerontology meetings, published lots of papers, and became something of a celebrity in the field. One of his closest colleagues and friends at Stanford was Alex Comfort, who was known to the world as the author of *The Joy of Sex* but who, within academia, was widely respected as a leading expert on the biology of aging.

But not everyone was enamored of Hayflick's science. "Len's talents lie in recognizing the importance of his observations and being extraordinarily skillful in proselytizing for replicative cell senescence as a model of aging well worth studying," said Woodring Wright, the University of Texas researcher who was a graduate student in Hayflick's lab from 1970 to 1975.

"But he was not well respected at Stanford, because he did not have a long history of mechanistic studies that would support that view . . . Hayflick has made some critically important contributions, but they were few and far between, and he didn't move the initial observations further along. After the initial observations, the field kind of passed him by." In fairness to Hayflick, he didn't move much beyond the initial observations for two reasons. First, the technical tools didn't exist to take the next steps in experimentation. And second, his academic and personal life blew up over his beloved cells.

Hayflick had continued to distribute WI-38 to other researchers, but with a twist. After the NIH stopped funding distribution of the cells, he and his wife, Ruth, formed a company in 1972 called Cell Associates, which, he says, adopted the same shipping and handling fees charged by the American Type Culture Collection. One of his regular customers, in fact, was the NIH. But he also started selling the cells to private industry. "It became apparent that Len was selling ampules to pharmaceutical companies," said Vincent Cristofalo, a Wistar researcher sent to California by Koprowski to try to recover some of the cells. "In those days, in that environment, when you did research with government support, it was in the public domain. When it came out that Len was selling these cells, a lot of people were appalled."

Not surprisingly, Hayflick was unrepentant. "I pretty much defied people to challenge my decision," he admitted. But the ownership of those cells was always an issue in the back of Hayflick's mind, and it came to the fore in 1975. The National Institutes of Health had decided to form a new research institute dedicated specifically to the biology of aging; Hayflick was interviewed to be the first director of the National Institute on Aging, he said, and, as testament to his rapid rise in this new field, was subsequently offered the job. During a meeting in Bethesda in the spring of 1975 with the acting NIH director, Ronald Lamont-Havers, Hayflick told me, he tentatively accepted the prestigious job, but with one caveat: he wanted the controversial WI-38 ownership issue resolved. "I *asked* for that investigation," he said. "I asked them to send a lawyer."

The acting NIH director promised to send someone out to Palo Alto to review the history of WI-38. When the man arrived in California, Hayflick claims, he gave him unrestricted access to all his files on WI-38 while he continued his laboratory work. The relationship was probably more adversarial than Hayflick suggests, for the visitor was not a lawyer; he was an auditor from the NIH's Office of Management Survey and Review, the group responsible for investigating grant fraud. "So he did his investigation and came to the conclusion that I had stolen government property and was selling it for personal gain," Hayflick said. On a night in October 1975, government investigators entered Hayflick's Stanford laboratory without his knowledge, confiscated both files and cells, and conveyed to Stanford officials the distinct sense that Hayflick was headed for monumental legal problems. The Santa Clara County district attorney's office launched a criminal investigation, and Hayflick scurried to find a criminal lawyer. Stanford not only refused to stand up for him, but Hayflick says the administration warned him he was about to be "showered with negative publicity." At the end of February 1976 he resigned in protest.

Soon he was in federal court, seeking an injunction to prevent the NIH from releasing details of the investigation to reporters until he had had a chance to respond to the government's allegations. The court refused, and within days details of the investigation began to appear in the press. One influential and damaging article appeared on the front page of the New York Times on March 28, 1976, detailing how Hayflick had earned profits of $67,482.33 by selling cells "that were the property of the Federal Government." Nicholas Wade, then a writer for the journal Science, obtained a copy of the NIH's investigation report through a Freedom of Information request and wrote a long, detailed, and damning article describing the allegations that Hayflick had stolen government property, personally profited from its sale, and even jeopardized the future of vaccine preparation because some of the WI-38 cell lines had become contaminated. "A personal tragedy which also raises problems of international diplomacy and public health has suddenly crashed down about the head of the distinguished biologist Leonard Hayflick," Wade's article began.

Once the story broke, Hayflick went from fame to infamy faster than the doubling time of one of his cells. Woodring Wright, who had finished his Ph.D. in Hayflick's lab at Stanford just before the scandal broke, speculates, "I think the reason he was a groundbreaker that got shafted is the particular way in which he did it. Among other things, it was his decision to resign. And then there was the fact that his science wasn't highly regarded." The head of the NIH investigation, James W. Schriver, was, according to press accounts, "widely respected for honesty and fairness." But many observers, even sometime antagonists like Paul Moorhead, Hayflick's former collaborator at Wistar, felt the NIH was settling an old score with Hayflick. "There was a cabal at the NIH that was definitely out to get him," Moorhead told me. When NIH investigators solicited help from Maurice Hilleman, the legendary vaccine researcher at Merck Research Laboratories, he refused to cooperate. "I was asked to be a principal witness against him, and I said that if there was an intent to convict him, I would make a campaign on my part that two top-level government officials would spend time in jail with him," Hilleman said in an interview. "He should have been celebrated as a scientific hero instead of being persecuted." Perhaps his greatest sin, as David Kritchevsky put it, was that "he was born too soon." In today's feverishly free-market biology, Hayflick's "crime" looks like a tempest in a dollhouse teapot.

The ramifications of the controversy were anything but small for the Hayflick family, however. "I went from full professor at Stanford to the unemployment line in one week," Hayflick said, his voice brittle and remote. "My wife and I lived on $104 a week for the next year." Newspapers were full of damning headlines. His lab group dissolved; his family was forced to abandon its faculty housing; he spent an entire year unemployed, before landing a marginal appointment at Children's Hospital Medical Center in Oakland. Even when he managed to get subsequent grant requests funded, federal authorities seemingly sabotaged the payments. At one point, the entire membership of the Advisory Council of the National Institute on Aging refused to conduct any business, including grant review, until NIH officials explained why a $562,000 grant awarded to Hayflick had not been paid out, after two years of inexplicable delay. But for all the heartbreak the case brought, Hayflick never abandoned his principles — or his stubbornness. After taking a position at the University of Florida in 1982, he immediately alienated university officials by refusing to sign a document in which he would agree to turn over all current and future fruits of his research to the university.

Hayflick has butted heads with plenty of people during his career, and not everyone has viewed his troubles with sympathy. "He's enjoyed playing the martyr," said Moorhead, his former collaborator, "but he's done very well." Stanley Plotkin, who worked closely with Hayflick at the Wistar in the 1960s, told Science in 1976, "I think that in the really classical Greek sense it was a tragedy, because it is a man who at the height of his powers brought about his own downfall." (When I asked Plotkin recently if he still felt that way, he declined to comment.) The personal and professional cost has been enormous, and Hayflick emerged from the debacle with "a very strong sense of bitterness," said Wright, his former graduate student. Indeed, Hayflick began publishing a series of personal, self-exonerating papers with titles like "A Novel Technique for Transforming the Theft of Mortal Human Cells into Praiseworthy Federal Policy."

What ultimately saved Hayflick, what ultimately rehabilitated both his reputation and his career, was those news articles. To William A. Fenwick, a lawyer in Palo Alto, the public dissemination of private NIH investigative information identifying Hayflick by name, before he had even had a chance to respond, appeared to represent a deliciously glaring violation of a recently enacted law, the Privacy Act of 1974, which prohibited the government from disclosing private information about citizens. Fenwick had helped create the legislation, and so Fenwick & West agreed to represent Hayflick on contingency and sued both the NIH and what was then known as the Department of Health, Education, and Welfare for violating the privacy act; Hayflick's lawyers filed a second suit against the government, seeking title to the cells and the ability to profit from their sale. Several months later, the government countersued. In dispute was an amount of money that, even in 1976 dollars, was kindling compared with the bankrolls of today's bio-moguls — roughly $67,000, which sum Hayflick immediately put into escrow (and later recovered, with interest).

To make a long legal story short, after nearly six years Hayflick's lawyers reached an out-of-court settlement in September 1981 in which the government conceded Hayflick's ownership of the cells and agreed that the money he'd earned from the sale of the cells (plus accrued interest) belonged to him. Dozens of prominent scientists, including Lewis Thomas and Alex Comfort, sent a letter of support to the journal *Science,* reporting news of the settlement (which the journal had neglected to do) and noting that "the effects of the negative publicity Hayflick received probably can never be fully reversed, because it is most unusual for the media to give the same coverage to the exoneration of individuals as they do to unsubstantiated but sensational allegations." William Fenwick told me the physical toll of the litigation on the Hayflick family was enormous. "I think he aged about ten years during the time I worked with him," Fenwick said. "I was worried because I thought he was going to die, and that his wife wasn't going to be too far behind. The stress was just horrendous."

By allowing Hayflick to keep a portion of the cells, and sell them as he wished, the government settlement arguably established the precedent for a biologist to transfer biological materials of his or her creation to the private sector, and for this reason, Hayflick today insists — perhaps a little too strenuously but not entirely without justification — that he bulldozed the legal landscape regarding intellectual property in biology to smooth the path for biotechnology. "Had it not been for my lawsuit," he says now, "the biotechnology industry would not be what it is today." If not quite the father of modern biotechnology, he might well be considered its St. Sebastian. But the long battle also had the effect of undermining Hayflick's scientific legacy. Because WI-38's status was so contested, vaccine makers increasingly turned to MRC-5, a human fetal–derived cell line developed by British researchers, to prepare their vaccines, even though Hayflick was, in Hilleman's estimation, "the real pioneer."

Hayflick's reputation — and finances — began to recover in the 1980s. He began to consult for the burgeoning biotechnology industry — an industry founded upon, and whose IPOs are energized by, the transfer of technology and materials not unlike WI-38 from academic labs to the private sector. His work as a graduate student at Penn on an obscure clot-dissolving enzyme known as streptokinase, as well as his knowledge of cell culture, led to a steady consulting position with Genentech in the mid-1980s, when the company was preparing its clot-busting drug TPA for market approval.

Perhaps more significantly, the impact of the Hayflick limit on aging research slowly began to transform the field of gerontology. Molecular biology and aging had begun to converge. The Hayflick limit forced researchers to think about what could account for cell senescence. Was there something inside the cell, in the chromosomes, that played a part? That was precisely the question that a young Russian researcher asked himself one night, sitting in a Moscow subway station, after hearing a lecture about Leonard Hayflick's work.

2

"A Circle
Has No Ends"

In the fall of 1966, a young postdoctoral biology fellow named Alexey M. Olovnikov found himself sitting on a bench, running an imaginary train over an imaginary track made out of the double helix of DNA. He had just attended a lecture at Moscow University given by Alexander J. Friedenstein, a well-known Russian cell biologist. In a vivid example of the scientific lecture as a form of intellectual pollination, Friedenstein had spoken at length about some exciting new research recently published by one Leonard Hayflick of the United States, showing that normal human cells have a limited capacity for replication. It was not the first time that Olovnikov had heard of Hayflick and his limit. Colleagues at work had mentioned it before, but he hadn't really paid much attention. This time, for some reason, Olovnikov was "simply thunder-struck by the novelty and beauty of the Hayflick Limit," as he later wrote, in a historical reminiscence both scientifically poignant and literarily florid. "I thought about this as I returned home from the University and walked along the quiet Moscow streets that were paved with gold-covered leaves on that early evening in late Fall as I made my way to the subway station. The Theory of Marginotomy came to me in that Moscow subway station."

There is a poetic symmetry (more poetic, alas, than Olovnikov's English) to the fact that the scientific insight of a lifetime came to him via railway tracks. At the time of the Friedenstein lecture, he had just turned thirty and was about to take a great scientific leap — a leap, however, largely obscured by the fog of the cold war. The Cuban missile crisis had occurred just a few years earlier, and Soviet-American relations were at their chilliest. Because he didn't belong to the Communist party, Olovnikov was forbidden to travel abroad for scientific meetings; indeed, the director of

the Gamaleya Institute, where Olovnikov conducted immunological research, was rumored to be a general or colonel of the KGB. Olovnikov was
very lucky even to be sitting in that subway station, daydreaming about
trains and tracks. Shortly after his birth in Vladivostok in 1936, his father, a
journalist, narrowly escaped imprisonment and almost certain death during Stalin's purges. While on a trip organized for journalists to the Soviet-
Chinese border in late 1936 or 1937, Alexey told me, his father "somehow
contrived to read a kind of hidden threat" in the faces of Soviet officials accompanying the journalists. Within an hour of his return to Vladivostok,
Olovnikov's father had hustled his wife out of their home, with neither luggage nor possessions ("As though they simply decided to go for a walk at
sunset," Alexey said), led her to the train station, and boarded the next train
to Moscow, never to return. The only item the couple brought with them
was their infant son, Alexey, bundled up in a blanket. The family later
learned that virtually every other journalist who had gone on the trip to the
border was arrested as a "Japanese spy" and sent to the gulags.

The ideas that came to Olovnikov that night in the subway station,
and which he elaborated and refined over the next five years, brilliantly knit
together the process of cellular aging, the immortality of cancer cells, and
the biological mechanism that could explain both. Olovnikov took Hayflick's notion of cellular mortality and reduced it — theoretically, of course
— to the movement of molecules through an infinitesimally small three-
dimensional space. He imagined the choreography of DNA and enzymes as
they went about their minute-by-minute business of cellular replication.
Marginotomy, as the word implies, refers to the biology of margins, of a
particular kind of biology at the ends of things. In this case, Olovnikov was
thinking of the biology at the ends of chromosomes. This is a genetic wilderness that has fascinated scientists intensely for more than half a century.

In each of our 100 trillion or so somatic cells (that is, the cells that
make up our bodily tissues, as opposed to the germ-line cells of egg and
sperm), there are 46 chromosomes (23 pairs), each one composed of tightly
spun spools of DNA. The double helix is often described as a kind of gossa-
mer thread of life, with genes arrayed along the thread, but in its natural
habitat, the chromosome, the thread gets wrapped around pebbly little bits
of protein called histone, so it's much more like small balls of twine, one after another, packed together in a bumpy amalgam. When viewed under the
microscope, the chromosomes can be arranged in pairs that look like

stubby, sausagelike X's, with a tiny knot of genetic material in the middle (the centromere) and then a short ("petite," or p) arm and a longer (q) arm. Now consider the implications if the Hayflick limit applies to the cells in your body as well as to cells grown in a dish; all those trillions of cells that make up your self are destined to replicate, or divide, no more than fifty or so times in your lifetime, and that's it. What, Olovnikov wondered, could possibly keep the cells from dividing more and renewing themselves after hitting Hayflick's wall? And why did cancer cells never hit that wall?

So he tried to imagine things from the point of view of a molecule, and he speculated that the answer lay at the very ends of the chromosomes — a region known as telomeres. He was hardly wandering into virgin territory here; two exemplary twentieth-century biologists, Barbara McClintock and Hermann Muller, had pointed to the importance of the chromosome ends decades earlier (Muller coined the term "telomere" in 1939), but they didn't possess the knowledge (or research tools) of a molecular biologist in the 1960s. Olovnikov knew that during cell division, the double helix of each chromosome must magically unravel and disentangle itself for a short time, as if the DNA, loosely speaking, let down its hair; this unspooling allows a Xerox machine of an enzyme, known as DNA polymerase, to run along each single strand of the double helix much the way a monorail runs along a track, copying the DNA underneath it as it goes along. By the time the parent cell heaves and contorts and tears itself into two daughter cells (a process called mitosis), a fresh, intact, newly copied double helix has already migrated into each daughter cell, providing a full complement of genes. And life goes on. Every time a bacterium replicates, or a protozoan, or a cancer cell, or simply the skin cells in the trough of a dimple, the DNA — a long string composed of the four chemical bases adenine, cytosine, thymine, and guanine, usually rendered as the letters A, C, T, and G — is dutifully copied in exactly this fashion before the cells divide. Without this process, there would be no growth, no development, no life. But this process also comes with a biological price — one that Olovnikov believed was quite relevant to aging and disease.

In bacteria, which can reproduce indefinitely, the chromosomes are circular, and as Olovnikov has noted, "A circle has no ends," so nothing can be lost at the end. But in all higher organisms, including humans, the chromosomes are linear, like sticks. And here, at the very tips of these chromosomal stalks, size matters. Olovnikov saw that the ends of our chromosomes can never be completely copied by the DNA-copying enzyme as it performs its duties; indeed he was the first biologist to recognize the importance of this seemingly arcane fact, known as the "end-replication problem." He realized that every time a cell divides, a little more of the end of the chromosome fails to get copied, and a little more of its DNA gets nibbled away. Why this is, and how it happens, dawned on Olovnikov as he sat waiting for the subway train after hearing about Hayflick's work.

"I heard the deep roar of an approaching train coming out from the tunnel into the station itself," he recalled. "I imagined the DNA polymerase to be the train moving along the tunnel that I imagined to be the DNA molecule." In a spectacular leap of physical imagination, he also realized that if the track represented DNA, and the subway train represented the enzyme, the locomotive of the train would represent the front, "locomotive" segment of the enzyme, which pulled the copying machinery of the molecule behind it like a railway car but was itself incapable of copying the DNA directly underneath it. In other words, every time a cell doubled, the very end of the chromosome — the part that lay beneath the "engine" of the enzyme — could not be copied and would therefore be lost. After fifty or so cell doublings, Olovnikov reasoned, the end of every chromosome would be chewed back quite a bit, to a point where it might begin to affect the stability of the chromosomes and therefore the capacity of the cell to replicate at all. "After the loss of a critical portion of the telomeric DNA," Olovnikov wrote, "the cells will change their normal, young phenotype to an old phenotype." Olovnikov argued that shortened telomeres were the cell's version of gray hair, an indisputable sign of old age.

Galvanized by this "serendipitous underground brainstorm," Olovnikov fired off a letter to Hayflick, sharing this important news, requesting additional details from the master. Hayflick, still at the Wistar Institute but increasingly embroiled in the politics of his controversial cell line, was frankly befuddled by the theory of marginotomy. "I didn't know what the hell he was talking about," he admitted. But Hayflick sent some more unpublished data to Olovnikov, and Olovnikov continued to mull the problem.

For all Olovnikov's ingenuity and imagination, his theoretical speculations proved another, more painful truth about scientific pollination — communication may promote the formation of new thoughts and theories, but the seeds of those new ideas will never flower if they fail to land in hospitable soil. In 1971, Olovnikov published his "theory of marginotomy" in a Soviet journal called *Doklady,* the equivalent of our *Proceedings of the National Academy of Sciences.* But this was a marketplace of ideas almost barren of foreign shoppers; scientists outside the Soviet Union paid virtually no attention to it. As Olovnikov has repeatedly reminded the world of biology ever since, his paper did not merely propose a biological mechanism that could explain both the Hayflick limit and the way cells grew old and senesced. It went much further. It tackled the paradoxical problem represented by germ-line cells (that is, sperm and egg cells) and cancer cells and later by embryonic stem cells. These cells, unlike all other cells in the body, are effectively immortal. There is no limit to their capacity to copy themselves; something in their biology grants them an exemption from the Hayflick limit. Olovnikov speculated that these cells achieved this exemption with the help of a special enzyme — one that helped repair and maintain the ends of telomeres. In its entirety, the theory of marginotomy was a remarkably audacious venture into theoretical biology; even more remarkable, it turns out to have been largely correct. But hardly anyone outside the Soviet Union seems to have been aware of it, and Olovnikov's few subsequent chances to publicize the theory were dashed by an unremitting string of bad luck.

The most auspicious opportunity undoubtedly occurred in the summer of 1972, when the International Congress of Gerontology held its annual meeting in Kiev. None other than Leonard Hayflick himself, by now a major figure in the world of gerontology, planned to attend. Here was Olovnikov's first opportunity to meet the famous Hayflick, his grandest opportunity to unveil the theory of marginotomy to Western scientists. However, merely getting to Kiev was not easy. The director of Olovnikov's institute in Moscow refused to permit his scientists to attend any international meetings, even one in the nearby Ukraine. Olovnikov refused to be deterred. He contrived to take his vacation during the week of the meeting and just happened to spend it in Kiev. But cold war hostilities managed to intrude even further.

Planning to attend that same meeting, Hayflick's good friend Zhores Medvedev, the prominent Russian biochemist, gerontologist, and dissident, "vacationed" under similar circumstances in Kiev. He met briefly with Hayflick before the start of the gerontology meeting and then suddenly disappeared. The story quickly circulated among attending scientists that

Medvedev had been kidnapped by Soviet authorities; the disappearance prompted coverage in the Western press. As it turned out, Medvedev had been followed, detained by officials, and then forcibly escorted back to Moscow, where he was based. But his abrupt disappearance, and the uncertainty about his fate, dominated affairs at the gerontology meeting, to the point that Hayflick devoted nearly all his time to locating Medvedev — and none to marginotomy. Olovnikov was as shocked by the kidnapping as anyone; like many Soviet citizens, he had read a typewritten, secretly shared samizdat version of Medvedev's outlawed book about Lysenko, the Soviet official who had corrupted science in the name of Communism. But despite these concerns, Olovnikov felt crushed. Although Hayflick sat in the front row when he finally gave his talk, hardly anyone paid attention to marginotomy because of the Medvedev crisis. When Olovnikov attempted to buttonhole Hayflick and solicit his opinion, the American scientist begged off. "Later," Olovnikov wrote, "I learned that he did not speak to me at length because he was heavily involved in attempts to learn the fate of his kidnapped friend, Zhores Medvedev . . ."

To the surprise of many Soviet friends, however, Olovnikov later received approval to submit a paper on his theory of marginotomy to a Western journal; it appeared in English in 1973 in the *Journal of Theoretical Biology*. But in a final dollop of cruel fate, Olovnikov now found himself with very stiff competition for scientific attention. The American Nobel laureate James D. Watson, codiscoverer of the structure of DNA, had published similar speculations on end replication in 1972, and of course the international scientific community held its breath when Watson read from a menu, much less published an important scientific speculation. It is heartbreaking to read Olovnikov, in his broken and desperate English, trying to establish the priority of his work. "I was glad that I, a young researcher, was ahead of the Nobel Laureate in several central positions," he wrote to me. "But this was only part of my emotions. Though, I never, not for a minute, hesitates about that my ideas, but it was very pleasant that I am not alone on this way. And in such good company! But even this is not a whole truth. As I already wrote in Experimental Gerontology, after the invention of that ideas in 1966, I several years (!) did not published them, rethinking them until 1971. Here is why I have had also another kind of thoughts: I could be late, would I further delayed with publications."

Olovnikov's paper, and theory, languished in relative obscurity for twenty years. It fell to a collection of brave and resourceful women scientists, working at the margins of molecular biology in the 1980s, to bring telomeres front and center as an area of tremendous scientific importance — leading very directly to the formation of the first biotechnology company dedicated solely to the notion of treating aging through enlightened molecular biology.

⚘

The telomere story, which will probably result in a Nobel Prize for its most distinguished protagonists, is a classic tale of modern biological inquiry. Several dedicated and incorruptible intellects toil in what appears to be, at least to the uninformed, a hopelessly abstract and, in this case, literally marginal precinct of biology. They study obscure organisms, like single-celled protozoans and yeast, that have no apparent relevance whatsoever to human biology or health. Then the scientists make a stunning observation that unsettles every corner of biology as surely as shaking a tablecloth unsettles all the china. Soon, by land, sea, and air, the field is invaded with researchers scrambling for a piece of the action, pursuing their intellectual curiosity with all the decorum and dignity of those nineteenth-century gentlemen geologists who pursued their curiosity about rumors of gold in California. Federal research monies pour in, venture capitalists begin to sniff something interesting afoot, companies begin to form, and, in an attempt to keep those fledgling companies in the public's awareness, giddy claims are floated about dramatic medical benefits just around the corner. Inevitably, these claims come with a tag line that is as indispensable to keeping a popular account of basic science aloft and in the public eye as is the tail to a kite, a line that invariably reads something like ". . . and might even lead to a cure for cancer." All that, and much more, unfolded in the telomere field between 1985 and 2000. Perhaps the field's most important legacy, however, has been linguistic, or at least metaphoric: telomere research reintroduced the term *immortality* to serious scientific discourse, and did so in such a way that the word became the coin of the commercial as well as the scientific realm.

There are several heroines in the telomere story, but none more central than a cheerful yet steely intellect named Elizabeth Blackburn. A native of Australia, Blackburn is a stately woman who nonetheless possesses an almost sheepish, schoolgirl repertoire of nervous laughs, modest shrugs, and circuitous verbal assertions. When I visited her several years ago in her laboratory at the University of California at San Francisco, I was struck by a kind of gentle, principled tenacity with which she explained the science and sociology of a field that had changed so dramatically, largely because of work that originated in her labs, first at Berkeley and then at San Francisco. She wore a sandy, earth-toned dress, sensible shoes, and white stockings; her fine crown of blond hair was tinged with gray; and her rimless glasses gave her a wise, grandmotherly cast. J. Michael Bishop, the chancellor of UCSF and a Nobel laureate, spoke admiringly of the spirited style she brings to her science. "She uses her intuition," he said, "but she's very meticulous and very rigorous. It's a great combination. And on top of that, she's very self-effacing." That was rare indeed in such a high-powered field.

Blackburn grew up in Launceston, a small town in Tasmania, the island southeast of the Australian mainland, where her parents were both physicians. One of seven children, she manifested an early (and fearless) love of the natural world, picking up poisonous jellyfish and serenading stinging ants. She considered pursuing music as a career because she loved playing the piano, but she ultimately gravitated toward science. At a time when many teenage girls had posters of the Beatles on their bedroom walls, Blackburn had a poster listing amino acids. "I loved the names and structures of amino acids and proteins," she once told an interviewer. She ended up pursuing biochemistry at the University of Melbourne. In 1971 she went to England to do graduate work at Cambridge with Frederick Sanger, the only biologist in the last hundred years to receive two Nobel Prizes for science, and then, in 1975, came to the United States to do her postdoctoral work with Joseph Gall, a prominent cell biologist at Yale University. It was Gall who introduced her to pond scum.

To be more precise, he introduced her to a single-celled creature that thrived in pond scum, a colorless flagellate called *Tetrahymena*. In the late 1970s, when Blackburn was working in New Haven, biologists were just learning to read the ABCs of DNA. Sanger, for example, had invented a powerful technique for reading the chemical letters, or sequence of bases, of DNA, which could unravel the secret messages of individual genes. Biologists all over the world scrambled to read any bit of DNA sequence. Nowadays there are machines that do this automatically, the same machines that allowed biologists to complete the sequence of the human genome to such fanfare in 2000. But in the 1970s, DNA sequencing required a hugely laborious biochemical procedure. Blackburn became interested in looking at the DNA sequences at the very ends of chromosomes for that most compelling of scientific reasons — "Technically," she told me, "it looked like it was feasible." In other words, it was doable. No one imagined it would have anything to do with aging, cancer, or cellular immortality.

Tetrahymena provided such a good starting material because, for some reason, the creature possessed thousands upon thousands of tiny chromosomes. That meant there were thousands of chromosome ends to work with. And to the few people paying attention, the ends of chromosomes were, as Blackburn put it, "behaving very strangely." For one thing, she and Gall discovered that the chromosome tips repeated the same short sequence again and again, the DNA equivalent of a broken record. The tips were also of variable length. It wasn't clear why, but it increasingly seemed that length and repetitive sequence were important. A few years later, after she had moved to Berkeley and set up her own lab at the University of California, it began to dawn on Blackburn that these short stretches of repetitive DNA at the ends of chromosomes somehow governed a very basic rule of grammar in the book of life: the very stability of chromosomes. Every living creature with a nucleus — which is to say, every creature other than bacteria, including us — obeyed this ancient rule of genetic grammar. Violate it and you were thrown out of evolution's most fundamental classroom: life itself. There seemed to be information, or structure, in the telomeres that preserved the integrity of chromosomes and kept them from unraveling, or falling apart, or sticking to each other in what would be a fatal nucleic embrace.

In fact, Blackburn and her colleagues noticed in some experiments that when telomeres from *Tetrahymena* were stitched onto the shorter telomeres of yeast cells, the yeast chromosomes suddenly stabilized instead of falling apart. The addition of DNA at the end of telomeres provided stability. Again, size mattered at the ends of chromosomes. And then the group noticed that with no help from scientists, chromosomes could sometimes add little bits of DNA at their ends. In biology, adding something requires some sort of glue, and for a cell biologist like Blackburn, that glue had to be a kind of enzyme, whose function was to monitor and groom and maintain the length of the tips of chromosomes. At the time, Blackburn was unaware of Olovnikov's theory — as was nearly everyone else, except Leonard Hayflick and a few others — but she, too, was becoming a disciple of marginotomy. Like Olovnikov, she gravitated toward the belief that a very special enzyme was at work here. Unlike Olovnikov, she was an experimentalist, and she suddenly found herself in a position to do something about that hypothesis. "At a personal level," she explained, "I got tenure and decided I could get brave. And so we'd just go and look for telomerase."

When the hunt began, telomerase (teh-LOM-er-aize) was a unicorn of an enzyme — theorized to exist by the footprints it left at the tips of chromosomes, but purely hypothetical. What's more, the world was not exactly awaiting its discovery with bated breath. The ends of chromosomes still seemed like a biological wilderness compared with other, hotter areas of molecular biology, and few researchers understood the potential significance such a discovery might hold. Fortunately for Blackburn, she was joined in this seemingly quixotic search by a graduate student named Carol Greider. Feisty, enthusiastic, smart, and with the involuntary strong-mindedness that comes from losing a parent (her mother) at the age of six, Greider grew up in nearby Davis, where her father was a physicist at the University of California. She graduated from UC Santa Barbara in 1983 and decided to join Blackburn's laboratory for graduate school after a single interview. "We just had a great conversation," Greider told me. "People click, but scientific styles click, too. She was very logical and very broad-thinking, and wasn't just looking at small, incremental-type questions. She wasn't afraid of proposing [an idea] that people wouldn't like, so she also has a certain amount of intellectual courage." Those, in short, were the founding principles of the telomere field.

Blackburn was similarly enthusiastic about Greider. "Carol joined the lab as a student," she recalled, "and was keen to do this — a brave, great student to do this. Because most people would say, 'Well, you know, you want me to look for an enzyme that you just *think* might exist?' But she was game." One of the reasons Greider was game was that *Tetrahymena*, this frisky single-celled swimmer, essentially lived forever. "It has plenty of telomerase," Blackburn said, "and it keeps just dividing and dividing, essentially immortally. Even when it's not making new telomeres, it just maintains them. And because it has a lot of telomeres, it has a pretty good level of this activity." So, just as you'd go to Newcastle to look for coals, Blackburn and Greider went to *Tetrahymena* to stalk this enzyme.

Not many people believed the enzyme existed. "At that time, there were two models," Greider recalled. "One was that there was a recombination-based mechanism for maintaining chromosome ends, because it was known that there was a problem with losing sequences from the end." Recombination refers to the way bits of DNA can occasionally be "given" by one chromosome to another. "And Liz Blackburn had, within a year or so of when I arrived, proposed an alternative model" involving telomerase, Greider continued. "Of course, most people sort of favored the recombination-based model because it was a known mechanism. It was pretty clear how that might occur. But being a young, naive graduate student, I decided to take on the other project! It seemed like it would be fun. And Liz and I hit it off really well, so we would bandy ideas about. So I just started grinding up cells and looking for biochemical evidence that there was such a thing as an enzyme that would add things onto the ends of chromosomes." In what she called "kind of a shot-in-the-dark experiment," Greider developed an assay — a lab test — that would reveal the presence of this enzyme, and had incredible good fortune. "The discovery was actually made on the very first gel I developed," she said, "on Christmas Day of 1984."

After spending nearly a year confirming the result, Blackburn and Greider published their findings in 1985 in the journal *Cell*. Even then, the discovery didn't exactly create a sensation. "I think a lot of people thought it was something specific to *Tetrahymena*," Greider said. It would take another twelve years before researchers would find the human gene for what ultimately would become known as the "immortalizing enzyme," but that *Cell* paper confirmed the existence of telomerase and opened the door to one of the most productive areas of research in modern cell biology. It also confirmed Olovnikov's original hypothesis, but his train still didn't arrive in the station, because Blackburn and Greider failed to mention his earlier theoretical work in their paper. "We weren't slighting him," Greider told me. "We just didn't *know* about it." In fact, she later went back to compare the number of times those early papers had been cited by other scientists. Between 1972 and 1990, she found, Watson's end-replication paper was cited 225 times. Olovnikov's "marginotomy" hypothesis was cited exactly 4.

༄

While Greider pawed through millions of cells, telomere research began to assume greater visibility because it suddenly looked as though it might have some medical relevance after all. In 1985, Howard Cooke, a biologist working at Britain's Medical Research Council laboratory in Edinburgh,

Scotland, put telomeres and aging in the same biological room, as it were. In a talk at the Cold Spring Harbor Laboratory symposium that summer, Cooke reported that the only normal human cells considered to be immortal — that is, sperm and egg cells — possessed longer telomeres than did somatic cells. As one symposium participant later put it, "Howard Cooke was the first to see these changes in lengths, in sperm and egg and in blood cells, and he speculated — and he was completely right — the whole story, everything about aging."

By this point, the research all funneled toward a familiar bottleneck in biology during the 1980s: finding the precise DNA sequence that marked the telomere to see if the sequence offered any clues as to why telomeres were so important. There were many efforts in the late 1980s to identify the sequence of telomeres, and for some reason this area of research, this field of "marginotomy," attracted a group of exceptionally talented women biologists. Greider attributes part of that to Blackburn's mentor at Yale, Joseph Gall, who "had a track record of being really supportive of women scientists very early on, when it wasn't politically correct." Besides Blackburn, two other prominent telomere researchers, Virginia Zakian of Princeton and Mary Lou Pardue of MIT, came out of the Gall lab. Greider entered the field by way of Blackburn's lab, of course, and these women were soon joined by Victoria Lundblad of Baylor University, who did a postdoc with Blackburn in which she studied the telomeres of yeast, and Titia de Lange, a Dutch-born researcher who began to look at the role of telomeres in cancer cells.

De Lange recalled the early days of the telomere field with great fondness, and a little remorse, when I visited her in her spare office at Rockefeller University in New York. With short dark curly hair, blue jeans, and a droll sense of humor, she projected a kind of hip urban sensibility, yet she also cultivated a serene, old-world atmosphere in her workplace, too; a beautiful oil painting hung above her desk, a still life aglow with a surface of burnished varnish, a painterly celebration of several bulbs of garlic. The scientists who studied telomeres, she said, had the familiar intimacy of a sect. "The telomere field was a quite naive, very small and friendly field, maybe fifteen or twenty people who loved telomeres," de Lange told me. "We had the greatest time with each other. We talked about our results *years* before we published. It was science the way it ought to be." It would stay that way for a while, but then the research simply got too interesting, and the academic and commercial stakes too high, for that familial atmosphere to survive.

What made it interesting was the continuing link between telomere length and cell senescence. Vicki Lundblad, for example, had created a fascinatingly ill-starred yeast strain in 1986 while a graduate student at Harvard. The cells possessed a devastating mutation — they couldn't maintain the length of their chromosomes. Saddled with short telomeres, these yeast cells tumbled into premature senescence and ultimately died. Lundblad dubbed the phenomenon Ever Shorter Telomeres, and christened this first benighted mutant EST-1. That suggested that a gene was related to the phenomenon of short telomeres and senescence. As Lundblad later said, "The field is rich in hypotheses, but very poor in genes." Now the telomere field had at least one gene in hand to study.

Back in 1988, a small army of researchers at no fewer than five separate labs simultaneously discovered just what the telomere looked like — at least in the syntax of human DNA. Each telomere, each chromosome tip in each cell of our body, turns out to be a kind of genetic stutter; it is a repetitive strip of DNA, the same six letters — TTAGGG — repeated again and again and again, sometimes as many as 2,000 times. "It's really the world's most boring sequence," de Lange told me with a laugh. "I also happen to think it's the most exciting part of our genome." To step ahead in the story, researchers ultimately discovered that the "boring" six-letter sequence formed a kind of repetitive physical motif that induced special proteins to coat the ends of chromosomes, much like the plastic caps that prevent the ends of shoelaces from unraveling. What's more, the telomeres seemed to function as a kind of buffer zone at the chromosome ends, an extended string of DNA padding that held the chromosome together and could be chopped back with each round of cell replication without cutting into the vital part of the DNA that contained crucial genetic information. By 1990, de Lange, working in the lab of Harold Varmus at UCSF, had also published work suggesting that telomere length could limit the life span of somatic cells. All these experiments, in yeast and *Tetrahymena* and even in humans, swirled around the notion — suspected, but by no means proven — that telomeres had something to do with the Hayflick limit, and by extension with aging.

"Telomeres are a wonderful example of what to fund," said de Lange. "Elizabeth Blackburn became interested in the ends of chromosomes in

Tetrahymena, which doesn't harm a lot of people. The telomere was just an entity of the chromosome. Nothing in the research was even remotely connected to human disease. Even when she found telomerase with Carol Greider in the 1980s, there was no way that discovery pertained to human disease. That little investment the NIH made in a small grant to a beginning investigator at Berkeley has had a huge impact." When de Lange moved to Rockefeller University in 1990, she recalled, "people here had never heard of telomeres, didn't know what I was doing and, more important, why I was even doing it." The age of anonymity was about to end with a bombshell experiment.

⌇

In the late 1980s there was a kind of distaff diaspora from UCSF among telomere researchers. De Lange went to Rockefeller, Lundblad went to Baylor, and in 1988, Carol Greider got her Ph.D. from Berkeley and moved to Cold Spring Harbor Laboratory on Long Island to continue her research on telomeres. But Greider spent an inordinate amount of time in Ontario. "My boyfriend was up there," she explained, "so I would go up and visit him at McMaster University." Indeed, she remembers traveling to Canada at Christmas in 1984, clutching the first gel that proved the existence of telomerase to show her boyfriend, Bruce Futcher.

Futcher's laboratory space was "one big room," she recalled, shared with another biologist, named Calvin Harley. "They had labs next to each other, so that's how I met Cal," Greider said. She began to sit in on the Harley lab's group meetings, give talks, swap ideas. Within a few months, Greider, Futcher, and Harley began collaborating on the next generation of telomere experiments — experiments that would nudge the work ever closer to aging research, and toward confirmation of Olovnikov's theory.

With his long, austere face and high forehead, Harley could easily be the model for one of Grant Wood's midwestern ascetics; he possessed the gentle, patient, dutiful manner of a clergyman's son. Among all the established figures in the telomere field, Harley seemed most enamored of the notion that telomeres had something to do with senescence and, possibly, human aging; indeed, he seems to have been the only scientist in the field to have been aware of Olovnikov's theory, much less taken it seriously. He was by all accounts an excellent and exacting scientist, and he and Greider, flung together by happenstance, launched a very productive collaboration.

"Cal was the one who told me about this paper of Olovnikov's that suggested that the Hayflick limit might have something to do with chromosome ends," Greider said. "But it wasn't possible at that time to do anything in human cells because nothing was known about the sequence of human telomeres or the structure of human telomeres, if they looked anything like *Tetrahymena* telomeres." Every six months or so, Greider and Harley would talk about doing experiments that might prove the link, but there was never quite enough information to take it beyond talk.

Now, telomerase, at least the version found in *Tetrahymena*, was turning out to be one of the weirdest molecules around. Part of it behaved like what is called a reverse transcriptase; this is the kind of enzyme typically used by viruses — the AIDS virus, to name the most conspicuous example — to copy ribonucleic acid (RNA) back into DNA. The other part of it was a short sequence of RNA itself, a portable template for making a short sequence of DNA. In total, the enzyme might be likened to a kind of roving tool-and-die machine containing both the mold of a physical widget and the machinery to stamp out one copy after another — copies, in this case, of DNA padding that could be stitched by the enzyme machine onto the ends of chromosomes. And it was an extremely powerful protein. Indeed, researchers later discovered that nature kept telomerase's two component genes on two separate chromosomes in humans, somewhat like the two-key requirement for nuclear weapon activation. In fact, the genes were almost always turned off in normal human cells.

By 1988, a number of laboratories had identified that repeating sequence, the stutter of DNA, at the ends of human chromosomes. Greider and Harley finally had enough to go on. "It was pretty clear what the sequence was, so I started to talk to Cal more, because now we could do an actual experiment on this," Greider said. "He had the expertise in human cells, and I had the expertise in telomeres." They applied to the NIH for funding in January 1990. The grant was turned down, but it was sufficiently interesting to one of the scientists on the NIH committee that considered Greider and Harley's request that he instantly became a convert to telomere biology. This was none other than Leonard Hayflick's former graduate student, Woodring Wright, now running his own lab at the University of Texas Southwest Medical Center in Dallas. Greider claims Wright told her years later: "He told me over dinner one time, 'Yeah, I reviewed your grant and thought it was really interesting and started working on telomeres.'"

After a delay of several months, Greider and Harley received funding, got their experiments to work, and published a much-cited paper in *Nature* in 1990 that seemed to connect most, if not all, of the dots linking the length of telomeres to the age of cells. They showed that when you looked closely at normal human cells grown in glassware, the telomeres seemed to grow shorter with each cell division, so that the more the cells had divided in culture, the shorter the telomeres became. To put it another way: the older the cell, the shorter the telomere. This came roughly fifteen years after Woody Wright and Leonard Hayflick proved that the counting mechanism, or "replicometer," that controlled the Hayflick limit operated in the nucleus; molecular biologists had now shown exactly how telomeres kept count of the number of times a cell divided. The experiment didn't say anything about cause and effect, Elizabeth Blackburn later pointed out, but it did suggest a powerful correlation between telomere length and aging. That correlation would soon take on a life of its own.

ᕲ

It is a sad axiom of modern biology that the hotter the field, the ruder the behavior of its practitioners. For Carol Greider, the axiomatic moment occurred in 1990 when, just as her *Nature* paper was coming out, she was waiting to give a talk on these dramatic new experiments at a meeting of gerontologists. She was preceded on the meeting's program by Woody Wright. "And at the end of Woody's talk, which was on something totally different, he brought up this whole telomere thing and started talking about telomeres and clocks," Greider said. "And he hadn't done any of the work! He gave us credit. He said, 'This paper just came out.' But he stole my thunder because I hadn't given my talk yet! It was just very unusual to turn around and have somebody talk about *my* data. He was very enthusiastic about it. He's a very enthusiastic guy. But I've never been in that situation before, where somebody gives my talk before me."

This minor breach of etiquette signaled the beginning of an unsettling change in the telomere field — a juncture at which the previously obscure research began to create ripples in a much bigger pond. With the Greider and Harley experiment, and with growing evidence that cancer cells could override the cell's replication-counting mechanism in their malignant version of immortality, it suddenly began to look as if telomere research could have a very significant impact on both aging and cancer. Moreover, the mere correlation implied by Greider and Harley's 1990 *Nature* paper exerted a powerful influence on the popular imagination, to say nothing of scientific enthusiasms. This quickly led to what Liz Blackburn calls a basic "misapprehension."

The simple version of this misunderstanding was that ordinary human cells couldn't make or use telomerase, the enzyme that maintained chromosome length, and therefore these cells were doomed to an inevitable mortality; a little bit of their DNA was amputated with each cell division, causing an inexorable countdown to senescence, decrepitude, death. The power of the metaphor was seductive; Blackburn recalls a visit from a French television producer working on a documentary about aging. "This idea seems to have taken on a life of its own," he told her, "because it has a sort of metaphoric appeal to people, like a candle burning down." Among a multitude of journalistic variations on this simplistic theme, here's another, later version that perfectly captures the public enthusiasm for this once-obscure field. "Find a chemical that can block the immortalizing effects of telomerase and cancer, and maybe find a cure," one writer noted. "Find a way to rejuvenate the short telomeres of aging cells, and perhaps discover a microchemical 'Fountain of Youth.'" Cure cancer *and* reverse aging by manipulating one simple enzyme — who could resist that?

"That was the very simple idea," Blackburn explained, with mirthful eyes and unforgiving logic. "It would be really lovely to say, 'Well then, the telomeres run down and then you don't make it to your ninetieth birthday, right?' That was a lovely idea — so simple that nobody but an idiot would believe it, unfortunately. But," she added, "we're all idiots, because everybody *wants* to find that there's a great simplifying principle, right?" Here she paused, and then continued with a quick, nervous laugh, "Sorry, folks! Life doesn't work that way. And in fact biology doesn't work that way. But it was very appealing."

For all her merry personality, Blackburn does not suffer fools, or foolish simplifications, gladly. But it's a common conundrum when popular culture encounters an intriguing idea tossed up by a dense experimental science like molecular biology. In its aversion to complexity, society craves the molecular version of a headline or sound bite, a short, sweet snippet of biochemistry that even the nonscientist can understand. Real biology doesn't often comply, but an oversimplified, undercooked message nonetheless gains currency, and the public becomes enamored of an illusion.

There were plenty of people who wanted to believe that shortened telomeres explained everything you needed to know about aging. The discovery of an enzyme like telomerase, which lengthened telomeres and seemed to confer immortality on cells, only raised the level of interest, and the stakes. "I think everybody saw that right away," Titia de Lange said. "The possibility of 'immortalizing' human cells — that was such an obviously attractive commercial application."

And, as so often happens, attention in the popular press began to drive a wedge through researchers in the field. Scientists who championed the view that shortened telomeres "caused" aging were courted and quoted by science journalists; those, like Blackburn, who expressed a more complicated, equivocal, or nuanced view tended to see their opinions appended, if at all, to the subterranean plumbing feeding the "fountain of youth" stories. Indeed, at the moment the notion of immortality became linked to the activity of the enzyme Blackburn and Greider had identified in 1985, the equally simple idea that telomerase could influence, and perhaps reverse, aging suddenly joined a long history of ideas that intrigued people with the possibility of tinkering with the aging process. There was no shortage of believers — scientists, to be sure, but also venture capitalists and investors and shareholders and, yes, science writers and other journalists, not to mention that sizable part of the population at large hungering for news that the process of aging could be blunted. And so, in the early 1990s, the biggest prize in the telomere field — and perhaps in aging, and perhaps even in cancer — was going to go to whoever could find the gene for telomerase, whether in a simple organism like *Tetrahymena* or in the ultimate destination, humans. The scientific race for that gene riveted the field of molecular biology in the nineties.

This competition even came with its own impresario, because around the same time, many prominent telomere researchers began to receive visits, often unannounced, from a dreamy-eyed young man who said he wanted to start a company that would address the problems of aging, a company that would hitch its star to telomere biology. His name was Michael West.

3
THE BORN-AGAIN
DARWINIAN

"I WORRY THAT I'M PATHOLOGICAL," Michael West mused on the phone, "because all I think about, all day long, every day, is human mortality and our own aging." It's not an altogether unreasonable preoccupation in our self-absorbed postmodern culture, but for anyone other than a mortician, it's an unusual inspiration for a business plan. Michael West's obsession with death, however, has left an indelible mark on the course of biotechnology, on the priorities of modern-day molecular biology, and to no small degree on some of the agonizing national conversations we have been having in recent years about the ethics of stem cell research, human cloning, and the social and medical consequences of altering, even a little, the normal course of human aging.

The first time I met West, in November 1999, he gave what I later realized was a vintage performance on the biology of aging. Like all great performers, he knew his material well, but what made the presentation captivating was the way he *felt* his material, tapping into a bottomless pool of scientific optimism and emotional longing. To West, the biology of aging is personal. At the time, he had just assumed leadership of a small company in Massachusetts called Advanced Cell Technology, one of those modest biotech start-ups that have the impermanent feel of a shop in a strip mall, with a generic reception area out front and, in back, a modest array of laboratory benches. He was also running late that day, which I had been warned was not unusual; temporal surprise — the unexpected visit, the unaccountable delay — seemed to be part of an idiosyncratic repertoire designed to command attention, no doubt refined over years of having doors slammed in his face and hearing that his ideas were crazy. His reputation preceded him, of course. He was, according to his many admirers, a visionary able (and willing) to peer deeper into our collective medical future than many others in science and biotechnology. He was, according to his equally numerous detractors, a Svengali of the double helix, an enthusiastic and mesmerizing entrepreneur peddling equal parts high-end molecular biology futurism and the oldest of old-fashioned snake oil. Both his critics and his admirers probably have it right: like the two strands of the double helix itself, the visionary is inextricably entwined with the salesman, the scientist with the merchant, inseparable, complementary, each strand of personality ineffective without its appositional thread.

When he finally appeared, about an hour late, he strode into the room in a dark green, but not conspicuously dashing, business suit and said, with surprising warmth, "Hi, I'm Mike West." He had a youthful appearance and a big grin, was of average height, his shoulders pinned back, but what you noticed right away was the round, open face and the dreamy eyes. And behind that benign and agreeable facade, you quickly became aware, too, of a gentle but insistent impulse to control. Although I had come to interview him for a magazine article, I never even had a chance to unpack my first question; as we exchanged pleasantries, he opened an Apple Powerbook on the conference table, snaked the power cord to an outlet while apologizing over his shoulder for being late, and immediately began projecting slides on the conference room screen. "I thought I'd show you some pictures," he said. And then he launched into a spiel that, in one form or another, he had been giving for ten years to venture capitalists, scientists, journalists, and just about anyone else who would listen. The message was simple and irresistible: aging might be arrested through better biology, and we might all live to be two hundred years old.

Click. "This is a brainteaser," he announced as the first slide appeared. It was a quote from the German philosopher Schopenhauer. "The task is not to see what no one has seen yet," West read on the screen, "but to think what no one's thought yet about that which everybody sees." He paused a moment for the words to sink in, just the right amount of time for a listener to connect Schopenhauer's infinitive *to think* to the central verb in the arsenal of a visionary. "And what everybody *sees*," he continued, "is aging."

Click, click, click. Soon there were graphs showing human life expectancy, the average age that members of any given population or culture can expect to reach, and a graph showing maximum attainable life span — that is, how old the very oldest members of our species live to be. In recent times, the oldest living human for whom credible records exist is said to have been Jeanne Louise Calment, a woman who died in France in 1997 at the age of 122. Few of us, of course, are likely to attain that age, but what medicine has been especially good at, particularly in the last hundred years, West continued, is altering the life expectancy curve. Whereas an American woman born in 1900 could expect to live on average to roughly age fifty (and men slightly less), a woman born in 2000 can expect to live nearly to age eighty (and men to about seventy-four). "But the point is that we have not, as you see, been improving *that* point," said West, indicating the maximum number of years a human can live. "It's remained inflexible. And my point, going back to the quote from Schopenhauer, is that everyone sees aging, but I think the thought no one's thought yet about that, which everybody sees, is that there may actually be a mechanism to actually change this [maximum life span]. And that's really what the focus of gerontology, of course, is."

Click. The next slide showed a bas-relief from an ancient tomb in Egypt, a tableau illustrating the myth of the Egyptian god Osiris. "This is really the image I wanted to show you, just by way of background," West said. According to the ancient Egyptians, he explained, Osiris was cut up into fourteen pieces by Seth, his jealous brother. Seth scattered these pieces all over Egypt, but Osiris's loving wife, Isis, painstakingly located the pieces, which were marked by growing clumps of papyrus, and reassembled them, making Osiris whole again, so that he subsequently became recognized as a universal symbol of regeneration. "There's a lot of evidence," West said, "that the Egyptians looked to the regeneration of plant life and tried to follow the path of Osiris in their own personal hope for regeneration, to see their loved ones again and so on . . . As you're probably aware, here's this barren desert and the Nile River, which is the source of life, and at every spring along the Nile you can see this new plant life spring up, and new crops. And what they saw in it was youth. Immortal life. Immortal, regenerating life."

As he spoke, his voice rose to a hushed, wondrous flutter; I had heard biotech executives wax poetic about their products and even their proprietary technologies, but never before about ancient gods. Here we were, just the two of us in a dark room (with a PR person sitting in), and it was clear that West bought into this story way beyond the level of corporate branding and bottom lines. "They built a lot of mythology around that," he continued, "and Osiris was the personification of immortal life, of immortal regenerating life . . ."

Click. The screen showed the standard image of Osiris, discovered in the tomb of Tutankhamen. With a kind of Gregorian gravitas, West recited what he called the Homage to Osiris from the Book of the Dead. It celebrates a god resistant to decay and rot, impervious to putrefaction and worms, destined for an everlasting existence. "That's the hope and the understanding of these agrarian cultures," West explained, "that somehow, basic in life, is a type of immortality. And there is the promise of regeneration. And they look to it somehow in trying to frame an understanding of biology and, of course, their own philosophical worldview as well."

By the time West had finished, he'd mentioned Hayflick, WI-38, Olovnikov, telomeres, and stem cells, knitting them all together in a river of thought that had gained depth and momentum. But as he spoke about Osiris and regeneration, with a conviction more religious than scientific or entrepreneurial, his voice quickened by true belief, the enormity of his kinship with the idea — the *possibility* — of immortality (if not literal immortality, at least dramatic extensions of human life span), and the degree of success with which he has shepherded and advanced that belief in the secular worlds of molecular biology and venture capital, suddenly became clear. He actually seemed to believe in some attainable version of immortality; moreover, this good-naturedly stubborn, enigmatic midwesterner had seeded the investment of tens of millions of dollars in venture capital into that notion. Apart from the actual science he has organized and overseen, which has been consistently impressive and highly imaginative, routinely published in top scientific journals, such as *Science*, West represents an amazing American success story, the Ponce de León myth reinvented in an age of genetic engineering and American entrepreneurialism.

West has attracted a feisty Greek chorus of critics over the years, and there are plenty of people who would dispute his paternity of these biomedical possibilities, to say nothing of the national controversies they have spawned. But he has popped up again and again, like some Zelig of modern biology, in the midst of this scientific and ethical thicket — in the laboratories where the cutting-edge research has been done, at congressional hearings, before presidential ethics panels, and frequently, in the press and on the air. From the technology of stem cells to the corporate ethics of using embryos for scientific research to human cloning, he has, as often as not, gotten there before anybody else. Indeed, part of his business and personal ethic is to get there first. On his curriculum vitae, he doesn't list his scientific publications first; he lists his patents.

What is most remarkable about all this is that not too many years ago, Michael West was neither a scientist nor a corporate visionary, but a creationist. He spent every waking hour — when he wasn't thinking about death, of course — trying to convince himself that the general theory of evolution was wrong and that the world began, as the Bible says, about six thousand years ago. Whatever else might be said about West, his personal journey from creationism to biotech prophet represents one of the most extraordinary intellectual transformations ever to influence a national debate in this country.

⁓

West's journey to the center of controversy has taken a most circuitous path. Born on April 28, 1953, he grew up in Niles, Michigan, just above the Indiana border near the eastern shore of Lake Michigan. His grandfather had established a chain of gas stations and fuel distributorships in the area, which eventually evolved into a profitable family-run truck and automotive leasing business. To put this business pedigree in vulgar terms, as some are wont to do, he came from a family of used-car salesmen. But a part of West's education was admirably experiential in an age of career-oriented learning. Throughout his life, he bumped into ideas that were of no practical merit but were too interesting to ignore, and he pursued them with unusual avidity. He had the philosophical equivalent of a wandering eye: he fell in love with fetching ideas that caught his fancy, and then would throw himself into their pursuit with something like infatuation.

Early on, one of those pursuits was science. He recalled in an interview how his father — "a great man," he said with reverence — helped him build a laboratory at home, in a storage area above the garage. "I was completely in love with science, as long as I can remember," he said. "But in my senior year in high school, I became just completely, profoundly interested in trying to understand really broad philosophical questions — the meaning of life and all these kinds of issues. And for about ten years of my life, I studied, pretty much on my own, philosophy and theology. Particularly delving into Mediterranean folklore — I'm really interested in Egyptian, Phoenician, Syrian, Babylonian, Palestinian, Greek, and, somewhat less,

Roman mythology and folklore." As part of this self-motivated quest, he said he learned to read Greek and Hebrew to try to get a better understanding of this mythology, to get closer to the roots of the Judeo-Christian tradition.

Following graduation from Niles Senior High School in 1971, West attended Rensselaer Polytechnic Institute in Troy, New York. He went there to study physics, but ended up with a bachelor's degree in psychology ("Really a mix of philosophy and religion and science," he explained). In 1976 he returned to Niles, and over the next six years he helped run the family business for his ailing father, serving as president and general manager of West Motor Leasing Company. But he never entirely abandoned his autodidact's journey through the foundations of Western thought. Then another element entered his thinking. "I was trying to understand the meaning of the soma," he explained, referring to the mortal, physical tissues and cells of our being, as if this were the most natural preoccupation for a young man in his twenties, "and of course the biological basis to it, and it took me about ten years to sort through the philosophical issues. And I finally came to the conclusion, back in 1978, 1979, 1980, that — well, my father died in 1980 of a heart attack — I really came to the conclusion that the most significant problem we have is the aging of our population, the aging of all of us. It's a disease — I use 'disease' advisedly, with quotes — it's a disease we all have. It's killing everyone. And the more prosaic aspect of it, of course, is that it's a strain on our health care system. Baby boomers are really going to strain the system, etc., etc. But on a more profound, humanitarian level, I felt it was the most significant thing. And secondly, I loved science and I loved puzzles, conundrums. I thought, What more fascinating area to spend my life in than trying to unravel the biology of aging!"

In one sense, West has never swerved from that mission in nearly twenty-five years — it is a philosophy, a passion, and a business plan all rolled into one, as high-minded as his exploration of ancient Mediterranean mythologies, as market-minded as his covetous glance at the graying baby boomers in his midst. And yet he came at this out of left field — the blue-collar Midwest background, the wandering sentimental education, the occasionally puzzling theological digressions on his pedigree. There is a murkiness, almost a fundamentalist haze, obscuring the early part of West's curriculum vitae. In 1979, while still working for the family business, he began to seek a master of science degree from nearby Andrews

University in Berrien Springs, Michigan, which has been described as "the flagship educational institution of the Seventh-Day Adventist Church." He took classes in paleontology, he recalled, in a building named after the founder of creationism in America.

Although he hadn't come from a deeply religious family, West found himself increasingly drawn to fundamentalist Christianity. As he put it during one conversation, "I spent a *terrific* amount of time exploring religion . . . My personal explorations were in a more orthodox Christianity. In a biblical, evangelical Christianity. But it was a personal exploration of it, going back more to the original, first-century Christianity." West's journey of spiritual exploration may represent a kind of extreme in the genre of self-discovery during one's postcollege years. West enjoyed a certain luxury in the pursuit of his curiosity. The family trucking business was sold in the early 1980s, and although West has never disclosed the amount of money realized in the sale, it's clear the money at least partially endowed his idiosyncratic quest for enlightenment and self-fulfillment. "I made a fair amount of money on that transaction," West told *Science* in 1999, "which has given me the freedom to do what I've wanted to do in science without having to worry about making a living."

As he described this phase of his life over the phone, West sounded uncomfortable; he didn't, as he put it, want to step on too many toes, ruffle too many public sensitivities. And while I found myself admiring his candor in speaking about something so personal, I felt an almost voyeuristic fascination with the substance of this confession. Because what he was confessing to was the fact that he had believed quite strongly in the biblical version of creation. He felt so awkward about it, in fact, that he lapsed briefly into the third person while talking about himself during that period of his life, as if creating even more distance between himself and the person he used to be. "I mean, basically the story was about a person who was a truth seeker. I'm speaking third person, me. I was a truth seeker, and for about ten years of my life diligently sought to see if there was truth, in particular, in the Christian religion. And, you know, a rather in-depth quest." Learning Hebrew and Greek, he said, allowed him to read the Bible in its original languages. He enlisted in the political rites of passage of Christian fundamentalism; at one point, he participated in antiabortion demonstrations, according to testimony he gave at a Senate hearing many years later. And he even traveled to Southern California to study the "science" of creation at the Institute for Creation Research. "I hung out with those folks a lot," he admitted.

The Institute for Creation Research, based in Santee, California, just outside San Diego, is "the most preeminent of the 'creation science' organizations," said Eugenie Scott, head of the National Center for Science Education. The group not only adheres to a literal interpretation of creation, Scott said, but argues that "you can support the 'young earth' idea with scientific data and theory." According to the group's Web site, human creation occurred, as described in the Bible, about six thousand years ago. The institute also believes that the story of Adam and Eve is all the explanation anyone needs for human mortality; because Adam and Eve sinned in the Garden of Eden, humans are condemned to die. West wrestled at length with these issues before he finally let Darwin into his life. "Unfortunately," he said, "kicking and screaming every inch of the way, I ended up having to admit the truthfulness of the general theory of evolution. Which, you know, when you think it through leads to the conclusion that aging and death, for instance, have been around since the — well, at least a billion years. And that's certainly before Adam and Eve."

Incredibly, West's rejection of creationism led to his present calling. "I spent some time studying the creationist movement out of San Diego, and, you know, I just had to admit, as much as I *wanted* to believe in [creationism], that the facts were otherwise. You have to dig for it, but there really are very solid foundations for the belief in general evolution, not just microevolution, but that everything, and all life on earth, is evolved from a single source. And that evidence is really — it's a bit of a pun, but it's rock solid. It's in the rocks, in the fossils. And the case against that, that there really is a scientific basis for belief in creationism, is, in my belief, pseudoscience . . . It's first and foremost a defense of the Bible. You know, these are Bible-loving people who want the Bible to be interpreted that the earth is only six thousand years old and that all the fossils were [from] Noah's flood and so on. The data are just overwhelming in favor of the evolutionary viewpoint.

"But with that," he continued, "comes the realization that the things that matter perhaps most to us — you know, the welfare of our fellow human beings, people we care for — are aging and death, and are facing us not for something that happened in recent history but because they're just part of the nature of life, and evolved with life. And inasmuch as that's a mechanistic process, we could use molecular genetic techniques — just like we understand now the molecular basis of the evolution of many species — potentially to understand the molecular evolution of that mortality. So it was a bit of a religious quest, gone into a new direction."

West paused here. I was almost breathless, struggling to take in and process this wild ratiocination that drew an unbroken line from creationism to the molecular biology of aging. Then he capped his remarkable monologue by saying, "I feel a bit like the apostle Paul in reverse, you know? The apostle Paul was the critic of Christianity become its greatest advocate, and I, in some sense, was a great advocate of Christianity who [had] now gone in the other direction." Indeed, armed with a master's degree in science from Andrews University, he proceeded directly to the lair of the infidels. He joined a laboratory of molecular biology. The year was 1982.

✍

West received his baptism into the world of science in the laboratory of Sam Goldstein at the University of Arkansas in Little Rock. Arkansas may seem a likelier destination for a creationist than a biologist, but Goldstein, in West's opinion, ran the country's leading laboratory studying the molecular biology of aging. He was among a handful of scientists looking at the molecules inside the cell, whether DNA, enzymes, or other proteins, that might help explain why bodies age and become vulnerable to disease. Goldstein, who died in 1995, was one of the pioneers of molecular gerontology; he knew Leonard Hayflick well and in a sense was attempting to extend Hayflick's work. Goldstein thought aging had to do with repetitive sequences of DNA (which, in the most generous of interpretations, fit into the emerging telomere story). Soon, West would be introduced to Alexey Olovnikov's subway train hypothesis of telomeres. He learned for the first time of the Hayflick limit. He crossed paths, at least by way of the kind of genealogy that counts in science, with Calvin Harley, who had earlier been a student in Goldstein's lab at McGill University in Canada and who was carving out a reputation as a leader in the field of cellular senescence. In one enormous, self-propelled leap, Michael West traveled from the world of creationism into the thick of molecular gerontology, from fundamentalism to reductionism.

Two important patterns emerged during West's apprenticeship in Little Rock. First, he immediately embraced the techniques of modern biology to attack the problem of aging. He was not the only person to see this, of course, but few if any of those who did shared West's deep, obsessive, philosophical hunger for answers. Second, behind his kindly, almost courtly manner, West has a stubborn and gently confrontational streak, and he apparently had an embarrassing clash with his lab chief. There was scientific substance to the row, according to West; he spent three years of his Ph.D. research trying to extend some notable findings about aging that the Goldstein lab had published in *Cell*. The findings had to do with repetitive sequences at the ends of chromosomes, tantalizingly similar to the telomere discoveries. But West says he ended up showing that the original research had been flawed, based on a laboratory mistake, or artifact. It clearly created tensions; at one point, West tried to find another scientific home, traveling to Philadelphia and pleading with Vincent Cristofalo to let him in his lab. "There was a technical error that Goldstein didn't see," Cristofalo told me, "and Mike's version is that Mike tried to tell him and Goldstein didn't want to listen. I'm convinced that what they published wasn't true, and that it was an honest mistake." "It was a real mess," West recalled. "And Sam was a great guy, but it got out of control, and I decided I should really separate myself from that. And I started all over again at Baylor, did another Ph.D."

West received a Ph.D. in cell biology from Baylor in 1989. But his time in Houston was not uneventful, either. He worked in the laboratory of James Smith, who, in the still-small world of molecular gerontology, had done postdoctoral research in the Stanford lab of Leonard Hayflick. West satisfied the requirements for his Ph.D., but his work could hardly be considered memorable; Smith had trouble remembering the topic of West's thesis when I asked him about it not long ago. And for the second time in a row, according to several scientists, he had some sort of clash with his adviser. "Jim and Mike had a falling out," Hayflick confirmed, "and I tried to get them back together, but they ended up getting divorced." With Ph.D. in hand, West started thinking about medical school.

At the time he was finishing his doctoral research, West and his first wife lived in a Houston apartment complex primarily occupied by older people. But they soon befriended another young couple living on the same floor, a medical student named Judson Somerville and his wife. Somerville and West shared a passionate interest in paleontology, old books, and big ideas. Somerville recalls giving West a hard time about his desire to become a doctor. "He always wanted to go to medical school," said Somerville, now a doctor himself, "and I told him he was a moron to want to get both a Ph.D. and an M.D. He was already a smart guy, and I told him he'd be wasting precious time that he could spend on pursuing his research." West ignored the advice and headed to Dallas to enroll in medical school at the University of Texas. But he continued to stay in touch with Somerville; indeed, in little more than a decade, he would attempt to clone a human embryo, using cells from his erstwhile neighbor in Houston.

‍ᗢ

The philosopher's wandering eye caught up with West in medical school, too. He repeatedly took leaves of absence and eventually dropped out. But while pursuing his medical studies, West began to walk the halls of the University of Texas Southwestern Medical Center and drop in on research labs. Around 1989, with the kind of selflessness that comes easily to the independently wealthy, he walked into a large laboratory at UT Southwestern, thought the research sounded interesting, and offered his services for free. It was a joint lab, actually, run by two biologists, Woodring Wright and Jerry Shay. Wright, of course, had been a graduate student in Leonard Hayflick's lab at Stanford shortly before the WI-38 scandal exploded, and he'd never relinquished an interest in aging research. But the lab had primarily focused on the way cancer-causing viruses seemed to confer a kind of deranged immortality — a continuous cycle of replication that was seemingly exempted from the Hayflick limit — on cells in culture.

Typically, scientific walk-ons are no more often granted a place at the lab bench than football walk-ons make the starting lineup of the Texas Longhorns. But as Shay recalled, "We, not being totally stupid, said, Here's a Ph.D. who wants to work for free. So we let him work in the lab, and eventually we published a couple of papers together." Wright was immediately impressed, too, but not without reservations. "Mike has absolutely magnificent skills as a visionary," he said, "and as someone who is willing to take ideas and push them as far as they can go. But he's not terribly rigorous in how he does it." Wright also harbored some concerns about West's problems with, for lack of a better term, authority. "We discussed a history of run-ins with his supervisors at several labs," Wright remembered, referring to West's clashes with Goldstein and Smith, "and how he needed to be sure a similar thing didn't happen again. Twice is a strong pattern, and three times is a killer. And although he had that pattern, we had no problems with him while he was here. But," Wright told me in 2001, "his character has led to other run-ins since then."

West began to do experiments with Shay and Wright. The cells they were studying seemed to exhibit strange behavior at the ends of the chromosomes, the region of the telomeres; they became ragged, frazzled, disorganized, entangled in themselves. Shay and Wright were beginning to come around to telomeres as a possible explanation. In fact, West arrived at just about the time that Wright, as part of an NIH review panel, had read Carol Greider and Cal Harley's grant proposal, which rekindled his interest in aging and senescence. "That's why Mike West knew about this at a very early stage," Greider told me. And it's also why Wright and Shay plunged headlong into the telomere field.

To hear West tell it, as he often does, everything suddenly fell into place — telomere biology seemed like a key piece in the puzzle of aging. "The *whole thing* really sounded like telomeres," he said. "And it started to smell right." Actually, the initial odor was anything but sweet. Shay and Wright recalled browbeating West, struggling to persuade him to overcome his Arkansas-induced aversion to repetitive DNA sequences and see that telomeres seemed to have a great deal to do with aging. Indeed, it took months to convince West that telomeres smelled "right." "Mike will not do something just because you tell him to do it," Wright said. "He has to be convinced by an intellectual argument that persuades him that it's the right thing to do. And unless you can muster the intellectual arguments that convince him it's right, he is not going to be a team player. After about six months of arguing, we finally convinced him how important telomeres were to the whole story, and then he became a total convert." Shay remembers it the same way: "Mike wasn't too much of a believer [in telomeres] at that time. But I told him to go do some more homework, and then he, Woody, and I sat down and had a series of discussions on the order of: What if? What if this is true, and what if you could develop a technology to stop or reverse senescence in cells?"

When I mentioned Wright and Shay's recollections to West, and suggested that perhaps telomeres had not in fact "smelled right" at first, he amended his story. The experience in Goldstein's lab, he said, had created a deep-grained bias against the telomere theory. "I have to admit, that experience . . . *completely* burned me on the concept of the loss of repetitive sequences as being a mechanism of cellular aging," he said. Before long, however, West became the greatest apostle of the telomere theory. It was St. Paul all over again. He was St. Paul of the Chromosome, and this time St. Paul began to preach this bit of gospel to men of the pin-striped cloth — the moneylenders.

✑

"What I felt was, we could do two things," West said of that crucial turning point around 1990. "Start working aggressively to get telomerase in telomeres, and to get telomerase cloned, and so on. And I had a nice, comfortable relationship there with Woody and Jerry, had a little grant and everything, and I could finish my M.D. But as I thought about it, I felt that if it was true that these mechanisms are central in cellular aging *and* cancer, this really needed to be a *substantial* effort. And I felt I could do a more substantial project if it was done via a biotech company, because we could have a hundred people instead of four. I remember Woody advising me, 'Mike, finish your M.D. You're not going to be able to come back and do it over again.' One more leave of absence, the medical school told me, would be my last if I didn't come back. But I felt the timing was . . . There were some biotech companies that were sniffing around the story, Carol Greider's lab, and so on, in Cold Spring Harbor. And I just felt, Look, now is the time."

The timing, West nonetheless admitted, was "a bit awkward." He was thirty-seven years old and a third-year medical student, trying to launch a company. At first, he and a friend, Bob Peabody, sat at a kitchen table in Dallas, trying to figure out ways to raise an "angel round" of financing, just so they could pay the travel costs of trying to raise serious seed money. West recalled working up patients at UT's teaching hospital when his beeper would go off; he would excuse himself, go into the hallway, and haggle over financing details. "I think we probably visited three hundred people and got kicked out of three hundred offices," he said. At first, no one seemed interested in the idea of an aging-research company. But West kept at it and got his first break in the early 1990s.

Around that time, the world of biological commerce was intensely interested in death, but most people didn't think about it the way Michael West did. Everyone was debating the merits of sequencing the human genome; there were fights about Big Science versus small science; there was great optimism, as there always is, about a new generation of cancer therapies. Inspired by the examples of Amgen, Genentech, and Biogen, hundreds of small biotech companies scrambled to raise money to cure diseases. Compared with the manufacturing triumphs of that first heady wave of biotech innovation — the clot-busting drug TPA, the blood-building erythropoietin (EPO), insulin for diabetics, interferon for cancer and multiple sclerosis, all of which were en route to becoming genetically engineered, billion-dollar molecules — the notion of using molecular biology to cure aging (or, as the less kind put it, "to cure dying") seemed almost laughable. It was hard enough just to get a palliative therapy like EPO approved by the Food and Drug Administration.

A handful of people didn't think it was too far out, though, and they became the initial investors in Michael West's dream. It began, like many "fountain of youth" ventures, with the old and the somewhat gullible. These first-round investors were, in the words of someone familiar with West's early money-raising efforts, "some fairly eccentric people who were interested in living forever." First in line was probably a retired Houston geophysicist named Miller Quarles, who in 1989 founded an organization dedicated to life extension called the Cure Old Age Disease Society. He was interested, as he put it, in "starting a company that would cure aging by the year 2000."

Quarles, who told me he takes fifty vitamins and pills a day, according to a regimen published in *Life Extension* magazine, said, "I'm eighty-seven years old, and feel like I'm forty." He began soliciting memberships to the society among friends and acquaintances; his dentist mentioned it to a patient whose husband worked at Baylor with a bright young man interested in aging named . . . Michael West. Soon, West and Quarles began talking, and at one point Quarles set up a meeting with a group of Texas oilmen to hear West pitch the idea of starting a company based on a molecule that he claimed would cure wrinkles. "Nobody was much interested in curing wrinkles," Quarles recalled, "but I told him again, if you start a company that will cure aging by the year 2000, I'll buy $50,000 worth of stock. He called back again, about a month later, and said, 'Quarles, I think I know how to find a longevity gene . . .'" West apparently was not referring to telomerase, which had not yet been isolated, but that's a petty detail; here was the first time that the molecular biology of aging story had seduced an investor, and the telomere story would soon be part of its appeal.

In recounting this history, Quarles simplified the science of telomeres

in exactly the way Elizabeth Blackburn had warned about, yet this mis-
interpretation had delightful financial ramifications for West. Quarles
thought of telomeres as "strings" at the ends of chromosomes, and everything could be explained by whether the string was long or short. "If the
string is short, the cell acts like an old cell," Quarles said in an interview, explaining the appeal of the idea. "If the string is long, the cell acts like a
young cell. So if you can stop the shortening of the string, that would stop
aging. If you could add to the string, you could *reverse* aging." Quarles was
in for $50,000, and a woman in her sixties with similar interests in life
extension also agreed to angel funding. Soon after, West attracted the interest of several more like-minded investors; Judson Somerville remembered
them as "a bunch of phenomenally rich people in Houston who wanted to
live forever." "We finally found a small group of people," West said, "maybe
five or six. One was Arthur Altschul, who was in New York and was involved in Sugen [a well-regarded biotech company]. A great guy. He just
said, 'I like the sound of this.' So I got $250,000 worth of investment, which
allowed us to travel and put together the venture financing."

But most important of all the early investors was Alan Walton. A dis-
tinguished-looking, well-spoken British-born scientist, Walton had been a
molecular biologist at Harvard Medical School and Case Western Reserve
University in Cleveland; he had left academia for good in 1986 to try his
hand at biotechnology full-time, eventually running a nursery for new
biotech companies at a venture capital outfit called Oxford Partners. Every
three months or so, Walton and his associates would invite scientists to an
informal retreat at the company's Connecticut headquarters and ask them
to ponder a question that particularly intrigued him: Is the state of research
on aging sufficiently advanced to consider commercialization?

The idea was definitely in the air when a startling coincidence one day
in 1991 assumed the weight of augury. "Three people called me on the same
day, asking about Mike West," Walton recalled. One call came from Daniel
Perry of the Alliance for Aging Research, a Washington-based group that
lobbied for more research on gerontological issues; West had begun turning up at meetings of the Gerontological Society of America (where, Perry
later recalled, "he was not greeted as warmly as he might have been by the
academic community"). Another call was from a business associate of
Walton's who'd run across a reference to West in a magazine article. The
third caller was a scientist at the National Institute on Aging who'd also

learned of West's intentions to start a company focused on aging. "They all said, 'Apparently this guy Michael West is doing exactly what you've been talking about,'" Walton said. "So I figured I better go down to Dallas and talk to him." Walton brought along a scientific colleague from Case Western named Arnold Caplan, who would later play a major role in the adult stem cell story. Caplan was "impressed with the vision," Walton recalled, "but not with the research or the science. I found West to be very charismatic. I remember being particularly intrigued that he had learned some ancient language, like Farsi, by which he labeled his test tubes." (Now there was the perfect union of scholarship and salesmanship!)

Walton took West under his wing. The more they talked, the more Walton found West's ideas "pretty stimulating." He started to put together a venture capital syndicate around West's plan. A group at Aetna led by Alan Mendelson expressed interest in the idea; it looked like the initial financing could be arranged. Walton began to scout out locations, seeking a biotech-friendly place near a major academic institution on the East Coast. At one point, officials in North Carolina were very eager to get the new company, Walton recalled, but Duke University balked ("Duke didn't get it," he said). They looked for a home on the West Coast. The one thing they didn't need to worry about was a name. In 1980, while he was still running his father's business and attending graduate school, West had quietly created a company on paper and registered the name. It was a foreign word, Walton recalled, Greek for "old man," something like that. The name was Geron. (West later told the *New York Times* that he took the word from a telling passage in the New Testament in which Jesus describes how a man who is old might be born again.)

And then Walton had an inspiration for showcasing West's ideas. Each fall, Oxford Partners sponsored a meeting on the West Coast called the National Conference on Biotechnology Ventures. It was held at a hotel in Redwood Shores, California, about twenty miles south of San Francisco, and it was intended to enhance what Walton calls deal flow: the heads of young biotech companies made presentations to a crowd of venture capitalists. Although it had the trappings, the wardrobes, and the gravitas of a serious financial gathering, it was really a test-tube version of *The Dating Game* for would-be biotech entrepreneurs — whoever told the most convincing story was likely to walk away with a princely sum. For the 1991 meeting, Walton decided to schedule a session titled Molecular Genetics of Aging during the morning of November 6. For speakers, he invited Daniel Perry of the Alliance for Aging Research, and Caleb Finch, a professor at the University of Southern California, who would provide a review of academic research on molecular gerontology. He also wanted to have a representative from a company, and the only company, even though just on paper, was Geron. So he invited West to give a talk too.

The decision to put West on the dais created behind-the-scenes consternation among the investors who had already agreed to back Geron. "If you put Mike West in front of that audience," Mendelson warned Walton, "he's going to have so much money poured on him that you're going to be excluded from the deal. You better get a signed term sheet ahead of time." Usually a term sheet indicates a venture group's degree of commitment to a company, but in this case it was intended to commit West to Walton's group of investors before he could be courted by bigger outfits; West had already met several times, Walton recalled, with representatives of Kleiner Perkins, the premier West Coast venture capital group. "I told Alan Walton we should get Mike West to New York to sign a deal before the meeting," Mendelson said. So, several days before the meeting West flew to the East Coast to negotiate the deal, and Oxford Partners, Aetna, and Matuschka Venture Partners signed term sheets with West. It's a good thing they did, because, as Walton said, "Alan Mendelson was absolutely right."

When West took the podium at Redwood Shores, he dispensed the same dream he'd been nurturing and refining and sprinkling with scientific fairy dust since he'd been a senior in high school. He spoke of the problem of aging and how it might be attacked by modern biology. He claimed his company had already discovered the molecular and genetic events regulating cellular senescence, and had created "potent" inhibitors of cellular aging. He made the same points he had been making for years — indeed, the same points he made to me in that one-man slide show nearly a decade later — but this time the audience and the timing and the social moment were exactly right. In minutes, Mike West went from being one of hundreds of glassy-eyed supplicants, the biotech version of a Fuller Brush salesman, to a self-made man and certified visionary. "He gave an absolutely stunning talk," Walton said, "and within an hour and a half, I think he had been promised $17 million."

As often happens at these meetings, speakers retire to a "breakout room" after a talk to meet with interested investors and continue the conversation. Following West's talk, the breakout resembled "a shark feeding frenzy," Walton said, as investors clamored for a piece of the action. Kleiner Perkins waded into the crowd and ultimately emerged as the lead player in the deal; Walton may have escorted West to the dance, but West left on the arm of another, more prestigious partner. Venrock, the venture capital arm of the Rockefeller Foundation, crowded in, too. "People who I couldn't even interest in *talking* to West before Redwood Shores were telling me they had to have a minimum of 11 percent," Walton recalled with a chuckle. "I remember Venrock fighting for every 0.1 percent of the deal that they could get. Mike's breakout room was packed, and people were writing checks in the room, which doesn't normally happen. And I would say that within a month or two, they had [laboratory] space, they had $15 million in the bank, and they were basically rockin' and rollin'." The share of the initial round of Kleiner Perkins funding was $7.5 million, but in some respects it was the fact that Kleiner Perkins anointed the deal that was even more important; it bestowed upon West the priceless wealth of credibility.

Flush with success (and cash), Mike West, lapsed creationist and born-again Darwinian, spent a few months setting up the company and then hit the road, recruiting scientists with his usual mix of visionary science and down-to-earth salesmanship. He went to Ann Arbor and won over Bryant Villeponteau, who was doing research in aging at the University of Michigan. He paid visits to Calvin Harley, Carol Greider, and Elizabeth Blackburn in those early years, partly as pilgrimages of respect, partly to woo. Like a lot of early start-ups, Geron was nowhere near having permanent lab space, much less a product, but it had a catchy name and would soon have a fetching logo that would leave no doubt about the mission of the company: it featured an hourglass, around which curled an image of the double helix. Not only was aging now seen to be a legitimate target for molecular biology, but it was a legitimate business interest for biotech. And in Mike West, the field had a leader who brought something other than just scientific enthusiasm to the project. Thinking back on his spellbinding talk at Redwood Shores more than a decade later, one quality stuck out in Walton's mind. In his fervor, his enthusiasm, his conviction, he said, West was "like a missionary." St. Paul, after all, is considered the most zealous of the apostles.

4

"Money for

Jam"

With a big financial wind at his back and a blessing from the archbishops of venture capital at Kleiner Perkins, Michael West sailed out into the new world of commercial gerontology like Columbus (or perhaps more like the sixteenth-century adventurer Ponce de León), to chart and discover a new world of aging research. Indeed, he was getting paid to spread his personal gospel and bring to fruition an unusually personal vision, and he set about to do one of the things that he has always done exceedingly well: identify the best talent in a given field or technology, and get them on the boat.

Formally launched in March 1992, Geron listed West as vice president of science during the first year or so, while Alexander E. Barkas, chairman of Geron's board, served as president and CEO. Barkas, with a Ph.D. from New York University, had a good feel for science. First the two found temporary laboratory space in Hayward, California, in the East Bay, just across the San Mateo Bridge from the more high-rent districts of biotech along the Highway 101 corridor. They recruited a panel of outstanding biologists to form the company's scientific advisory board and consultants. The scientific advisory board (or SAB) is a curious ornament that adorns every biotech company; it has both substantial and cosmetic purposes. These outside scientists vet research initiatives and provide some long-range vision, but just as important, they lend credibility and scientific sheen to a commercial venture barely out of the box. Geron's board was a *Who's Who* of telomere research, with a little biological stardust thrown in. It included Carol Greider (who ran a lab at Cold Spring Harbor Laboratory now); Woody Wright and Jerry Shay, Michael West's former mentors at the University of Texas; James Watson, the famed codiscoverer of the double helix;

and, for historical, scientific, and even sentimental reasons, Leonard Hay-flick. Serving on Geron's board was a good deal for the scientists; in exchange for about $35,000, according to one former member, each scientist agreed to make himself or herself available for thirty-five days of consulting per year, including attendance at up to half a dozen annual advisory board meetings. The first meeting of Geron's SAB occurred in New York, high above Rockefeller Center, in a meeting room provided by Venrock. The mere proximity to the Rockefeller fortune brought out the middle-class wonder in Hayflick. "There were people in cubicles behind computer screens, people by the dozens if not hundreds," he recalled, "who I later learned were managing the Rockefeller assets."

With the scientific board in place, West — with occasional help from Barkas — set out to recruit full-time scientists. In 1992, Bryant Villeponteau was working at the University of Michigan. He was a self-described "token molecular biologist in the field of aging" and had become a convert to the telomere story after reading a paper in 1989 about Vicki Lundblad's mutant yeast and their Ever Shorter Telomeres. "Mike West came and visited me," he recalled. "If he gave me any notice, it was pretty short. He told me he had just started this company, and he told me all this stuff that hadn't been published yet, which made me even more excited." (That was another of West's tools of seduction: he loved to broker information and realized that sharing it — with scientists, with investors, even with journalists — was a way of drawing them into the fold.) It didn't take much to persuade Villeponteau. "When Mike West came with the idea not only of doing telomeres, but with aging as the idea for a company," he said, "that was very exciting." Villeponteau left Michigan almost immediately; he came to California as the company's first scientist and started on January 1, 1993.

West paid an obligatory visit to McMaster University in Ontario, home to a powerful telomere research team. He spoke with the lab chief, Silvia Bacchetti, Calvin Harley (Greider's collaborator and one of the best researchers in the field), and an Iranian graduate student, Homayoun Vaziri. Harley apparently fretted over the decision; he consulted with Hayflick, Greider, and others before agreeing to join Geron and moving to California in the spring of 1993. Vaziri soon followed.

Geron also made a pitch to Greider. She recalled that West and Barkas flew out to the East Coast one day, took a cab from the airport to Cold Spring Harbor, and, newly flush with all that venture capital cash, held the taxi at the lab, its meter running, while they took her out to lunch and asked her to join Geron's scientific staff. "I really did respect Alex Barkas, who had a very good scientific viewpoint," Greider said. "They came out and talked to me about telomeres and telomerase. I said the critical experiment at that point would be that if you could artificially lengthen the telomeres, you could artificially lengthen the life span of the cell. They said, 'Well, we've done that experiment, and it works.'" It turns out, Greider added, that they hadn't truly proven it, and that realization made it clear to her that a culture of exaggeration began with the company's very first acts of recruitment. "They didn't have any data. I should have been more critical, but I really felt strong-armed." Greider ultimately declined to join the company, but she agreed to serve on the advisory board.

Even while maintaining her independence, however, Greider became entangled in the company's research agenda because she was already embarked on a project with the now corporatized Cal Harley. Indeed, she had inadvertently been dragged, by prior collaboration, into the young company's highest-priority project: to find and clone the gene for telomerase. The more optimistic scientists believed that this enzyme could literally immortalize cells in culture, allow them to crash through the Hayflick limit and keep on going; from there, some scientists were even willing to take the short, metaphoric leap to suggest that "immortalizing" aging cells in the human body might achieve an analogous kind of molecular rejuvenation. "The whole principle of putting telomerase into normal cells and making them immortal was a crazy idea," Homayoun Vaziri told me, recalling some of the early discussions during Geron's first year, "but Mike West and I were obsessed with these ideas. So we sort of brought it out of this impossible task and made it an experimental task, took kind of a cowboy approach to doing it."

⌒

As a company, Geron seemed to crystallize a new energy state of biotech ambition: nothing, not even the inexorable process of aging, could stand in the way of the new biology. With enough money, enough smarts, and enough gumption, technology might even be able to repeal the oldest and most inflexible of natural laws, at least for a while. This commingling of corporate mission with demographic desire was no accident. West was shrewdly cognizant of the depth and strength of this social fascination, this cultural need; he also realized that there was a market waiting to be tapped for perhaps the first time by the biotech industry. "You know, in World War II, when the draft came, everyone was knocking each other down, trying to get into the army," he told me once. "When the baby boomers were confronted with Vietnam and not Hitler, their response was, 'Hell no, we won't go!' And I think, in regard to aging, that you're going to see a similar response of 'Hell no, we won't grow old!' I think you're going to see an enormous increase in the interest in aging." Yet that, of course, was not the explicit message of the company. Bluntly articulating that goal would make it sound crass, almost quackish, not serious and sober. The early genius of Geron — and, by extension, Mike West — was in stating that goal by inference and indirection, with a shrewd, promiscuous use of scientific language and a form of self-promotional breast-beating that left academic scientists, especially those naive souls in the telomere field, shaking their heads in disbelief.

Newspaper and television reporters, with a few notable exceptions, responded to West and his mission as if the Geron labs were located in Hamlin, not Hayward. To the befuddlement of many telomere biologists who watched this parade from the curb, the amount of press coverage lavished upon a company that had merely piggybacked and overinterpreted its own research continued to stun them — a cultural thermometer in its own right, measuring the temperature of public interest in aging research. The coverage did not necessarily reflect the fact that Geron's scientists were better than anybody else's (although they were actually very good) or that the company's science was providing definitive answers to long-standing biological questions (although it certainly contributed significantly to the telomere story), but it did, at some level, convey just how successful Geron was at getting its company "story" across and just how smitten the public seemed to be with that story. In the early to mid-1990s, several other biotech companies were formed with philosophies or technologies similar to Geron's — Osiris Therapeutics and StemCells, to name two surviving examples, as well as start-ups like Jouvence, Molecular Geriatrics, Life Span, and Apollo Genetics. Most people have never heard of them, and that's the point. Like the generation it sought to serve, Geron planted itself in the limelight and then dragged the light along with it as it grew up, matured into a formidable research and development enterprise, and then aimed itself at real products.

Geron had by no means cornered the market on molecular aging research. Then as now, a number of competing theories aspired to explain the molecular degradations that characterized the process of aging, and nothing had been settled. It had been known since the 1930s, for example, that a severely restricted diet extended the life span of laboratory rats up to 50 percent; that research ultimately found stunning and precise ratification, as we'll see later, in the molecular biology of yeast. Bruce Ames, a distinguished biochemist at the University of California at Berkeley, had painstakingly accumulated a body of persuasive evidence over the years showing that oxidative damage to human DNA and other cellular components, caused by the chemical rowdies known as free radicals, inflicted enough dents and dings in one's DNA over a lifetime to lead to the cellular dysfunctions associated with aging. Researchers who studied fruit flies, like Michael Rose at the University of California at Irvine and Seymour Benzer at the California Institute of Technology, had uncovered evidence that some aspects of life span in these simple organisms could be attributed to genes. And in 1992, Cynthia Kenyon at UCSF began to discover genes in a simple nematode that, when manipulated, could double its life span. So aging, and its possible manipulation by molecular medicine, was not simply a matter of telomeres, and perhaps not even principally a matter of telomere biology. Genes, cellular metabolism, caloric intake, DNA damage — they all seemed to play important roles, and there was no scientific agreement on what *caused* aging.

But the creation of a company is an explicit promise of products, sooner or later, and Geron's mere existence was itself a scientific statement that aging, and its treatment, was no longer a matter of dubious nostrums and unproven herbals, a fringe territory of pseudoscience that relied on customer hope, not scientific efficacy, to drive the market. Indeed, the in-house scientific staff gloried in the sheer bravado of the enterprise. "Geron had a lot of guts," said William Andrews, who joined the company in 1993 and served as director of molecular biology for five years. "Mike West had a lot of guts. It took a lot of guts to start a company on aging, when everybody thought that to work on aging you had to be a quack." In that insular, adrenalized environment, the company nurtured its own creation myths and rituals and celebrations. There were monthly parties called Geronimos, and, continuing the Native American theme, a deserving employee each month was given the "Sitting Bull award" for extraordinary service. A

Geron official once boasted, "All of our vice presidents are marathoners or mountain bikers or white-water rafters." "Tremendous energy," Andrews said of those early days. "Just incredible. We all knew that we were working on something that could change the world. And when we did something, we *celebrated.*"

Even as the company became a darling of the media and venture capital communities, it set teeth gnashing in the academic world. Such unhappiness is not uncommon in contemporary biology, fueled as often by envy as by substantive scientific issues, but in this case it seems to have been pronounced, long-lasting, and occasionally even bitter, a situation in which the stresses become especially apparent because the stakes for the research — in terms of science, money, prestige, and potential medical impact — are so high. Indeed, the suspicions (not to say frictions) that arose between Geron and the academic community in the mid-1990s served as a template for the relationship that has characterized the stem cell field, leading to lawsuits, restrictions on the use of research materials, and, according to some, an unseemly constriction of free academic inquiry in the name of intellectual property. In other words, the telomere story paved the road, bumps and all, upon which all of biology (and society) would later travel during the stem cell story.

The seeds of discord were sown early on. West paid a visit to Elizabeth Blackburn, probably in 1993, hoping to arrange a research contract with her highly productive and respected lab. In cases like this, which have become routine, a biotech company pays an academic lab anywhere from $50,000 to $150,000 a year to conduct research, usually over a three-year period. In exchange, the university typically grants the company exclusive commercial rights to develop whatever discoveries are made and patented by its researchers. Geron began to arrange sponsored-research agreements with leading telomere labs, but the agreements themselves became controversial. As a perfectly legal, even shrewd negotiating strategy, Geron's lawyers often tried to specify the division of labor on research projects in these agreements. As one technology transfer lawyer involved in these negotiations explained it, Geron constantly pushed to have its own people doing human-related work for patent reasons, while the company's academic collaborators would take the lead on animal-based research.

"Very early on," Blackburn recalled, "Michael came by. *Very* early on. When they hadn't even moved to their current quarters. And said, Did we want to do this sponsored project with them? And first of all, I said, like every respectable scientist I know, 'Money for jam? Sure!' You know, this is easy money . . ." Blackburn leaned forward with a self-mocking laugh, as if to suggest that anyone who hesitated about accepting such corporate largesse was, to use one of her preferred bouquets, an idiot. "And then I started thinking, 'No, I don't think I want to do this.' We just wrote a little, simple proposal, but I realized there is no such thing as a division in our lab, and I just realized this was buying trouble. This wasn't even worth it, because it was going to lead to problems of demarcation down the line. And so I pulled out. Our university contracts-and-grants [officer] said, 'You don't usually turn down grants for $400,000!' I said, 'In *this* case, you do.'"

In one sense, Blackburn could afford to take the high road. Her lab was, and continues to be, very well funded, and consistently productive. But academic labs typically scramble for money, which can lead to industrial collaborations that are, in a word, awkward. And Geron quickly became an elephantine presence in the field. "It was a rude awakening," Titia de Lange told me, "to have a company in our midst." In addition to sponsoring academic research in at least half a dozen leading telomere labs, the company organized big telomere meetings, at glamorous resorts in Hawaii or Northern California. It reproached academic researchers who sounded too negative in the scientific literature about clinical applications of telomere research. "They have been corporately very generous," said the Whitehead Institute's Robert Weinberg, a latecomer to the telomere field, "but they also displayed an unseemly aggressiveness in their attempts to coopt and dominate the entire field of telomere research, by hiring on all the important research players in academia, securing their intellectual property, and then consuming their research for the sake of the corporate agenda."

The telomere field still had a friendly, cottage feel to it, but Geron's involvement charged the atmosphere. "The commercial interests brought us a lot of good things," de Lange said. "They advertised our field, and that benefited us tremendously. It's made it easier to get published, to get funded, to get people to come to your lab. But it's a different style. It's much more pushy. We were used to dealing with very basic research, and nobody would talk about human applications to cancer and aging. Then all of a sudden, in your midst, there are people saying, 'We are going to cure cancer, we are going to cure aging.' It's a very different message. We weren't interested in hearing that message," she added, "but a lot of other people were."

⸙

Of all the research collaborations arranged by the new company, probably none was initially more important — or ultimately more problematic — than the agreement with Carol Greider to clone the telomerase gene. "Cal Harley and I were collaborating on it, and Cal moved his lab to Geron, so I was collaborating with the company," she recalled. "I was nervous about it. As it turns out, I should have been much more nervous."

Greider had tremendous stature in the field. She had been the first author on the 1985 paper that demonstrated the existence of telomerase. But Greider — and this was not well understood outside the immediate field — had not actually found the complete telomerase molecule. What she had done in the mid-1980s — and it was no trivial accomplishment — was to design very exacting experiments that conclusively demonstrated the *activity* of that molecule, and in 1989 she had isolated a small RNA portion of it. Still, nobody had plucked the entire molecule out of the cells of any organism, much less those of humans, and nobody had discovered the gene that encoded the protein structure of this fascinating enzyme. That was the ultimate prize, because if you found the telomerase molecule in protozoans or yeast, or in a hippopotamus for that matter, the general thinking was that having one version of the molecule would make it immensely easier to find the analogous enzyme in humans. And once you had the gene in hand, you could finally do experiments proving what it could — or couldn't — do. More to the point, discovering the animal form of telomerase was key to building an intellectual property estate around the technology. Patents begot subsequent rounds of funding for more research, more research begot more patents, and Michael West was acutely aware of the power of patents. So there was a keen desire, in both academic and commercial circles, to identify the telomerase gene.

As a matter of scientific strategy, the obvious place to look for the gene was once again in that hearty denizen of pond scum: *Tetrahymena.* As a matter of scientific protocol, the obvious person to lead the search was the person who showed that the enzyme existed in the first place: Carol Greider. But as if to prove there's no sentiment in industry, that obvious choice proved surprisingly controversial. According to Bryant Villeponteau, he wrote a grant that he planned to submit to the NIH in November 1992, proposing to clone the telomerase gene, but when he showed a copy of the grant application to several members of the Geron scientific advisory board, all hell broke loose. "Carol Greider was incensed because she didn't want Geron stealing her project, and she went to Jim Watson about it," Villeponteau recalled. "She wasn't happy, and then he wasn't happy, and they basically gave the board an ultimatum that if we sent in the grant, they were going to pull out." So Geron's first and most visible project triggered an in-house turf war.

It took a peace conference in Dallas to iron out the differences. "We were meeting to talk about how to work out a 'collaboration' between my lab, Geron, and the Wright-Shay group," Greider recalled. "They all wanted to 'help' on the project going on in my lab to clone the RNA component of telomerase. I saw it as treading on my graduate student's toes; they saw it as putting more hands on an important project. So to work out the issues, Cal and I flew to Dallas and met together with Mike West and Alex Barkas." The meeting aimed to mollify Greider, but it ended up stirring old resentments about the Greider-Harley grant that Wright had voted to turn down in 1990. "The thing that gets me," Greider said later, "is that Woody had the gall — over dinner at his house — to tell me that he rejected the grant, but he thought it [the telomere story] was so interesting that they started working on telomeres. That was at a time when Mike West was in the Wright-Shay lab." Inspired by the promise of telomere research, Wright worked on the ideas academically, and West founded Geron on them.

In time, outside competition made the situation more tense and difficult. The hunt for telomerase became an international race. Geron and Greider would ultimately be joined by the lab of Thomas Cech, a Nobel Prize winner, at the University of Colorado; Vicki Lundblad at Baylor; Roger Reddel in Australia; several groups in England; and, unbeknownst to anyone, a group at the Whitehead Institute in Cambridge, Massachusetts, in the lab of Robert Weinberg. And that was just the academic competition. Amgen, the largest and most successful biotech company, and other biotech companies would ultimately join the chase too. It wasn't a race; it was a mosh pit.

What kept the competition unusually open is that telomerase was a kind of apple-and-orange molecule that came in two parts. The smaller, nucleic acid portion is called the RNA component; the other, larger part was the so-called catalytic protein. Each component part had its own gene. Each part sparked its own race.

ॐ

There is a certain giddiness in the way Geron reported its progress in these endeavors, and also in the way the press quickly grasped the metaphoric possibilities. In April 1994 a *New York Times* business story ably captured the larger implications. "Throughout fiction and mythology," it began, "seekers of eternal youth pay a heavy price for finding it, as life refuses to be anything but finite. Cells that aspire to immortality face a desperate end as well: they become cancerous, replicating without ceasing until they kill their host. Now scientists have confirmed the presence of an 'immortalizing enzyme' in tumor cells, and they hope it may offer a pre-eminent target for anticancer drugs." The *Times*, barely two years after Mike West's performance at Redwood Shores, was referring to Geron's quest for the "immortalizing enzyme," whose presence had been detected in many types of tumor cells. Soon, like a nouveau riche family with houses in both city and country, Geron enjoyed watching each update in the telomere field, each publication or press release, find a home in both the financial and science sections of the press. In that same *Times* story, a Geron scientist famously predicted that the company would begin clinical trials with a telomerase inhibitor by the spring of 1997. (As of this writing, in the fall of 2002, that still had not happened, although one promising compound was on the verge of testing.)

Geron's first big media splash, however, occurred in September 1995, when the company announced that its researchers had cloned the smaller, RNA component of telomerase — the rubber stamp portion of the molecule. This news made a big splash in the scientific press, as papers that appear in *Science* tend to do, but its ripples reached the popular press. In a company press release, Geron noted that "telomerase is believed to be an 'immortalizing enzyme'" and that the company had "established a leadership position in the biology of cell senescence, which appears to play a causal role in aging." Once again, popular renditions of the science connected the dots, sometimes with considerable elegance. Writing in the *New York Times*, Nicholas Wade incorporated ancient mythology to explain the latest in molecular biology: "The span of human life is determined by fate — in Greek mythology by the three Fates, Clotho, Lachesis, and Atropos.

Clotho spins the thread of each person's life, Lachesis marks off its length, and Atropos cuts it. Molecular biologists have recently come up with a surprisingly similar explanation for the mechanism of mortality. Their thread is a tassel of DNA tacked on to the ends of chromosomes and known as a telomere, from the Greek words for 'end section.' And their counterpart to the three Fates is a strange enzyme known as telomerase."

The erudite press accounts created a mythologizing sheen and feel-good aura about this kind of research. They overlooked, however, the messy scientific story that had unfolded behind the scenes during the previous two years. Turf battles notwithstanding, Greider's group had had problems cloning the RNA part of telomerase, according to Villeponteau, and she had resisted all initial offers of assistance from Geron's in-house scientists. "In the fall of 1993, about nine or ten months later, she relented," Villeponteau recalled, and allowed Geron's in-house biologists to join the effort. "She hadn't been successful, and so she said that, under these limitations, you can help. There were three different approaches, and my approach worked. We pulled out the RNA component in about five months. It was a coup. That was really Geron's finest hour. That was a very important result, scientifically and financially." The work was indeed important. It gave Geron scientific credibility, and was the first of many big hits by the company's talented research staff. Several years later, when the patent on the discovery was finally issued, four Geron scientists — Villeponteau, his wife, Junli Feng, Walter Funk, and William Andrews — won recognition from the United States Patent and Trademark Office as runner-up Inventors of the Year for their work, second only to the discoverers of the protease inhibitors used to treat HIV-infected individuals.

Ariel Avilion, a graduate student in Carol Greider's lab, had worked on the telomerase project as her doctoral thesis, but when it came time to write it up, Geron officials sought to prevent her from including the DNA sequence of the gene in her dissertation, with the argument that it might undermine their patent application. This tension between academic and corporate interests is not unusual, but in this case it became such a contentious issue that even the director of Cold Spring Harbor Laboratory, Bruce Stillman, had to weigh in on Avilion's behalf. Ultimately, Avilion included the information in her dissertation and Geron obtained its patent, but this skirmish adverted to the kinds of squabbles that could, and would, occur in connection with other regenerative medicines, especially research on stem cells. "It does get very complicated even when you have an academic collaboration," Greider said wearily. "So no collaboration is easy, and that one was a little bit difficult."

Greider was being a little bit circumspect. John Maroney, the technology transfer lawyer at Cold Spring Harbor Laboratory who negotiated with Geron, detailed a long and tempestuous interaction with the company over both the terms of its sponsored-research agreements and the terms by which Greider's lab could share the telomerase gene with other researchers. "It was our impression," Maroney said in an interview, "that Geron was taking the juicy parts of the results and then going after them intensely, without giving appropriate opportunity and deference to Carol Greider's group. This created a real tension. The other part was that they chose to assign people to the tasks that lay before everyone in such a way that only their employees, Geron's employees, would participate in the key inventions, and so excluded [academic researchers] from work that would lead to patent applications. It didn't happen purely as a logical assignment of work. It was part of a strategy. It was not an accident that we got the mouse and they got the human."

Another area of tension concerned the legal terms by which the telomerase material was shared with other researchers — the so-called material transfer agreement, or MTA. The moment Cold Spring Harbor sent out its RNA-component gene with the legal language that Geron desired in the MTA, Maroney began to hear cries of complaint from very high places — in one instance, from the lab of then NIH director Harold Varmus. "I think the problem with Geron's MTA," Maroney said, "is the problem with all MTAs — the reach-through rights," that is, the company's insistence on commercial rights to any discoveries made with its material. The Geron MTA caused such a ruckus among academic researchers that Maroney "felt very strongly that Cold Spring Harbor Lab should make a unilateral patent claim on telomerase. That would have occasioned an explosion, but I was prepared to do that. Normally, we would have." But a "combination of factors" at the Long Island lab, Maroney said, prevented such a dramatic confrontation. He declined to mention any specifics, but the presence of James Watson, the lab's longtime president and elder statesman, on Geron's SAB certainly must have complicated matters.

All in all, these were precisely the problems that Liz Blackburn had sniffed in the jam pot when she turned down Michael West's offer of research support. Just as important, this would be the same jam pot that, several years later, would make the intellectual property issues surrounding stem cell research so sticky.

༄

When scientific papers finally come out in journals, researchers have usually been working for months on the next step. As Cal Harley put it, "The RNA component was a big achievement, but it wasn't the brass ring." The brass ring was the so-called protein component of telomerase. In recognition of his success on the RNA component, Villeponteau was assigned to lead this next phase of the cloning project, but it was his turn to play Sisyphus. "I was on that project for two and a half years, and it was eventually successful," he said. "But unfortunately, here the problem was pulling out a very rare gene." The Geron team had in-house arguments about the best way to proceed. "We also had wrong information from Carol Greider," Villeponteau recalled, "who had supposedly found the *Tetrahymena* protein, but really hadn't."

At first, Geron's aggressive program to court academic researchers appeared to have paid off handsomely. In 1995 — before, actually, the *Science* paper on the RNA component came out — Greider and a graduate student, Kathleen Collins, reported the isolation of crucial components of the second, protein part of the enzyme. They didn't have the whole beast in hand, but to everyone who read the paper, it looked as though they had identified two key parts — an arm and a leg, as it were — called p80 and p95. These parts came from *Tetrahymena*, but this was nonetheless a critical step, because it could form the basis of a very important patent and could seemingly provide decisive information that would allow any good molecular biologist to search for and find the human gene. Greider and Collins published their results in *Cell*, and although Greider insists that another year or two of confirmatory work still lay ahead, it appeared to some people that the hunt for the second telomerase gene was already over. "Everyone thought it had been done, and they'd all been scooped, and too damn bad," Titia de Lange recalled with a laugh.

One of those people was a Swiss-born researcher working in a laboratory in Colorado. When Joachim Lingner learned that telomerase had already been tracked down by Greider and Collins, he felt stunned. He'd been trying to do the same thing for nearly two years, and now he'd clearly been beaten. But Lingner had managed to make a fair amount of progress of his own, and when he took the time to compare the Greider data with his, he realized, as would researchers at Geron and elsewhere soon, that Greider and Collins's paper was wrong. This development exasperated Michael West, of course; Geron was in the process of preparing an initial public offering on the stock market that would make West and all the other pioneering people at the company very rich. But West was not as upset as you might have expected, because at about the same time, he had developed another case of wandering eye.

5

CONTROLLING THE

HEADWATERS

PATHOLOGY LABS ARE USUALLY DREARY and sobering lessons in the frailty of the flesh. Medical students poke through organs ravaged by disease — a heart from someone who ate too many cheeseburgers and french fries, a smoker's lungs, an alcoholic's liver, precious sweetbreads and viscera distended by tumors. One day during medical school in Texas, Michael West stared down at an unusual tumor in his dissection pan. The tumor is known as a teratoma or, when malignant, a teratocarcinoma. It is a rare form of cancer that typically develops in a cell of the reproductive organs, appearing adjacent to either an ovary or a testicle, and it is arguably the most bizarre amalgam of run-amok tissue that can occur in human beings. It behaves something like a cancerous embryo, capable of forming any of the body's two hundred or so tissues, but without any of the embryo's exquisitely honed biochemical checks and balances that herd cells toward normal development. "Oftentimes, the way [these tumors are] diagnosed is that a woman, for example, will have some abdominal pain and discomfort, and she'll go in to her doctor, and he'll do an x-ray of the abdomen and he'll see teeth!" West told me, recalling the episode. "And he'll ask her, 'Well, do you have dentures, and did you swallow them?' And she says no. Then he says, 'Well, you have a teratoma.'" As they spin out of control, these primitive cells recapitulate the process by which embryonic cells hurtle themselves down particular developmental pathways to form adult tissues.

West was fascinated by the teratoma in his dissection tray that day. "I opened it up, and sure enough, here's an incisor and a molar," he said, in an anecdote he has frequently recited in interviews. "Beautiful pristine teeth. And the first thing I thought is, Can we make that happen in a dish? People need those cells. I asked my professor, 'Do we know what causes these?' And he said, 'Well, you know, it's reproductive pathology.' And I said, 'I know, but what cells form these tissues, and how does it work?' No one had a clue. I looked in textbooks, and there was hardly anything on it. And I just stumbled across some stuff on mouse embryonic stem cells, which pointed out that if you put a mouse embryonic stem cell in perfect conditions, it'll form a teratoma. And all of a sudden, it all made sense — that of course these are very primitive cells gone awry. Normally they travel into the gonads and form reproductive cells, but they kind of miss the mark and form this abnormal growth. And I thought, If we could just culture human embryonic stem cells, we could potentially make teeth and lots of other things for people."

It didn't take a genius or a visionary to reach this conclusion. Medical students have been pawing through teratomas for generations, and many no doubt have marveled at the stunning, grotesque little clumps of tissue containing not just teeth, but hair, skin, occasionally eyeballs, and even partially formed organs. Some students have perhaps even marveled at the miracle underwriting biological development, where a single cell, in the form of a fertilized egg or a wayward germ-line cell, can, for lack of a better word, unfurl — no single verb can truly capture the immensity and complexity of this biological feat of sheer *becoming* — into a mosaic of mature, distinct, functional, and cooperating tissues. Indeed, a haywire mosaic is the mark of one of these patchwork cancerous tissues, but a perfectly composed mosaic is what we call an active, living human being. So there was nothing new about teratomas. Unlike the tens of thousands of medical students who have examined similar tissues, however, West immediately began to think about how that remarkable power might be harnessed for medical purposes.

West began to familiarize himself with the history of stem cell research, the major players, and the state of the art around 1990. Telomeres were still in the foreground of his thoughts at this time; Geron had just committed its resources (not to say its reputation) to the race to clone the telomerase gene. But West's mind increasingly became a private battleground between telomere biology and stem cell biology. In time, that internal tussle would spill out into a corporate battle for the very soul of the company he had so recently founded.

⁂

Strictly speaking, the origins of embryonic stem cell research date back to the early 1980s, when researchers in Cambridge, England, led by Martin Evans isolated the first mouse embryonic stem cells. About that same time, a researcher at the University of California at San Francisco named Gail Martin isolated a version of the same cells. But there were intriguing early hints about this same class of regenerative cells in cancers known as embryonal carcinomas during the 1970s and, in a broader sense, in the purely observational annals of nineteenth- and early-twentieth-century medicine. Mark Pittenger, a researcher at Osiris Therapeutics in Baltimore, points out references to "wandering cells" in old medical journals, regenerative cells that appeared to gravitate to wounds and facilitate healing. Some of the first clues, Pittenger notes, were observed by battlefield surgeons, who operated on grievously injured soldiers who were not expected to survive their wounds; when some patients recovered, the doctors became curious and began to take a closer look at the healing process. Julius Cohnheim, a famous nineteenth-century German pathologist, was among the first to assert that the "wandering cells" seemed to play a role in tissue regeneration.

Since the early 1960s, stem cells — from adults, not embryos — have been a useful, accepted, and indeed lifesaving part of conventional medical practice throughout the world. Every successful bone marrow transplant succeeds because of stem cells — a transplant that "takes" does so precisely because hematopoietic (or blood-forming) stem cells from a donor recolonize the patient's depleted marrow and totally rebuild the blood and immune systems from scratch. In fact, bone marrow transplants have increasingly evolved into stem cell transplants, as doctors have learned to isolate these potent cells either from umbilical cord blood or even from circulating blood drawn from a patient's arm. It's just that they were rarely called stem cells for most of the forty years they have been in medical use.

What exactly is a stem cell? That is a more controversial question to answer now than it was just a few years ago, following a host of preliminary but intriguing reports suggesting that brain stem cells can form blood, and blood stem cells can form neurons. This degree of developmental versatility has forced scientists to rethink the definition of the term. For practical purposes, however, everyone would essentially agree on two key criteria. One, embryonic stem cells have the capacity to self-renew. In a sense they are immortal; unlike Leonard Hayflick's normal human cells growing in a dish, stem cells theoretically never reach a limit to replication (although populations of cells may "crash," or stop acting immortal, after being maintained in vitro for a number of years). Two, they are "pluripotent" — that is, they possess the potential to become any of some 220 or so different cell types in the body. They can become not just brain cells, but three different classes of brain cells (neurons, astrocytes, and glial cells); not just blood cells, but the full hematopoietic rainbow (red blood cells, platelets, antibody-making B cells, T cells, eosinophils, and so on); not just bone marrow cells, but the gristle and brick of the human edifice (muscle cells and cartilage and ligament and bone and fat and stroma, the spongy upholstery lining the inside of bones where the blood-forming stem cells nestle to work their magic). They were truly protean, all powerful, and all potent ("totipotent," in the lingo) when they could also give rise to the reproductive cells, egg and sperm, as well as nonembryonic tissue like the placenta. Of all the miracles that modern biological research has unearthed and laid upon humankind's hearth, this cellular aptitude for change and specialization may be the greatest. "With stem cells," said Ronald McKay, a researcher at the National Institute of Neurological Disorders and Stroke in Bethesda, Maryland, and one of the earliest scientists to grasp their potential, "you tickle them and they jump through hoops for you. It's as if we have the power to build the machine. It's not that we can take it apart and count the parts and try to put it back together again. We're saying, 'Take this cell, and this cell will build it for you.'"

In 1992, even while he scrambled to set up Geron's first labs and get the telomerase cloning project rolling in Northern California, Michael West began to scratch the stem cell itch. Every now and then, he would drop in on a professor at UCSF named Roger Pedersen. Pedersen ran the university's in vitro fertilization clinic and was an expert in human reproduction. He had long been interested in the basic biology of human development: how fertilized eggs turn into complex, multiorgan creatures. Pedersen had conducted research in this area for many years, and he would have attempted even more ambitious research had it not been for the constraints placed on fetal experimentation by a series of administrations, beginning with that of Jimmy Carter and continuing through those of Ronald Reagan and the first George Bush.

When I visited Pedersen in his office several years ago, I was immediately struck by his punctilious mannerisms and appearance. He was dressed in slacks and a brown pullover sweater, had neatly trimmed dark brown hair and a dark mustache, and wore dark-rimmed glasses; he sat at a small desk, his legs crossed almost into a tourniquet. He didn't seem especially at ease talking about reproductive biology and stem cells; in fact, he insisted that a representative of UCSF's public relations office sit in on the conversation. Despite his reserve, despite the almost postural way he sought to make himself invisible and anonymous, Pedersen would ultimately play a critical, if largely reluctant, role in transforming stem cell science from a purely academic pursuit, done primarily with mice and other lab animals, into a thriving, controversial area of biotechnology devoted to human cells and tissues.

Pedersen's interest in stem cells dated back to the 1970s. He studied early development, and had been interested in growing cells culled from mouse embryos — the feat first achieved by Martin Evans in England in 1981. In this procedure, researchers would flush an early embryo from the womb of a pregnant mouse and then isolate the stem cells for research. Evans's work inspired many biologists to use the laboratory mouse as a kind of living, breathing genetic blackboard. They would remove a single stem cell from a mouse embryo, erase a single gene in it, reinsert it into the very early embryo known as a blastocyst, and then allow the animal to develop to see what would happen to animals lacking that single gene (these were called knockout mice because you knocked out a gene). Similarly, you could insert single genes into a stem cell and repeat the whole process, seeing what effect this genetic addition would have on the animals (these were called transgenic mice). This experimental addition and subtraction of interesting genes has become a critically important research tool in recent times because it's almost the only way, short of ethically unacceptable human experimentation, to make sense of all the genes spit out by the Human Genome Project. But a handful of researchers saw different possibilities in the technology. They were interested in going back to the very beginning, to the Big Bang of fertilization, and trying to understand how early development unfolded. What factors — what chemicals, what maternal prompting and physical cues — knocked primordial stem cells out of their immortal spin cycle and caused them to peel off toward one fate or another, down one path to become neurons, down another to become liver or muscle or blood?

"I'm a student of the embryo," Pedersen told me that day in his office, "and I look to the embryo to teach me how it controls differentiation. And what I see is that the embryo can take a pluripotent cell, including an embryonic stem [ES] cell put into it, and make *anything*. So in principle, we should be able to make anything if we create the right microenvironment." Decades of government and foundation funding had been thrown at that mystery, but there is a fundamental paradox at the heart of embryology: brilliant minds have whet their keen intelligence on the stone of lowly animals — a menagerie of mice and fruit flies and zebra fish and rats and nematodes and a flabby frog called *Xenopus*— but they have never been able to extend that research to human biology. This despite the fact that the ramifications of such research in humans — in preventing birth defects, in treating infertility, and in generating basic knowledge about one of the signal events of life on the only planet known to possess it — are as great as in any province of science. Developmental biologists were all dressed up but had no place to go. While the recipe and choreography of vertebrate development had not been entirely figured out, the animal research created what Pedersen called a "trajectory of knowledge," and that trajectory intersected a forbidden target. "What we ultimately want, right?" he said. "A molecular explanation for development. And that trajectory converges on the existence of human embryonic stem cells." And therein lay the problem: all the funding stopped when you added the word *human* to the phrase "embryonic stem cell." By tacking on that one qualifying adjective, tens of millions of dollars in potential research funds went up in smoke, consumed by the political and ethical fires that have burned, unchecked, for thirty years around the issue of human embryo research.

That is why, to weary and beleaguered embryologists like Pedersen, Michael West appeared at the door of their labs like a godsend, like a can-do capitalistic angel. West shared their scientific enthusiasms; he understood the political and ethical obstacles they had endured and that still lay before them; he appeared to echo the moral gravity with which they wanted to approach any research involving human embryos and fetuses; and he seemed to offer a logistical and financial alternative that would allow human embryonic stem cell research to proceed.

"I started trying to make it happen," West said. "And I talked to Roger Pedersen at UCSF about the ethics of it. The first thing that concerned me was really not, Can we get away with it? but my own personal concern: Was it ethical? I wasn't sure that it was, and I certainly wasn't convinced that the public would feel that it was, for the same reasons as other people had squawked at. But after I thought it through with him, he told me he actually had a long-term interest in trying to get a human ES cell. And it had also occurred to him that that would be a nice project. And so we started talking about it. I said, 'Well, look, maybe I could sponsor work in your lab to get them.'"

"He did approach me in early days about the possibility of funding," Pedersen confirmed. *Very* early days — it was probably in the fall of 1992, within a year or so of West's rainmaking performance at Redwood Shores. But the potential of stem cells utterly captivated West, an enthusiasm he quickly transmitted to his team of backers. "He said he had investors who would be interested in supporting that kind of work," Pedersen continued. "Because I was talking at that point about the desirability of studies to derive all the differentiation of human embryonic stem cells, starting before 1995 for sure. And Mike somehow got wind of that. I don't know how. But he came and talked to me."

Pedersen, however, was not ready to accept industrial support, and for a very provocative reason. "I said, 'No, I'm not interested,'" Pedersen recalled, "'because I think that this area of investigation is something that is so at the headwaters that it's not appropriate for private investors to control the headwaters of the river. It's something that could benefit all of the people, and therefore it should be developed by the people, by the federal government.' And I actually proceeded to — how shall I say this?" His voice fluttered with a kind of naïveté. "I proceeded from there to an era of hope that that would be the case."

Twelve years of Republican administrations in the White House had coincided with twelve years of a de facto ban on embryo research, but Bill Clinton had just been elected president in the fall of 1992. Pedersen, like many other scientists, was convinced that the political environment was about to become much friendlier to research on human embryonic and fetal tissue. "I was pretty sure that, from things Clinton had said, his position was such that he would make a change," Pedersen said. Because of this, Roger Pedersen politely told Mike West that the timing wasn't right. In a few short, politically traumatic months, Pedersen would discover just how quickly the weather could change.

ᔕᔐ

In the spring of 1993 human embryonic stem cell research became formally, and fatefully, entangled in the long, controversial history of reproductive medicine, by which is meant not only that it moved into the orbit of abortion politics, but also that it fell under the gravitational pull of federal laws regulating fetal research and of America's conflicted social attitudes about in vitro fertilization. This age of social unease began on July 25, 1978, when Louise Brown, the first so-called test-tube baby, was born in England. From that moment forward, the science of in vitro fertilization — facilitating the union of a sperm cell with an egg cell (or oocyte) in a lab dish, and then implanting the resulting embryo into a mother's womb — began to revolutionize opportunities for infertile couples who wanted children. Since that first brown-haired prototype rolled off the line, an estimated 100,000 children in the United States, and more than a million worldwide, have been born through in vitro fertilization, or IVF. But at a critical moment in the late 1970s and early 1980s, American research into human reproductive biology hit a political fork in the road that, a generation later, determined the direction, itinerary, and speed limit of stem cell research, too. "You can't really understand the present controversies," said John C. Fletcher, a retired bioethicist at the University of Virginia, "until you understand how all this began with fetal research after *Roe v. Wade*."

There is no better exhibit of America's social and political ambivalence about reproductive biology, embryo research, and related issues than the late Pierre Soupart, a fertility expert at Vanderbilt University, who remains infamous among reproductive biologists to this day as the scientist who died waiting for a federal research grant that was approved but never funded. In 1972, well before Louise Brown's birth, Soupart became the first American scientist to show that a human egg could be fertilized in a petri dish, outside the body; several years later, he asked the NIH to fund a three-year research project assessing the potential safety of in vitro fertilization. The NIH approved the grant in 1977, but Soupart's proposal was caught up — ground up, really — in the government's deliberations on the ethical conduct of reproductive medicine.

Prior to the early 1970s, unregulated embryo and fetal research occurred in the United States and elsewhere. Some of it was grisly — neurological researchers in Finland decapitated aborted fetuses, for example, and researchers in the United States once immersed fifteen living fetuses in a salt solution to learn if they could absorb oxygen through the skin (one fetus survived this experimental treatment for twenty-two hours). All of it was unregulated and unmonitored. Recognizing the potential for abuse, the government began to explore the feasibility of conducting embryo and fetal research under the larger bioethical tent of protecting human research subjects. In 1975 the National Commission for the Protection of Human Subjects of Biomedical and Behavioral Research issued a landmark report that concluded, among other things, that a human fetus deserved protection as a human subject in such research but was in fact a legitimate object of scientific inquiry, and that, indeed, fetal research was vital to improving human health. "That was the first time that was ever said in biomedical ethics," said Fletcher, who was a bioethicist at the NIH at the time and has chronicled this early history in a paper written for the Clinton-era National Bioethics Advisory Commission. But in sanctioning the notion of research on human embryonic and fetal tissue, that early commission added a balancing regulatory component, insisting that a governmental Ethics Advisory Board had to approve research that exceeded "minimal risk" to the human subject, and that in special circumstances — including embryo and fetal research — the secretary of the Department of Health, Education, and Welfare, as it was then known, had to waive the "minimal risk" provision. With congressional bidding, HEW established the Ethics Advisory Board (EAB) in 1977 under Secretary Joseph A. Califano, Jr. It considered not only Pierre Soupart's proposal, but the larger question of whether the government should even fund in vitro fertilization research.

By law, no embryo research — or any research into in vitro fertilization, for that matter — could receive federal funds without the approval of this board. In 1978 and 1979, during Jimmy Carter's administration, the EAB held meetings, discussed the ethics of embryo research, and ultimately approved Soupart's proposal; by then he had been waiting two years to begin his research. Reflecting the mores of the day, the EAB went so far as to say that the government could fund in vitro fertilization research, including the study of embryos created through IVF that might ultimately be sacrificed — as long as the embryos came from lawfully married couples. Since embryo, fetal, and in vitro research potentially involved more than minimal risk to subjects, however, funding still required one final approval: a waiver from the secretary of HEW. Fletcher and others believe that Califano would have signed off on the recommendation. But by the time the ethics board finally reached its decision, Califano was gone, forced to resign from the Carter administration in September 1979 for his aggressive campaign against the tobacco industry. He was succeeded by Patricia Harris, a Carter appointee who opposed the funding of IVF research, though not on ethical grounds. "She said infertility was a middle-class and upper-class problem," Fletcher recalled. "The official view was, and probably still is, that infertility wasn't a disease." Harris refused to approve Soupart's proposal, or IVF research in general. Then, when the EAB's charter lapsed, "it was summarily disbanded" by Harris in September 1980, according to former EAB member Albert Jonsen, who writes in *The Birth of Bioethics*, "The Ethics Advisory Board still hovers as a ghostly presence in the Federal Regulations, charged with mandatory review of certain types of research, but it exists nowhere in reality."

Torn by the ethical implications of embryo and fetal research, and possibly intimidated by antiabortion protests, HEW took no action, and the political winds grew chillier still for government-financed research after the 1980 election. Beginning with Ronald Reagan's first term in 1981, the ethics board became a bureaucratic pawn in a strategy to block all forms of embryo and fetal research. Technically, the legal need for the panel continued to exist. It was just never rechartered or allowed to convene and thus could never satisfy the regulatory requirement to approve potential research. Although fetal tissue research was still possible without the EAB's approval, all research involving human embryos disappeared into a regulatory rabbit hole worthy of Lewis Carroll. Pierre Soupart died in 1981, having waited in vain for years for funding that had been approved by every pertinent body, including the NIH and the EAB. More important, the notion of public oversight for embryo, fetal, and in vitro research died too.

"It was a kind of catch-22," explained Fletcher, who sees today's cloning and stem cell debates as having been foreshadowed and shaped by these previous controversies. "As a result, nobody sent in any proposals on embryo research or fetal research, and they still don't." This is not to say that the research ceased; to the contrary, it flourished, but in the unregulated environment of the private sector. In the United Kingdom, by contrast, a parliamentary committee headed by Dame Mary Warnock, a moral philosopher and member of the House of Lords, issued a report in the mid-1980s assessing the legal, scientific, and ethical issues at play in reproductive medicine and established guidelines for a government agency that could oversee and regulate such research in both the public and private sectors. In 1990 the British Parliament approved the creation of the Human Fertilisation and Embryology Authority, which has subsequently provided oversight to fertility research and clinics, fetal tissue experimentation, and later, both stem cell and cloning research. As a result, ethical monitoring of research in Britain is considered much tighter than in the United States, and the government maintains precise statistics on the number of frozen embryos stored at clinics, the number of in vitro births, and the success rate of the procedures.

Exactly the opposite situation prevailed in the United States, and that, almost everyone agrees, has been a disaster. The lack of oversight has led to ethically dubious medical practices and a well-chronicled horror show of abuses, many of which are described by Lori B. Andrews, an expert in reproductive law, in her book *The Clone Age.* Beginning in 1980, when 20,000 attempts at in vitro fertilization resulted in only 3 successful pregnancies, fertilization doctors often misled women about the success rate of the technology, at the same time minimizing health risks to women undergoing the procedure. Embryos were stolen or given to the wrong couples. Ricardo Asch, a fertility specialist at the University of California at Irvine, was forced to flee the United States after illegally bestowing dozens of embryos on couples for whom they were never intended. Physicians often misrepresented the source of sperm used in artificial insemination, and nearly 150 cases of AIDS among inseminated recipients were later traced to poorly screened sperm donors. In the absence of federal oversight, the field of assisted reproduction quickly became the black sheep of academic medicine, a discipline that has sired an appalling amount of personal heartbreak, sensational litigation, cowboy-style medicine, and abysmal success rates — "truly the Wild West of medicine," according to Andrews. Between 1980 and 1997, according to a recent editorial in *Science,* there was a 400 percent increase in the number of triplets born to women in their thirties, and a 1,000 percent increase in babies born to women over forty, when the risks to a developing fetus are much higher. The American practice of assisted reproduction has unfortunately become a textbook case of the way medicine develops when it proceeds without the scientific oversight or moral suasion of federal funding agencies. And that was exactly where stem cell research seemed to be heading if the government failed to support it.

But, paradoxically, the rise of in vitro fertilization conveyed another message: that market economics can sustain a shadow technology, and even allow it to thrive. Assisted reproduction, despite its uneven success record, has become wildly popular and profitable. In the past two decades, an estimated 600,000 Americans wanted babies, wanted them immediately, and were willing to try assisted reproduction technologies that cost about $10,000 per IVF attempt. When baby boomers want something, they create an enormous economic vacuum of consumer need. Modern medical research — regulated or not, reduced to practice or not — abhors that vacuum and rushes in to fill it. Despite the mixed-up embryos, despite lives and marriages ruined by the failure of in vitro fertilization, there was a market for it, and that market was driven by the first wave of boomers. In the 1990s that same cohort, a decade further along the conveyor belt of life, would hear a different kind of biological clock ticking, and would articulate a different kind of biological need. Appropriately enough, it was the first baby boomer president who addressed the problem.

The discovery of mouse stem cells in the early 1980s certainly upped the ante in the high-stakes game between developmental biologists and antiabortion activists. These cells provided a Rosetta stone for deciphering the biology of early life, and as Pedersen noted, the trajectory of research headed in an unmistakable arc toward human stem cells and human embryonic development. That trajectory had of course been frozen in flight for well over a decade, but the election of Bill Clinton in the fall of 1992 seemed to augur a change. "He immediately supported, upon election, a change in the regulations," said Pedersen, who monitored the changing political landscape with increasing excitement. Indeed, in January 1993, during his very first week in office, Clinton issued an executive order rescinding the federal ban on fetal tissue research. Following the president's lead, Congress, in a form of addition by subtraction, passed the NIH Revitalization Act in June 1993, which finally dissolved the regulations requiring the problematic Ethics Advisory Board and explicitly forbade executive branch interference into fetal tissue research. Even before Harold Varmus arrived as the new director of NIH in the fall of 1993, the institutes had taken steps to convene an expert panel that would examine the potential of human embryo research and suggest ways that federal funding might be implemented.

For the moment, Mike West was frozen out of the action — there was no need for researchers to go to a fledgling company for funding when it looked as though the NIH, with much deeper pockets, was about to get into the game. But West could bide his time. Stem cells were still a goal in the middle distance. And there was plenty to do closer to home. Get Geron's name out there, generate some buzz, placate the academics, who were growing increasingly restive at what one journalist derisively called the company's "fountain of hype," and most of all, get the human gene for telomerase, and get it first. Balancing the airy scientific vision that West routinely dispensed was a very practical side as well: getting there first meant patents, and patents meant control of intellectual property. So he was content to paddle around the headwaters by himself, watching the situation develop, until, irony of ironies, the same antiabortion forces with whom he'd once marched in picket lines would get in the way, and provide him with his main chance.

6

"THE WHITE HOUSE
WAS NERVOUS . . ."

ONE OF THE FIRST MISTAKES Harold Varmus made as the new director of the National Institutes of Health, he reckons, occurred on the morning of February 2, 1994, when he walked into a conference room at the Pooks Hill Marriott Hotel in Bethesda, Maryland, and greeted nineteen members of a special government advisory panel too effusively. This group of experts — doctors, ethicists, developmental biologists, lawyers, and others — was being asked to handle a particularly hot potato: how the federal government should fund research involving human embryos. Varmus, the tall, dark-haired, briskly efficient molecular biologist who shared a 1989 Nobel Prize with J. Michael Bishop for their work on genes related to cancer, had intended simply to open the meeting, outline the general task, and thank these high-end volunteers for tackling a very difficult issue.

"I made the mistake of saying how close I felt to many of the members," Varmus recalled with a small laugh one summer afternoon. He was wearing a T-shirt, shorts, and a baseball cap with the insignia of the minor-league Brooklyn Cyclones as he sat on the esplanade overlooking the East River near his New York City home. "One of the members was my personal physician when I was in San Francisco, Bernard Lo. And a couple of them were personal friends. You know, I never should have said that! Really dumb! And I got beaten up by some of the fiercest pro-lifers, who said, 'This whole thing is a sham. The new director has set up the committee that he wants.'" In terms of Beltway gaffes, it was nowhere near the level of Monicagate — although, in an odd way, that Clinton scandal, too, may ultimately have exerted a subtle effect on the politics of stem cell research — but the reaction to Varmus's heartfelt greeting instantly educated the new NIH director, the panelists, and the scientific community at large to the hair-trigger emotions that animate the politics of embryo research.

One of the problems that arises during national debates on thorny ethical issues like embryo and stem cell research is that well-reasoned, in-depth public-policy reports by expert panels, even in the best of times, generally go ignored; for all the lip service we pay to expert advice, our political culture devotes little attention to the actual conversation and has time only for the executive summary or the vote count. To the extent that they are read at all, the reports merely provide talking points (or points of attack) for the professionally organized foot soldiers of our national debates, all carefully coached and deployed by lobbying groups on either side of the issue, who write the scripts, hone the message, dispatch the troops, and pull their invisible strings. Add to this the heightened emotions and legitimate philosophical differences that have always swirled around the abortion debate, to say nothing of the metaphysical uncertainty about when developing human life is deserving of protection, and it becomes clear that serving on a national commission designed to provide answers to these fundamental biological and social questions is a hugely important but thankless, and sometimes fruitless, task.

The Human Embryo Research Panel of 1994 was merely one of several in the past three decades to confirm this dreary truth about public service. The panel's report opened the door ever so slightly to embryo research in this country; in fact, well before news of Dolly, the cloned sheep, and even more bizarre biological creations jangled society's nerves, the document presciently addressed the acceptability of both embryonic stem cell research and research cloning, anticipating two issues that later reached the front burner of national debate. But at a critical juncture in the process, the door didn't open quite soon enough, and there wasn't sufficient political will to hold it open long enough, to see whether this avenue of research could be undertaken in a responsible fashion under strict federal oversight. The failure was due to bad timing, electoral happenstance, and, perhaps, a failure of political nerve in the very administration whose rise to power had so heartened Roger Pedersen and other developmental biologists only a few months earlier.

Committee work of this sort is about as exciting as hanging wallpaper, and about as sticky and unforgiving of mistakes, too, but the 1994 panel set in motion a train of events that, by promptly running off the tracks, may affect the science and commerce of regenerative medicine for decades to come. The 1994 committee was headed by Steven Muller, president emeritus of Johns Hopkins University, who had no background in the issues under discussion. The real power resided with the panel's two cochairs: Brigid Hogan, a prominent stem cell researcher at Vanderbilt University, headed the scientific aspects of embryo research, and Patricia King, a lawyer at Georgetown University, headed up policy aspects. The two-headed leadership reflected the two conflicting agendas, scientific and social, that the panel hoped to reconcile. "Essentially, the NIH wanted, with the revitalization of the NIH with the arrival of Harold Varmus as director, to see if they could move into assisted reproductive genetics," Hogan recalled. Varmus brought real scientific stature to the job of NIH director when he arrived in November 1993, following Bill Clinton's election in the fall of 1992, and although the NIH leadership had begun to consider the feasibility of embryo research even before his arrival, Varmus galvanized the Bethesda campus, creating an institutional ethic to reassess every aspect of biomedical research — not just embryonic and fetal experimentation, but clinical research and even the bureaucratic organization of the NIH. "Our charge," Hogan explained, "was to see what was science here, what could be done, and then characterize ideas and experiments that people might want to do — which were feasible, which might require further discussion, and which might be beyond the pale for the foreseeable future."

In short, the committee had the dubious honor of drawing a map for navigating a political minefield, providing a scientific rationale (and bureaucratic cover) for researchers who might want to embark on federally funded embryonic research, while indicating areas of inquiry simply too socially or ethically problematic to traverse. There were, in fact, many compelling reasons to pursue such studies, beginning with one that is routinely overlooked in debates that dwell on medical utility: pure basic research. Well-conducted research on early embryonic development, for example, could finally unveil the score of the biochemical symphony controlling human development, a score as beautifully intricate and split-second and interwoven as a Bach concerto. It could reveal the activation of genes during the earliest moments after fertilization, genes that might otherwise remain silent throughout postnatal development, genes that could help explain the development of many congenital health disorders. In practical terms, this kind of knowledge could dramatically expand the diagnostic capabilities of prenatal tests like amniocentesis and, in a more distant but nonetheless attainable future, suggest ways to correct congenital conditions.

The committee worked against unforgiving deadlines, real and symbolic. Upon completion of its report, the panel was technically required to present its findings to the Advisory Committee to the Director of the NIH (or ACD), and this committee met only twice a year, in June and December. The timing here is critical. On the February morning when he greeted them, Varmus urged the panel to try to complete its work and make its recommendations in time for the June meeting of the ACD. From the very first day of the panel's convening, however, that request was probably unrealistic; the amount of thought, discussion, and writing was overwhelming. But Varmus realized, as surely many panelists did not, that speed is often the father of accomplishment, at least in a bureaucratic setting in a political town. The June deadline soon became impossible, and the target for a draft of the report slipped to the end of the summer, with formal presentation to the ACD at its December meeting. That the June deadline occurred before the midterm national elections in 1994 while the December deadline occurred after may have determined the future of regenerative medicine in the United States for a generation.

The Human Embryo Research Panel did what these commissions do — it held public meetings, heard expert and public testimony, drafted position papers, took votes, and ultimately issued a report. During that long spring and summer of 1994, panel members wrestled with many of the issues that have become numbingly familiar to many Americans: When does life — or, as some prefer, "morally significant" life — begin? When does an embryo become an individual that deserves societal protection and has a right not to be killed? When does a ball of embryonic cells become sentient? And does an early embryo have equal moral status to a living human being? For all their diligence and good intentions, however, the committee essentially reinvented a moral compass that already existed elsewhere, because as panelist R. Alta Charo, a law professor at the University of Wisconsin, later said, "The panel made a series of recommendations that, for the most part, were totally unremarkable, and very closely followed the Warnock report recommendations in England. On one particular point, we went beyond the Warnock report." On that one point, as we shall see, the entire enterprise foundered.

The scientific and moral issues attached to human embryo research are obviously complicated questions, but the ground had in fact been well prepared for the NIH panel several years earlier, and Hogan, born and raised in England, was particularly familiar with this history. In the mid-1980s, Dame Mary Warnock had chaired a parliamentary committee that confronted exactly the same issues. Britain's leadership in the world of reproductive medicine, signaled first by the birth of Louise Brown and then by a series of subsequent successes, made in vitro fertilization a high-profile domestic issue in England, and indeed the Warnock report paved the way for Britain's Human Fertilisation and Embryology Authority, created by an act of both houses of Parliament in 1990.

Hogan had been struck by cultural differences that marked the public debate on the two sides of the ocean — differences that would be apparent during the later social debates about stem cells and cloning. In Britain, she said, legislators were brought into the process very early, the level of discourse was high (in part because doctors, she said, "stopped working for a year to lobby members of Parliament and the public"), and the government willingly accepted responsibility for establishing regulations controlling embryo research in the country's private IVF clinics — something legislators in the United States, in the face of fierce public division, have never had the stomach to do. During the U.S. debate, Hogan added, "we didn't hear much support from the medical community. They kept their heads in the trenches." But there were other significant differences, too. "First of all, the British are not nearly as polarized on abortion as we are," Alta Charo said. "Second, the U.S. does not have a good system of party discipline." By this she meant that in contemporary American politics, elected representatives are "subject to local issues and local pressure groups. The British don't have the phenomenon of large, well-organized interest groups like we do."

Those interest groups — right-to-life organizations, conservative foundations, the Catholic Church — wheeled into action as soon as Harold Varmus uttered his warm greeting to the panel at its first meeting. Anti-abortion activists instantly seized on the gaffe, citing it during a months-long campaign designed to establish that the panel had been handpicked to reach a favorable consensus on embryo research. That was the environment in which the Human Embryo Research Panel began its work.

☙

In most respects, the NIH panel faithfully reiterated the recommendations of the Warnock report in England, which had operated as law for three years. It recommended allowing federal funding for research on so-called leftover embryos stored in fertility clinics, including the isolation (or "derivation") of embryonic stem cells ("If human ES cells could be obtained from blastocysts," the panel wrote, "they would have enormous potential in many clinical fields and the long term impact would be very high"). It recommended allowing federal funding for somatic cell nuclear transplantation, or cloning, as long as the resulting embryos were not implanted for birth. It recommended against federal funds for nuclear transfer between species, such as between cow and human. In short, the report served as a remarkably limpid crystal ball revealing the near-term future of biomedical research in reproductive technologies, with an ethical road map to all the potential hazards.

In one significant respect only did the NIH panel go beyond the Warnock report — it concluded that the creation of human embryos for research purposes might also be federally funded under certain stringent, and in some ways unforeseeable, circumstances. "We did not have stem cell research in mind at the time," said Charo, who pushed for this recommendation, "although we heard testimony about it. It was more like, if you were trying to develop a new contraceptive vaccine and if it failed, you would create an embryo. So the recommendation was that under very limited circumstances, federal funding would be allowed to make embryos for research purposes."

The creation of human embryos for research alone is an enterprise thick with ethical thorns, and the committee did not make this recommendation lightly, nor easily. Ronald Green, a religion professor at Dartmouth College, prepared an excruciatingly detailed and reasoned position paper for the committee, describing the moral status of an embryo in the context of the accepted biology of early embryonic development. Here are a few salient facts from that discussion. In normal sexual reproduction, embryos begin to develop, of course, following the union of sperm and egg cells. The fertilized egg, or zygote, begins to divide, first into two, then four, then eight cells, and so on. About four to seven days after fertilization, the cells, now numbering about one hundred, organize themselves into a hollow ball called a blastocyst. Clinging to the inside wall of this hollow, fluid-filled ball of cells is a bumpy sconce of separate cells, known as the "inner cell mass"; these are the embryonic stem cells, numbering roughly twenty to thirty. The outer part of the ball, called the trophoblast, ultimately becomes the placenta; the stem cells ultimately specialize, or "differentiate," into all the cell types, tissues, and organs of a mature, viable human being. In humans, much of this occurs in transit, as the early embryo migrates from the ovary through the fallopian tube toward the womb, where it implants about a week after fertilization. Stem cells are a vanishing miracle; they arise during this dark migration and persist in their pristine, pluripotent state for about a week after implantation. Approximately fourteen days after fertilization, the cells begin to build the specialized tissues of a human, after the embryo has safely nestled into the uterus.

The embryo at fourteen days contains about 2,000 cells. It has no brain, no limbs, no feelings, no sentience, no consciousness, no organs — indeed, no differentiated structure or morphology of any sort (neural development begins at about day eighteen). Although its cells obviously contain the genetic program to orchestrate subsequent embryological and fetal development, that program is by no means equal to a guaranteed or inevitable destiny. Genes activate that program in response to environmental cues, with perhaps the most important early cue being implantation in the womb. If those cues are not present, or are insufficient, or arrive at the wrong time, or are complicated by environmental stresses or toxicities, development may shut down or veer off into aberration. That happens more often than not. "Reproductive embryologists report that human embryos have a very high natural rate of mortality during the first few weeks of development," Green writes. "Estimates vary greatly, but some studies suggest that in normal healthy women, between two thirds and three quarters of all fertilized eggs do not go on to implant in the womb." In a purely biological sense, there is no right to life for any embryo, if at minimum 50 percent of fertilized eggs in healthy women fail to implant, to say nothing of the miscarriages and aberrant development that can occur later. "Humans are not particularly good at making babies," the writer Jon Cohen recently observed. "More often than in any other species, embryos we conceive have an abnormal number of chromosomes, a condition called aneuploidy." Humans may well be more respectful and protective of embryos than nature seems to be.

Green argued that an individual life may not actually begin until about two weeks after fertilization. This is when the first hint of structure, or morphology, known as the primitive streak, appears in the embryo. The primitive streak inscribes a kind of equatorial reference point on the embryo, around which the poles of eventual body geography — head and feet, front and back, left and right — organize themselves. In addition, Green argued that it is still possible, within the first two weeks of development, for an embryo to split into two identical embryos, destined to become twins, which undermined the argument that "personhood," in conventional terms, began at fertilization. Indeed, the crux of the committee's overall argument here was that on the moral scales upon which bioethicists weighed the relative values of competing interests, an early embryo — "early" meaning up to fourteen days old — did not possess the same moral status as a fully formed human being, especially an ill and suffering human being who might conceivably benefit from medical research conducted on an embryo. This was precisely the line British society had drawn in the Human Fertilisation and Embryology Act — fourteen days and no further. It was a recommendation that offered cold comfort to those who believe that morally significant life begins at the moment of conception. Rather, it was a pluralistic solution, befitting a pluralistic society, but it ran into opposition from those who believe absolutely that life begins at the moment of fertilization, that the creation of embryos solely for research is immoral, and that the destruction of embryos in the course of research is tantamount to murder. Richard Doerflinger, representing the Pro-Life Secretariat of the U.S. Conference of Catholic Bishops, equated a fertilized egg with an adult human being; it was owed the same respect, the same degree of reverence, and the same biomedical rights. As an aside, he impugned the integrity of panel members by referring to the commission as "a special interest group" and charging that members had "financial or other personal interests" in embryo research.

Despite suggestions that the panel was stacked, the recommendation to allow the creation of embryos for research purposes caused severe divisions within the group. "That was an issue that was very divisive within the panel," recalled Thomas Murray, a panel member and currently head of the Hastings Center, a bioethical research institute in the Hudson River Valley of New York. "We had to vote on various proposals, and some of the votes were damn close. So the report was by no means a consensus report." Charo recalls pushing for inclusion of the recommendation to allow the creation of embryos for research, realizing only later that that might have been a miscalculation. "Pat King was convinced that that recommendation was politically damaging," Charo said, "and I admit that I pushed and I misjudged . . . The whole committee misjudged, because if we had rolled out the recommendations individually, they might have been more acceptable to the administration. That was a missed opportunity." Indeed, King, a committee cochair, ultimately wrote a dissenting opinion on the issue of creating embryos, saying, "The fertilization of human oocytes for research purposes is unnerving because human life is being created solely for human *use*. I do not believe that this society has developed the conceptual frameworks necessary to guide us down this slope . . ."

As the committee wrestled with these controversial and difficult positions, members of the panel became targets of anonymous campaigns of physical intimidation and even threats of violence. Alta Charo received a chilling message on her answering machine at work: the sound of a gunshot, and then a click. Another panelist, who asked not to be named, received a warning visit from police after his name appeared on a list of doctors and researchers that had been discovered by investigators in the possession of a person "who had committed crimes." The wife of a third panelist, I was told, became intensely fearful for his physical well-being. "Many of us got hate mail," Charo said. "I got quite a lot, and I developed a form letter to respond." "They had to have extra security guards at the public meetings, and the committee members had to be escorted into and out of the bathroom," recalled Kathi Hanna, a consultant who worked with the panel. "It was really ugly." Ronald Green, the bioethicist at Dartmouth, who described his experience on the committee in his book *The Human Embryo Research Debates*, wrote that even several years after serving on the embryo panel, he received mailings with his name listed among a group of "assassination targets."

For a nation that prides itself on the quality and openness of its democracy, the public debates involving abortion and embryo research have repeatedly regressed into this kind of heavy-handed, ideological fundamentalism, enforced by anonymous thuggery, that we correctly (though self-righteously) denounce in other countries and other cultures. Indeed, in the aftermath of September 11, when hundreds of abortion clinics received mail purporting to contain anthrax, the tactics of certain antiabortion fanatics could be seen — and were investigated — as a form of domestic terrorism. This is not to suggest that anyone opposed to abortion or embryo research is a terrorist; rather, it is to acknowledge that the culture of absolutism that so often characterizes people who hold those beliefs provides hospitable growth conditions for even more extreme and violent expressions of political opinion. The sense of personal, physical threat was palpable to virtually every member of the 1994 panel. Tom Murray remembered eyeing a disheveled middle-aged man who attended one meeting clutching a suspicious bag. "I kept thinking, 'What's in that bag? How long would it take for me to get to the guy if he pulled anything out of the bag?' I even remember backing my chair away from the table." (The bag turned out to contain position papers.) I was told that by the end of their service, several panel members felt so traumatized by the anonymous threats that they vowed never to participate again on a similar commission. Hardly an atmosphere conducive to ethical reflection.

∽

To ask private citizens to accept the risk of physical harm while serving the public good is, of course, a duty we normally associate with soldiers, not moral philosophers and ethicists. That sacrifice was repaid by several acts of gross political ingratitude as the panel finalized a draft version of its recommendations toward the end of the summer of 1994, with long-standing and paralyzing consequences. As he read the committee's draft report, Varmus was impressed by both its scope and its prescience. All the issues that have come to dominate headlines in 2001 and 2002 — embryonic stem cells, cloning, parthenogenesis, creation of embryos for research purposes — were addressed in the report. "*Very* forward-looking," Varmus said, citing specific details seven years after the fact. "Of course, stem cells were going to be discussed. We'd all been working with mouse stem cells for years. But the fact that somatic cell nuclear transfer [cloning] was discussed really surprised me, because that really wasn't much of an issue. Nobody even believed at that point that it would work. But it was discussed. I mean, *everything* was discussed. But the thing that was attracting my attention was those situations in which you'd have to create embryos for research purposes." That was clearly the most controversial part of the report. "And I think there was probably interest in keeping it under wraps," he added.

The first hint of trouble became apparent on August 19, when, on a layover at Logan Airport on his way to a scientific meeting in Maine, Varmus picked up a copy of the *Boston Globe* and noticed a front-page headline: "U.S. Panel May OK Human Embryo Study; Bid to Allow Funding Seen Drawing Fire." The *Globe* reported that the NIH panel, in its draft report scheduled for release in late September, was "expected to make the controversial proposal" of public funding for the creation of embryos for research purposes. This, and a similar story leaked to *Science* around the same time, represented the first public suggestion that the panel had taken its one fateful step beyond the Warnock report.

The most disquieting negative reaction of all, and the most surreptitious, came not from antiabortion activists but from the Clinton administration. "There were some press leaks that functioned as a kind of DEW line," recalled William Galston, who served on the Domestic Policy Council, "telling us that this was something we were going to have to deal with, probably sooner rather than later." That same summer, while the report was still being drafted, Varmus began to receive "quite a few telephone calls" from distressed domestic policy officials at the White House, including Galston and Joel Klein, a lawyer, questioning him about the drift of the report. He talked administration officials through the report's findings, making stops at the White House, the presidential science adviser's office, the Office of Management and Budget. "In general, I got a pretty good reception at these chats," Varmus told me. "But it's clear that the White House thought the report went too far, let's put it that way. And there was a difference of opinion about how much too far." But when the draft report was released for public comment on September 27, the committee was almost instantly deluged with negative comments, and not just by right-to-life activists. A *Washington Post* editorial called the creation of research embryos "unconscionable" and charged that the panel "took a step too far." That got the White House even more concerned.

The Clinton administration wanted Varmus to denounce the findings of the panel. "There was interest in . . . *not* going along with the report," Varmus told me carefully. "I mean, it was definitely the case that the White House, that I was given directions that I shouldn't agree with all of this and, you know, either say that I don't agree with the report or that it should be put on the shelf . . . But they wanted me to cool it, there's no doubt about it. At one point, Leon Panetta called me up and ordered me to do it. And I refused." The reason the White House wanted its NIH director to repudiate the report, Varmus believed, was political. "They were nervous," he said, about the November midterm elections. With good reason. That November, the "Republican revolution" swept a tide of conservative freshmen congressmen into the House of Representatives, stunning the administration and, in effect, cementing Bill Clinton's misgivings about embryo research into hardened public policy.

Still, Varmus refused to go along. Moreover, he never breathed a word of the White House's discontent to anyone on the panel or its staff, mindful that even the slightest political breeze could, and probably would, influence the committee's final deliberations. As a scientist, he believed embryo and fetal research — with proper federal oversight — was essential to medical progress, even before human embryonic stem cells were on the table. But the issue was no longer purely scientific; it had become political as well. Indeed, the White House formed an ad hoc working group, including Leon Panetta, the chief of staff, his deputy Harold Ickes, and policy adviser George Stephanopoulos to, in Galston's words, "ensure that the issue would be carefully monitored." In refusing to go along with the White House, Varmus worried that his job might be in jeopardy. On November 9, several days after the stunning election results, Varmus briefed officials at the White House about the embryo panel report, and according to one of the participants, William Galston, "It was pretty clear to all of us that he intended not only to receive the report, but would act on its recommendations." As a result, Galston said, the White House prepared a "strategy of preemption."

A year earlier, Bill Clinton had looked like a beacon of progressive spirit to the scientists interested in human stem cell research. By the fall of 1994, following the humiliating collapse of his health care reform program, and with the disaster of the midterm congressional elections fresh in everyone's mind, the administration couldn't create enough distance between itself and embryo research. Galston insisted in an interview that the reason wasn't political; he cited Patricia King's dissenting opinion, which suggested that the public was not ready for the creation of embryos for research purposes. Yet in a scholarly paper he later wrote about the episode, he approvingly cited a peculiar remark by the panel's chairman, Steven Muller, who said, "By a huge majority, the public has no idea what ex utero or preimplantation human embryo research means or what it involves. But it does, to most people, sound terrible"; Galston added that "for this reason, among others, the White House working group feared a political backlash." That not only confirms that politics was a factor but seems to equate public ignorance with democratic wisdom and also sounds like an eerie endorsement of the "wisdom of repugnance," the conservative moral philosophy of bioethicist Leon Kass. When Galston spoke about the embryo report at Harvard's Kennedy School some months later, a number of bioethicists in the audience were surprised at how much the "wisdom of repugnance" philosophy appeared to inflect the Clinton administration's thinking.

The timing was excruciatingly bad: if ever there was an issue that begged for intervention from a bully pulpit, as even George W. Bush realized eventually, it was the complicated moral, scientific, and social dimensions of modern biology, as exemplified by stem cells, cloning, and related embryo research. No one seemed more poised or more qualified to assume that forward-looking, future-shaping role than Bill Clinton in 1993, by political inclination, progressive temperament, and personal ambition, and few other politicians could have made a more persuasive argument in favor of the basic research. But not only did the White House repudiate the report's recommendation on creating embryos; it appeared to some panelists that by throwing the car in reverse and stomping on the gas to retreat from the controversy, the Clinton administration essentially ran over the entire report.

The political concerns of the White House were not misplaced and even resonated among some members of the embryo panel. Tom Murray thought the recommendation supporting the creation of research embryos went too far, and that the White House acted reasonably in rejecting it. "I thought it was politically a nonstarter," he said, "and I think it was a savvy move by the White House." Members of the embryo panel, however, did not know that the White House had been actively working behind the scenes for months to undermine their findings. In late September, when the completed report was ready to be circulated for comment, Ron Green said, "We had every reason to believe our report would be well received and that most of our recommendations would soon go into effect."

Nothing could have been further from the truth. For panel members, the reality became sickeningly clear on December 2, 1994, when the group filed into a room at NIH headquarters and formally delivered its report to the Advisory Committee to the Director. Varmus appeared in the room and made what one panel member called a "profiles in courage" defense of the integrity of the process by which American health policy underwent review. The advisory committee promptly approved the findings of the report. By this time, however, Varmus himself was out of the loop. Without warning Varmus, the White House cut off everybody at the pass — its NIH director, the embryo panel, and, arguably, the short-term feasibility of tightly regulated, ethically vetted, publicly funded research on human embryos, including experiments on leftover embryos in IVF clinics and research involving stem cells and cloning.

Sitting in the same room with the panel members was the White House liaison official, Rachel Levinson. Moments after the final report had been presented to Varmus, barely after the NIH director had left the room, Levinson began showing people a copy of a White House statement — an executive order stating that the Clinton administration prohibited the government from funding the creation of human embryos for research purposes. This preemptive strike caught everyone by surprise, including Varmus. "I didn't know it was coming, actually," Varmus conceded. "I remember going back to my office after the meeting and starting to write some statement about the report. And getting a call from the White House, saying, 'Here's the executive order that the White House is issuing.'" The only consolation was that he still had a job.

People who saw him that day recalled that the usually unflappable Varmus looked stunned by the developments. "He was *really* not happy with what happened," recalled Kathi Hanna. "The White House hit him broadside, without warning. There wasn't a moment, a *second* for Varmus to accept the report. It was very upsetting to members of the panel." "We were stunned that the White House disavowed the report before we even presented it," said Murray, adding, "I have no complaints with what the White House did, I just wish they could have waited a day." Charo agreed. "I was surprised," she said, "particularly because Clinton didn't let the process run its course. In fact, he stepped in so quickly that it had to mean that the White House had been planning its intervention before the report was even out. Because of Clinton's action, the steam was kind of taken out of the whole thing."

"That was not the intention," Galston said. "It may have been the result. I know it was the view of the political people in the White House that if we were going to do something, we had to do it quickly. We couldn't just leave the report out there, because it would begin to rot, like a fish left out on the table. It seemed abrupt and ungracious, and I regretted that. But we wanted to send a very clear message that from the president's point of view, one part of this discussion was not going to continue, but rather was going to end. My sense is that, one way or another, whether the president had

done it or the new Republican Congress pulled the plug, the results would
have been the same."

✑

Several months after the elections, the penny dropped — fell, rather, like a
manhole cover on the entire field. At that time, Varmus was juggling a delicate set of priorities. The NIH budget was not as politically sound as it
would eventually become during his tenure as director; there was concern
that defying the White House and making an issue out of embryo research
would jeopardize funding for the entire NIH. Varmus also understood that,
technically speaking, human embryo research using spare embryos could
still proceed. Since Congress, at Clinton's behest, had previously eliminated
the need for the problematic Ethics Advisory Board, there was no regulatory impediment, and Varmus interpreted the executive order as "relatively
narrow," allowing some room for maneuvering. "The White House wasn't
saying 'You shall not do embryo research,'" he said. "It was saying, 'You
shouldn't create embryos for research purposes.'" But he also knew that any
research involving embryos would require guidelines, which meant more
committee work, more public comment, and more delay in the face of a
conservative Congress on the attack and a backpedaling White House.
Once the Clinton administration backed away from the fight, the Republican revolutionaries in the House led a final charge.

The foot soldiers of Newt Gingrich, who now fancies himself a friend
of science, presided over the final sacking of the embryo panel's report. In
August 1995, two Republican congressmen, Roger F. Wicker of Mississippi
and Jay Dickey of Arkansas, inserted an amendment into a government appropriations bill related to the federal budget for fiscal year 1996. The rider
expressly prohibited the Department of Health and Human Services (including the NIH) from funding work that would create human embryos
for research purposes, or work "in which a human embryo or embryos are
destroyed, discarded, or knowingly subjected to risk of injury or death
greater than that allowed for research on fetuses in utero" outlined in previous law. The legislation was passed in January 1996. Renewed each year
since, the Dickey-Wicker amendment has successfully blocked any public
funding for human embryo research, including stem cell research, even
though Dickey was voted out of office in 2000 and Wicker has long since
attempted to dissociate himself — at least by name — from the legislation

(he later told Varmus, "I prefer that it be known as the Dickey amendment"). So, following fifteen years of a de facto moratorium on embryo research, and a brief window of opportunity to change the terms of the debate in 1993 and 1994, the Clinton White House stared the future of biomedical research in the eye, and blinked. It's far from certain that the Clinton administration could have persuaded Congress to support embryo research, but a public discussion might well have prevented antiabortion politicians from resorting to a stealth amendment to change the law, without even a moment of publicly accountable debate. In any event, the window slammed shut in 1995 — slammed tighter even than during the Reagan and Bush years, and by 2001, a born-again Christian embracing the "culture of life" occupied the White House. The following year, George W. Bush's handpicked bioethics council recommended a further, four-year moratorium on research cloning — in effect, a recommendation to extend the federal moratorium on embryo research to a total of twenty-five years, with the argument that more public discussion was necessary.

The Clinton administration's decision to undercut the NIH panel represented a turning point in the history of stem cells, therapeutic cloning, and regenerative medicine. Even allowing for the administration's uneasiness about the recommendation to create embryos for research purposes, the reluctance of the Clintonites to articulate an argument for other embryo-related research in 1994 — or, more important, to fight to bring such research under strict federal oversight — has had tremendous implications for the speed and quality of stem cell research, the control of intellectual property in the field, and the hope that this science might bear timely fruit for millions of American citizens desperately awaiting new treatments for mortal diseases. Moreover, after the Lewinsky scandal broke in 1997, Clinton's bully pulpit on matters reproductive and moral shrank to the size of a footstool. Even if the timing for discussion had improved during his second term, he had effectively squandered any moral authority for conducting such a discussion.

"That was the moment," one Washington observer said, in retrospect, of the fall of 1994. "We'd barely come out with the report," Brigid Hogan said, "and Clinton came out against the use of NIH funding for fertilization [experiments]. The general feeling is that not enough support was gained, that the balance was not strongly enough in favor in Congress that it would be fought. I think they all agreed not to fight over this." Moreover, the

White House's decision to back down from a fight over embryo research may have had a little-appreciated domino effect on the psychology of public servants who later tackled controversial new biomedical technologies like cloning while serving on Bill Clinton's National Bioethics Advisory Commission. "On NBAC, there were three members from the embryo panel — Tom Murray, Bernie Lo, and Alta Charo — who wanted to say it would be okay to clone for [nonreproductive embryo] research but were gun-shy," said Kathi Hanna, who worked on both panels. "It was like, 'We already know how he [Clinton] feels about this. We already feel burned by this.' So they were deliberately very ambiguous about that in the 1997 cloning report, and then it came back to haunt them."

These developments also had long-range implications for how embryo research would subsequently be pursued in the private sector. The cold reception for the embryo panel report and then passage of the Dickey-Wicker amendment, for example, marked the collapse of Ron Green's faith in public-policy discussions related to reproductive technology and regenerative medicine; he cast his lot with private industry. "Some of my colleagues in the field say, 'Why don't we have something like the Human Fertilisation and Embryology Authority, as they have in Britain — this one overarching entity that supervises all reproductive research, all use of embryos, all cloning endeavors, and actually extends to clinical oversight?'" said Green. "And though I personally feel that would be ideal, I would resist it in the United States at this time, because we are far more divided on these issues than is Great Britain. And my experience in Washington tells me that any centralized body of that sort would quickly become hostage to the most strident political groups. And that's what's happened at the federal level with regard to embryo research. A minority opinion has prevailed, in effect. So this is why I'm here." By "here," he meant Advanced Cell Technology, the biotechnology company whose ethics advisory board he heads and where he helped promote private research into embryonic stem cells and therapeutic cloning. "I'm here because I think there's room for the private sector, and the aim is to help the private sector do it properly."

Varmus tried valiantly to sustain momentum for publicly funded human embryo research. The debate changed dramatically four years later, when scientists succeeded in isolating human embryonic stem cells. Varmus obtained a legal opinion in January 1999 from Harriet Rabb, the legal counsel at Health and Human Services, stating that the NIH could legally fund studies of human embryonic stem cells, as long as the agency did not fund the actual derivation of them (that is, once researchers somewhere else destroyed an embryo in the process of isolating human embryonic stem cells, other researchers could work with those cells and conduct experiments with government support). "But what frustrated me tremendously," Varmus said, "was the time it took to get money out the door. And one of the things I'm unhappy about, as I look back on my having left the NIH with a year to go in the Clinton administration, is that they didn't get money out the door. They should have. It would have helped. Because it's much harder to stop research than to keep starting it."

⌒

And so the moment passed. After a brief flirtation with ethical and regulatory legitimacy, researchers interested in federal funds to conduct human embryo experimentation and in vitro fertilization research found themselves back out in the cold; the Republican revolution of 1994, drawing intellectual and moral sustenance from the conservative, antiabortion wing of the party, effectively drove prospective human stem cell researchers into exile. Drove them, in fact, into the arms of the private sector. And brokering this bitter marriage of convenience was none other than the now disillusioned Roger Pedersen.

Pedersen saw his dreams fall apart at close range. He was among the biologists who discussed the embryo panel's findings with journalists at a press briefing in the fall of 1994. He had allowed himself to become enthusiastic. "Things looked very hopeful," he told me, "until the fall of 1994, and you know what happened then." Crestfallen that the embryo panel's report had "provoked the opposition into successfully stopping the research," Pedersen realized the path to federal funding had once again become long and arduous. "And I was frankly dismayed," he said, "because I had begun to be very enthusiastic about the area of research. Particularly the prospect that cells could be derived from primate stem cells. And in fact, in 1995 Jamie Thomson succeeded in deriving such cells from primates, showing that in principle you could. So it seemed to me therefore only a matter of time and effort."

After the shipwreck of the NIH's embryo panel, Mike West's offer of Geron cash to support human embryonic stem cell research suddenly looked a lot more appealing to Roger Pedersen. "Some months later," West said, Pedersen called him up and said, "Well, you know, some things have changed, and I'd be open" to corporate support of stem cell research. West wasted no time scooting up to San Francisco to discuss the possibility. Disillusioned and impatient, Pedersen didn't need much convincing. "I did agree to work with Mike then," Pedersen admitted. "I submitted a proposal to them, and I also put them in contact with other people in the field that I knew were doing this kind of work. Because I'm in the field, I was aware of early leads as to at least who was doing the work. And I became a conduit to their funding by putting Mike in touch with these people."

While Pedersen was glad to take West's money, he passed on the "fountain of youth" rhetoric. He never viewed the promise of stem cells through the prism of gerontology, nor did the other researchers that West ultimately funded. But neither can it be forgotten that to West, the human embryonic stem cell was the microscopic, twentieth-century embodiment of Osiris's flesh — a seed of regeneration, a fount of primordial healing, the body's version of recuperative alchemy. If human embryonic stem cells ultimately live up to their potential (and provide even a penny of return for Geron), it will vindicate both a scientific vision and a personal mission that Michael West set in motion in 1995, almost entirely on his own initiative and with (at best) tepid encouragement from his colleagues at Geron, when he rushed to Pedersen's rescue.

West was, in Pedersen's perhaps double-edged appraisal, "quite an operator." No sooner had Pedersen handicapped the stem cell field for West than the creationist-cum-entrepreneur was combing the scientific literature, working the phones, booking flights. "There's a fellow in Wisconsin," West recalled Pedersen telling him, "that has just — it's not published yet, but he's isolated these cells from a rhesus monkey." That scientist was James ("Jamie") Thomson, a developmental biologist who worked in the Department of Veterinary Medicine at the University of Wisconsin. "And I think the next day I was in Wisconsin, at Jamie Thomson's lab," West said. "Because I realized, having them [stem cells] from a rhesus monkey is so close to humans, we could just learn how to work with them and grow them and differentiate them and file patents. But we'd have this head start on the whole world. So that's what I did. I licensed that patent and set up a collaboration with Jamie."

Not long after, West learned that John Gearhart, a researcher at Johns Hopkins and director of the university's IVF clinic, had publicly announced his intent to try to isolate human pluripotent stem cells from fetal tissue. Gearhart had cleverly realized that fetal tissue research, with proper oversight, was still entirely legal despite Dickey-Wicker, and that experiments in mice had demonstrated that stemlike cells could be isolated from the developing reproductive tissue of fetuses. "I can almost remember the day," West recalled. "It was right around my birthday. I'm en route to Michigan to visit my mother on a Friday night, I think, and I just heard about this. I was talking to somebody, and they told me about Gearhart having putative embryonic germ-line [EG] cells in culture." West called up a colleague at Geron and asked her to do a quick literature search of Gearhart's publications and to fax them to a Kinko's in South Bend, Indiana, not far from his mother's home. He picked up the papers over the weekend, then flew on to Baltimore. Without an appointment. That Monday morning, he called the lab to make sure Gearhart was in and then rushed over. "It was Mike West who appeared at my door one day," Gearhart recalled with a high-pitched laugh. "Unannounced! He just gets on a plane and *goes*. I found Mike to be a focused, charming kind of a guy who looked pretty far ahead in science, and obviously a big risk taker. And so it was through Mike that this evolved."

It would oversimplify things to say research agreements were struck on the spot, and there is a certain self-mythologizing quality to West's account of his wheeling and dealing. But staking claims in intellectual property has become an increasingly critical feature of modern biotechnology, and West had, in blitzkrieg fashion, managed to sew up collaborations with several key researchers in the field; the fact that Geron was funding experiments that were otherwise impossible for federally funded scientists to conduct gave these three collaborators a significant inside track on isolating human embryonic stem cells — and, not coincidentally, it gave Geron the inside track on rights to commercialize work that might emerge from this now privatized research. "I knew that Mike would make things happen," Pedersen admitted in retrospect, "and I thought it was a good strategy, just because of the redundancy of effort."

It is difficult to overstate the importance of Roger Pedersen's decision in 1995 to turn to the private sector for support, and of Michael West's decision to fund human embryonic stem cell research at several academic institutions. This two-man, free-market override of the congressional funding ban completely transformed the technological landscape. It unleashed the entrepreneurial side of West, who moved, with lone-wolf alacrity, to sign up key researchers and jump-start a field of research that had languished for years. It set up an ethical debate about stem cells — and, later, human therapeutic cloning — that has been driven, and bedeviled, by unanticipated announcements from the private sector, announcements that have often dropped like bombs in the midst of civil and thoughtful conversation. It positioned Geron to amass seemingly dominant commercial rights in the field. It allowed a select few researchers to embark on a new form of medicine that was directed not against one or two diseases, but that held the prospect of regeneration and rejuvenation of human tissue against a broad spectrum of diseases. Most of all, and most shockingly, it may have come unbelievably cheap. A prominent stem cell scientist later told me that Geron's initial investment in Thomson's research at the University of Wisconsin cost the company about $30,000. That sounds implausibly low, but even a typical sponsored-research agreement — something like $200,000 spread over three years — would have potentially purchased a fabulously large patent estate for a relatively modest sum.

In many respects, West had little in common with the three stem cell scientists he had courted. Thomson was quiet and introverted, quick to squint in the limelight; Gearhart, an excellent scientist with a long history in the field of developmental biology, had a tightly wound, almost pedantic propriety about him, with the pinched carriage of a church deacon; Pedersen seemed as mild-mannered as an accountant. And yet it bespeaks the long-standing frustration in the scientific community that when Mike West walked into their labs and their lives — spinning theoretical reveries about the unlimited potential of the research they wanted to conduct, plunking hard cash on the table as a down payment on those dreams, appearing a bit more fanciful and enthusiastic and maybe *flighty* than your average Ph.D. — when he walked into their world, with his open face and open wallet, they received him with open arms. He had, outside the notice of all but a handful of informed insiders, become the self-appointed impresario of one of the most celebrated scientific races in the past several decades of biological research. And he operated in the cracks of a convoluted logic warped by abortion politics, in which it was okay to destroy embryos in the private sector but not in the public; it was okay to work with fetal or embryonic tissue with industrial funds but not with government support; and it was not okay to trust Harold Varmus and the NIH to oversee human embryo research but was fine for that awesome responsibility to reside with a small, private biotechnology company and its founder, a headstrong and mercurial erstwhile creationist.

Not that the folks at Geron were thrilled with West's initiative, either. They were still, as far as anyone was concerned, a telomere company. "The irony was," West said, "I had floated this thing past the analysts for Geron. I was excited about this, but it was a complete yawner to them. Nobody was interested. 'Oh, stem cells, that's old hat.' 'Stem cells have been around forever. Besides, you know, the human ES cell doesn't even exist.' 'If you have some monkey cells, and if you use those, you'll get monkey viruses that kill everybody.' There was," he claimed, "hardly any interest at all."

7

CLONING IN
SILICO

DURING HIS THIRD YEAR as a postdoctoral fellow at the University of Colorado, Joachim Lingner picked up the June 2, 1995, issue of *Cell*, one of the flagship journals of molecular biology, and experienced the moment every postdoc dreads. The issue contained a paper from Carol Greider's lab, reportedly identifying two key parts of telomerase, the "immortalizing" enzyme. Lingner had spent the previous two years trying to do exactly the same thing. He'd traveled from his native Switzerland to Colorado in April 1993 to work in the laboratory of Thomas Cech. He had come with specific plans to clone the telomerase gene, one of the most coveted snippets of DNA known to biology at the time. And as testament to his dedication, his lab area had become a small, fetid metropolis of lasagna dishes, because a single-celled creature called *Euplotes aediculatus* seemed to like living in store-bought, deep-dish lasagna glassware more than in any other kind of receptable known to modern science. And if you were looking for the telomerase gene in *Euplotes,* as Lingner was, you used whatever housing seemed to keep them happy. Now, holding the journal in his hands, reading the grim news that he'd been beaten, it all seemed for naught.

Lingner was thirty-one years old at the time, a persistent and exacting biochemist, "a one-man Swiss band," in the admiring words of Titia de Lange. But he now saw his career begin to disintegrate in front of him. Postdoctoral fellows in biology typically have two or three years to produce an original piece of research, something sufficiently novel, difficult, and brilliant that it will land them a permanent job in an intensely competitive market. In going after telomerase, Lingner had selected a sufficiently novel problem, and it would indeed have been brilliant if he'd found the gene, but he had chosen an excruciatingly difficult way to go about it. Modern biologists typically choose one of two routes to finding a gene of interest, and both strategies derive from the well-known fact that genes are simply discrete lengths of DNA, which in turn contain the instructions specifying the sequence of amino acids that make a particular protein. With that truism in hand, you can use genetic tricks, such as creating mutants, to locate the segment of DNA that encodes the gene you are looking for. Or, like Lingner, you can use biochemical tricks to isolate and purify the protein you're after (in this case, telomerase), decipher the sequence of amino acids that make up the protein, and then translate from that protein sequence back to the gene. Only then does the "real" biology begin.

Given how much was at stake in this search — fame, patents, and, to certain true believers, the molecular fountain of youth — the quest at times resembled a biologist's version of blindman's bluff, Ph.D.'s staggering through a molecular landscape without quite knowing what they were bumping into. Here was Lingner, growing creatures in lasagna pans. There was Greider on Long Island, trying to isolate the same protein from *Tetrahymena,* but having identified two proteins that had no apparent cousins in the burgeoning digital warehouses of known proteins. There was Vicki Lundblad, at Baylor, with several intriguing genes in hand from her "ever shorter telomere" yeast cells; she just didn't know, according to other researchers, that one of them was telomerase. When they weren't bickering about the right approach to take in their own labs, Geron's scientists were scrambling to figure out what everybody else knew — sometimes buying their way into that knowledge with licensing agreements, only to discover that money still couldn't buy them scientific happiness. Then a dark-horse competitor, Robert Weinberg, came out of the blue and almost scooped everyone. In the end, it wasn't sheer scientific brilliance that led to the human telomerase gene. There was also the luck of who pushed a computer button at the right moment — "cloning in silico," as one scientist put it.

On almost a weekly basis, newspapers describe the discovery of important human genes with great fanfare — one that is said to control an activity like the metabolism of fat, for example, or a self-preserving function like tumor suppression. The work of Lingner, Lundblad, Cech, Greider, Weinberg, and the scientists at Geron was destined to land on the front pages of newspapers in August 1997. But newspaper accounts rarely have the space or luxury to hint at the ragged pageant of ingenuity and sweat and ego that is modern biology on the move, with its pitiless competitions, its inevitable mistakes and missteps, its persevering loners, its tribal celebrations. With apologies if this account seems a little like inside baseball, the quest to clone the human telomerase gene perfectly captures what contemporary research is really like, on a gene that may well have important medical ramifications.

ॐ

If the "immortalizing enzyme" ultimately deserves a definitive history, Lingner will be the unsung hero of this chapter of the telomerase story. He had chosen *Euplotes,* one of those whiskered (ciliated, that is) pond swimmers because, for some unknown reason, they were stuffed like sausages with telomeres. "I encouraged him to work on it in *Euplotes,*" Cech recalled in an interview, "because of the fact that there were about fifty million tiny chromosomes per nucleus. We'd been working on the telomeres in *Euplotes,* so we were aware of the beauty of this organism for exaggerating questions of telomeres. It had about a millionfold more of everything than a human cell does." But, as Cech admitted, he may also have sold Lingner a bill of goods (not for the first time in lab chief–postdoc interactions). "It was true that the telomeres were a millionfold more abundant, and that telomerase was also vastly more abundant," he said with a laugh. "But it turned out that these cells were about a millionfold harder to grow than anything else that had ever been grown. So it's not clear how much of an advantage it really was. But once Joachim had taken the bait on this, he was off and running. And these cells grow in Pyrex lasagna dishes that you get at the grocery store. Open to the air. With bacteria and algae in there, along with the *Euplotes.* They sort of crawl along the bottom. They are *hell* to grow."

It was a crushing amount of work. Lingner and a student helper had to maintain colonies of algae in separate tanks just to feed the *Euplotes.* They had to crack open millions upon millions of the one-celled creatures and then attempt to pluck a single protein out of the scum and mess. Lingner had made considerable progress, using a clever biochemical trick to snag telomerase out of the murky soup of molecules, when the dreaded issue of *Cell* arrived. "I remember clearly the reaction in the laboratory," Cech recalled. "Lingner was crestfallen. He said, 'I've been beaten. No one will ever care about this.' And I sort of had to buoy his spirits at that point and say, 'Well, yes, it looks like you've been beaten. You know, they'll get all the fame and glory for this, but the second example is often very interest-ing . . .'" That, too, may have been a bill of goods; as Cech, who won a Nobel Prize in 1989, knows only too well, second place in science is often no place. Nonetheless, he encouraged his postdoc to persevere. Lingner's premoni-tions about his career prospects were all too prophetic. Although he agreed to continue working on the project, potential employers didn't seem very excited about his research, including three laboratories in Germany and Switzerland that turned him down for a job. "I guess others felt probably more miserable about my situation of being scooped than I did," he said.

At first it appeared that Greider's work could not have come at a more propitious time for Geron. The company had staked its young reputation — to say nothing of its credibility with investors — on finding the telo-merase gene. "I think the company would have been history if we'd failed to get it," said Bill Andrews, Geron's former director of molecular biology. "At that time, it was so important that it was absolutely essential to the suc-cess of Geron." Despite Greider's rocky relations with the company, her iso-lation of the two proteins associated with telomerase, p80 and p95, ap-peared to give Geron a leg up in the search for the gene. When the Cold Spring Harbor scientists published their findings, however, Greider felt that another year or two of confirmatory work lay ahead. That follow-up work coincided with the year in which Geron made preparations to take the company public. The initial public offering had tentatively been planned for the summer of 1996, and progress in the lab had been promising — the cloning of the RNA component of telomerase had been published in Sep-tember 1995 to great acclaim, and a dozen Geron scientists were blasting away at the catalytic component, trying to use Greider's discovery to fish out the human gene. There was just one problem. Scientists in the com-pany, who at first greeted Greider's discovery of p80 and p95 with exulta-tion, quickly began to have doubts, largely because they couldn't find anal-ogous proteins when they combed through computerized databases of other organisms. "There was skepticism," Andrews conceded, "because we couldn't find a human homologue."

They were not alone. Lingner began to have his doubts, too. Con-tinuing to work on cloning a gene after someone else has apparently done it is molecular biology's equivalent of sweeping up after the elephants have passed through — anonymous, unglamorous, minimally rewarding work. But a lot was at stake — not just Lingner's job prospects, but an entire field of biology and, possibly, patents worth tens or hundreds of millions of dollars. Ultimately, Lingner decanted a tiny amount of the vanishingly rare enzyme out of millions upon millions of his ciliated creatures and began to determine the protein sequence of the enzyme. That was when he got a hint that it might be worth the effort to persevere. Well after the Collins and Greider paper came out, in the summer of 1996, he managed to isolate parts of telomerase from *Euplotes,* and he was immediately struck by a discrepancy: the sizes of Greider's proteins were "very different" from his. After Lingner had begun to assemble bits of the telomerase gene from *Euplotes,* it became clear that his gene and his protein didn't match Greider and Collins's. True, they came from two different organisms, but evolution is a parsimonious and penny-wise shopper; if she finds a biologically useful gene in one creature, the same gene — and protein — often gets passed along to other creatures, so that there is biochemical similarity, or "homology," between species. When Lingner didn't see the expected similarity with Greider's version of telomerase, he immediately thought, "One of us must be wrong." Then he got "really excited" when he saw that one of his protein fragments had the characteristics of an enzyme known as a reverse transcriptase; this is the class of enzyme that the AIDS virus uses to copy its lethal genes, written in RNA, into the DNA that fatally inserts itself into the cells of HIV-positive individuals. This made a great deal of sense, because telomerase essentially protects the ends of chromosomes by "reverse-transcribing," or copying, its little tagalong RNA segment into DNA that caps the chromosome.

"When he finally got the gene," Cech recalled, "he came running to me and said, 'It's completely different from the Greider protein! Plus, hers doesn't make any sense at all, and ours makes *perfect* sense.' So he knew at that point that he had made the right decision to allow me to talk him into continuing. The reason hers didn't make any sense was that the yeast genome had been completely sequenced, and there was no copy of p80 or p95 in the yeast genome. And to a biologist, that's a red flag. You say, if this is really the fundamental, ancient telomere-copying mechanism, of course yeast have to have one, too."

Lingner obtained these results during the summer of 1996, in a race against at least a half-dozen academic labs and also a team of twelve people at Geron, setting up the final, delicious irony of all this feverish activity. In August 1996, Tom Cech headed off to Hawaii for a scientific meeting, with

Lingner's stunning new results in hand — in mind, rather, because he had no intention of showing any of the precious data. By now, Geron scientists had also known for many months that Greider had the wrong protein. The Hawaii meeting was sponsored by Geron, which was spending tens of millions of dollars to clone the gene that Cech — and, as it turned out, one other researcher — had already found. And the entire meeting took place without Geron learning the one fact that was central to its corporate identity and existence.

∽

Vicki Lundblad set off for Hawaii with a secret, too. When she had set up her lab at Baylor in 1992, she had created an enormous pool of yeast mutants — but talk about piecework! She and three other colleagues had handpicked 35,000 yeast cells in search of mutants. They turned up about a thousand, and by the summer of 1993, she knew that four of the mutant strains had short telomeres. One of them, dubbed *est-2*, would turn out to bear its mutation in the gene for telomerase — but she didn't know it, and wouldn't know it for "an embarrassingly long time," on the order of three years. "They looked like, smelled like they were likely to be the component of telomerase," she said. "But we didn't have any good biochemical evidence that it was." The only thing she knew for sure was that they didn't look like Carol Greider's genes. By the summer of 1996, she had managed to sequence the *est-2* gene and had a paper in press, but as she headed off to the Hawaii meeting, she still didn't know what the gene was. Suddenly it looked like the race for telomerase might not be over after all.

Both Cech and Lundblad gave talks at the meeting, but neither gave more than a coy hint of having interesting new data, none of which they would reveal. So the patent estate of Geron suddenly teetered on two small, misidentified proteins (as Greider later put it, "We cloned two components, which turned out subsequently to be accessory components, but not the true catalytic component of the enzyme"), and several very clever and high-powered labs were closing in on the prize. And here, at Geron's own meeting, under the noses of Geron's own very interested scientists, the leaders in the race forged a secret alliance.

"After Tom's talk," said Titia de Lange, "there was a coffee break, and Tom and Vicki talked. They looked at each other and kind of smiled. Because both had this . . . this nugget of *gold* in hand." "Tom knew he was onto something," Lundblad recalled of the encounter. "You could tell." As de Lange looked on, Cech said to Lundblad, "We really ought to compare sequences."

In a highly competitive field like molecular biology, here was a moment of genuine high drama. Cech, whose boyish good looks belie a fine sense of scientific aggression, brought his Nobel credentials and a long-standing interest in telomere research to this encounter. Lundblad was one of the pioneering women in the field; she was also a Geron collaborator. Indeed, she had shared her research results with the company, but no one realized that one of the yeast genes she had identified was in fact telomerase!

"So this, of course, is a very tricky moment," said de Lange, who was standing beside the two during the coffee break. "Because you are giving away the most important information in your lab. But they got together and bingo, one of Joachim's sequences was the same as Vicki's and allowed them to see they had a reverse transcriptase. It was a wonderful merging of superb biochemistry and superb genetics, and the best thing about science is when people get together and talk about unpublished data." It was also a difficult moment for the Greider lab, especially for Kathy Collins, the researcher who was the lead author on the mistaken *Cell* paper. "It cannot have been easy," said one telomere researcher, "to have a Nobel Prize winner climb up your back and grab the prize."

All this, of course, occurred literally and figuratively out of Geron's earshot. William Andrews and his team of cloners, attending that same meeting, were still scrambling to identify the telomerase gene. The company's mission was so tied up with the enzyme that if someone else got the gene first, it would have been considered a crushing corporate blow. Indeed, the financial future of the company might well have played out in a different manner if the company's stumbling progress in the telomerase race had become public knowledge even two months earlier. Geron had scheduled its long-awaited initial public offering on the stock market for July 30, 1996, several weeks before the meeting in Hawaii. At that point, the company was still not close to having the right telomerase gene in hand, and one can only speculate how the news would have played on Wall Street, especially after all the extravagant press releases. As it was, the company continued to enjoy its charmed public life, for by the end of the first day of trading, investors had snapped up 2 million Geron shares offered at $8. The company had raised $16 million and had an instant market capitalization of about $175 million. Mike West says he is constrained from discussing details of Geron's business arrangements, but he did mention once in passing that the company had spent "something like $35 million" trying to clone the telomerase gene. If so, it cost twice as much as they raised in their IPO, inaugurating a burn rate in which net losses totaled more than $150 million in their first five years as a public company.

The upshot of the secret sharing between Cech and Lundblad came out at a second meeting, in November 1996, at the Banbury Center at Cold Spring Harbor Laboratory. "I was making the schedule about who would be speaking when," said de Lange, who co-organized the meeting with Carol Greider, "and I contacted Vicki, and they were very hush-hush but they had telomerase. I had to keep my mouth shut, and it was just so hard!" By sharing their data, Cech and Lundblad had quickly realized that they both had versions of the telomerase gene, one in *Euplotes* and the other in yeast. Theoretically, that information would allow any decent researcher to find the human gene, simply by searching for similar DNA sequences in the database of human genes being spit out and posted daily on the Web by the Human Genome Project. "That's the fish bait," explained one telomere researcher. "Any graduate student could do this."

It should by now be apparent that no one was interested, from a commercial point of view, in the version of an enzyme that could "immortalize" pond scum or yeast. The human gene was the key target. It is for that reason that, with a little well-earned schadenfreude, Carol Greider relishes the recollection of Cal Harley, Geron's vice president of research, looking "very nervous" as he wandered around at that November meeting. "It was very clear that Tom Cech had something and that Cal didn't have the rights to it," she said. "They knew about the results right there, right away," de Lange added. "It fell like a bomb. It changed everything. So Geron knew the sequence then. At the [Cold Spring Harbor] meeting, Tom and Vicki didn't give us all the data, but Geron already had the *est-2* sequence, and Vicki had a paper in press with all four genes. So that sequence was coming out." There was a snag, however. You couldn't find human telomerase using the yeast information. You needed the *Euplotes* sequence. And nobody would have that until the Cech lab published its results.

The Cech and Lundblad labs worked together to refine their results, and they submitted a joint paper on the isolation of telomerase, in both *Euplotes* and yeast, to *Science* on March 4, 1997. Once the *Euplotes* sequence was published, everyone and his brother would be gunning for the human telomerase gene. One of the competitors — a male, as you might surmise from the quote — referred to the gene as "the big bad boy. Everybody wanted it." "As soon as that sequence was out," de Lange said, "the race was on."

In a story full of false starts and unexpected turns, here was the final twist, perfectly emblematic of our digital age: Normally, finding the sequence of a gene in one species, like yeast, allows you to fetch the human gene quickly, by using a genetic computer program known as a BLAST search. In this case, the BLAST search didn't work; a single idiosyncrasy in the yeast sequence threw off the most sophisticated computer programs. You needed Lingner's sequence, and it turned out that after December 1996 you had to pay to get it, because the University of Colorado decided to patent Lingner's work. "We patented the discovery of the *Euplotes* telomerase," Cech said, "and we did so with the argument in the patent application that the human gene would look the same. I mean, we stuck our necks out a little bit," he continued, with a laugh. "But there's nothing to lose, because if the human gene didn't look the same, no one was going to be interested in the *Euplotes* sequence and patent anyway. So then the University of Colorado tried to sell that patent, tried to license it to a company. They sent feelers out to a number of pharmaceutical companies, and to Geron. And Geron came through quickly and aggressively and generously, in terms of the financial inducements." One Geron scientist recalled that the company paid between $200,000 and $250,000 for the initial rights, with more substantial benchmark payments, but what's surprising is that, even as the prize threatened to slip through the company's fingers, there was some internal disagreement about whether that was too much to pay. "I remember, before Christmas of 1996, we were arguing whether to pay that money to Cech," Villeponteau recalled. "But telomerase was so important to the company. We *had* to be the first ones to clone it."

So here was Geron, scrambling to license the patent of a gene it considered to be a corporate birthright. Now the company had Lingner's sequence, which presumably provided a kind of road map for finding the human gene. Cech reckons that Geron's biologists had a head start of at least four months on the rest of the world. Still, they couldn't snag the gene! Even with the *Euplotes* sequence, they got nowhere. "People at Geron were very demoralized," Villeponteau said. "We had a four-month head start, and we thought we had lost the whole race."

In April 1997 the Cech lab bailed them out — again. Lingner and Toru

Nakamura, a student in the lab, kept trolling for the human sequence, using their BLAST search to go through Genbank, an enormous database of human gene sequences that was updated daily. On April 2, the same day Lingner and Lundblad's paper was accepted for *Science,* someone working on the Human Genome Project deposited a scrap of human DNA in Genbank; about a week later, researchers at the Institute for Genome Research in Maryland made a long-awaited "data dump" of short human DNA sequences into the public domain. Soon afterward, both Lingner and Nakamura, working independently, stumbled upon the fish bait — a patch of human DNA roughly matching the *Euplotes* sequence. They e-mailed Cech at 3 A.M. with the news. "So then we called Geron," Cech recalled, "and we said, 'We've got this.' And they were excited about it. Then we decided to collaborate, because we didn't know how hot on our tails other groups might be."

Very hot, as it turned out. Not long after, Cech found himself attending a scientific retreat in New England for the Whitehead Institute, a citadel of molecular biology affiliated with MIT. "And I went to a poster session," recalled Cech, who was on the Whitehead's board of scientific advisers, "and talked to Chris Counter." Counter was a postdoc in the laboratory of Robert Weinberg, and he was among many skilled young greyhounds chasing the telomerase gene. "This was after we had found the human gene," Cech continued. "And he appeared not to have a clue. And I said, 'Well, you know, I'm sorry, but it's all over! We've done it already.' Which was true. I said, 'We've already got this, and you aren't at first base. And so we'll let you know when the sequence is done.'"

Normally, when a Nobel laureate tells a twenty-something gene jock to forget it, there's just not a lot of ambiguity in the message. But when Cech told Counter to "put down your pipettes," he took it as a challenge, not a deterrent. "I said, 'Well, thanks for the tip,'" Counter recalled, "and we went right back to the lab." "He and his colleagues in Weinberg's lab are very skilled and very ambitious scientists," Cech admitted, not without grudging admiration, "and they closed that gap *extremely* quickly. It turned out that Weinberg's group was much . . ." — and here Cech loosed a hearty but nervous laugh — "much faster than we thought humanly possible!"

✐

In the summer of 1995, Christopher Counter arrived at the laboratory of Robert Weinberg at the Whitehead Institute in Cambridge, Massachusetts, as a postdoctoral fellow. Counter had been a protégé of Cal Harley's in Canada, which is to say he was a disciple of the telomere story. "As is the custom with all entering postdocs, I told him to go spend a couple of weeks talking with everybody in the lab and then come back to me with a project that he would like to work on," Weinberg explained one day, while plucking leaves from an enormous indoor plant in his office. "And then we would negotiate back and forth, and haggle over what might be an attractive line of work. And he came back after about two weeks and said, 'Dr. Weinberg, what's going on in your lab is really quite interesting. But I really want to work on telomerase.' And I said, 'Well, that's a great idea. It's a very interesting field. But you really can't. Because you worked on telomerase as a graduate student, and it's my holy duty as your postdoctoral mentor to divert you in a new direction.' And he said, 'Thank you very much for your kind advice' and two days later began to work on telomerase."

Counter and a colleague, Matthew Meyerson, began trying to clone the yeast version of telomerase, and ended up finding the same telomerase gene Vicki Lundblad had found — est-2. Indeed, they were close on the heels of Cech and Lundblad. Counter said in an interview that the Whitehead group had submitted a paper on the work to *Nature* before Christmas 1996, which ultimately got turned down, but not until after the holidays. (They later published a more developed piece of research in *Proceedings of the National Academy of Sciences.*) But then they, too, ran into the computer snag that had hung up Lundblad and Geron — the yeast gene didn't lead directly to the human gene.

"They had gone into this with the pretense — or the promise, whatever — that this would lead directly, as day follows night, to the human gene, which did not happen," Weinberg recalled. Finally, on April 25, 1997, Lingner, Lundblad, and their colleagues in Colorado and Texas published a tour de force paper in *Science* describing the telomerase gene from *Euplotes* and yeast. Once the *Euplotes* sequence appeared in print, the race for the human gene entered its final, and now public, furlongs — the least interesting stretch, in the view of Cech, Greider, and just about any non-Geron scientist, because the big scientific hump had been Lingner's herculean effort to find *any* gene for telomerase. For Geron, however, exactly the opposite was true — the basic science was interesting, but the human gene was where the commercial prestige (not to say intellectual property license, products, and money) would ultimately reside.

By now, Geron had nearly two dozen scientists working around the clock to find the human telomerase gene. William Andrews, who headed the cloning team, told me he did not leave the site of Geron's labs in Menlo Park for three straight weeks — did not go home, did not go out for a meal, did not take a walk to clear his head. He slept on a couch in the office and, when that became a little too gamy for upper management, the company rented a trailer and put it in the parking lot so the scientists could have a place to sleep. Meals were brought in. Life was put on hold. The only thing that mattered was getting their hands on the human DNA sequence before anyone else. They almost didn't.

The Whitehead team took Lingner's published data and "ran and compared it with the yeast sequence," Weinberg remembered. "There was extensive homology" — meaning there were identical passages in the two genes' DNA texts — "and that enabled them to pull out the human sequence. It was cloning in silico. We had the yeast sequence. We had used it to scan human [DNA] by computer. Unsuccessfully. The moment the Cech sequence came out, within hours we popped it in, and it yielded homology to our gene, and homology in turn to the human gene. We were hungry for that sequence. On a day-to-day basis, we were looking to find ways of cloning out that sequence. Obviously it was a godsend when the Cech sequence came out." This was keyboard biology: you could type in the DNA letters of the gene for telomerase found in yeast or *Euplotes* just as you would type a word or phrase into a search engine like Google, punch a button, and send supercomputers electronically scurrying through vast digital tracts of human DNA sequences, looking for a proximate match — that is, for homology.

Using their still-unpublished *Euplotes* data, Joachim Lingner and his colleagues got there first, by a couple of weeks. That temporal advantage proved crucial. Toward the end of April, using the just-published *Euplotes* sequence, Chris Counter and many others found the same fish bait and raced to clone the same human gene. "Matthew was BLAST-ing like there was no tomorrow," Counter said. Around this time, Weinberg dropped in on Counter and offered some nonscientific advice informed by decades of wisdom and experience. "Send your wife some flowers," he said, "because you're going to be spending all your time here." By the end of May, they had cloned the human gene.

"They got some help from Merck eventually on the sequencing, so they had some Big Brother help," Cech said later. "But then again, we had all these people, a dozen people at Geron working like crazy to try to finish the sequence, too." With a four-month lead, Geron and the Cech group submitted a paper to *Science* on the human gene sequence on June 23; about three weeks later, on July 15, Weinberg's postdocs, who had not put down their pipettes, submitted a similar paper to *Cell*. Less than three months had elapsed between the time the human sequence first appeared in the Genbank database and the time a paper was submitted for publication. "Weinberg didn't get the Lingner data until two weeks later, and that made all the difference," Villeponteau said. "Cech really saved the day for Geron." "It's one thing to say it, but another thing to show it," said Counter, recalling Cech's admonition. After a pause, he added, "But he showed it, and kicked our ass."

In telling his side of the story, Weinberg savored the apparent inability of Geron's biologists to find the company's meal-ticket gene. "Cech sent the *Euplotes* sequence over to Geron," Weinberg told me, "and they fumbled around with it. And then Cech himself did some sequence searching and found the human sequence, pulled it out, and undoubtedly communicated it to Geron." Cech's lab, in collaboration with the Geron scientists, published its paper on the sequence of the human telomerase gene in the August 15, 1997, issue of *Science*. Weinberg's two postdocs, warned by Cech not even to bother, just missed beating the Geron team; Counter and Meyerson published exactly one week later, on August 22, in *Cell*. The finish was so close that for years afterward, people in the field weren't sure who exactly had made the discovery. But companies make sure: a day before the *Science* paper came out, Geron filed a patent application based on the work.

Perhaps the larger point is the public reaction to the news that the telomerase gene had been cloned. "We had people calling us up the day after," Gregg Morin, a former Geron scientist, recalled, "asking, 'How big is the pill going to be?'"

〜

That the race for the human telomerase gene ended in a virtual dead heat seems to have affected the sociology of the field as much as its intellectual property. The isolation of telomerase threw an unusually harsh light on Geron's relations with academic researchers and crystallized tensions between the academic and commercial camps that persist to this day, and that foreshadowed many of the problems regarding patent rights and freedom of scientific inquiry that later clouded the prospects of stem cell research. The friction erupted, as it often does, over issues of intellectual property.

Because of its licensing agreement with the University of Colorado, Geron received rights to commercialize the research that grew out of the university's patent on telomerase, based on Lingner's discovery in the ciliate *Euplotes*. Once the human telomerase protein was isolated, Geron and Colorado patented that molecule, and then Geron shared it with researchers throughout the world — but not without strings attached. Geron now claimed to "own" the rights to nature's immortalizing enzyme, and the company sought to control the way other scientists could use the enzyme in experiments, according to the terms of its material transfer agreement. Very quickly, Geron established a reputation for taking an unusually aggressive position in its MTA. Among other things, the company required researchers to submit a statement of experimental aims to Geron; although the company doesn't characterize it as such, that gives Geron tacit veto power over basic research. The MTA also grants Geron rights to subsequent commercialization of any discoveries. Those terms are neither unique nor unusual, but Geron's aggressive enforcement of them has alienated many researchers.

I first heard about the "Geron problem" from a friend of mine, a respected immunologist who, like many researchers, was intrigued by the role of telomerase in cancer (cancer cells, as noted earlier, use telomerase to keep replicating). We were walking in Times Square one night when I asked if he had ever tried to use Geron's telomerase; there was a hitch in his stride as he threw back his head. Trying to get the enzyme from Geron, he snorted, was "ridiculous." That was the word I kept hearing from academic scientists. "The Geron licensing agreements are ridiculous," Carol Greider told me at one point. She said that John Maroney, the lawyer at Cold Spring Harbor Laboratory, told her he had "never run across a company that was so difficult and pushed so hard." Maroney confirmed this. "I'm being nice when I say we learned some lessons from Geron," he said. "Their patent prosecution strategies were more 'independent' of us than our other corporate partners. We had less consultation with them than we either had or have experienced." Ultimately, Cold Spring Harbor amended Geron's conditions and distributed its reagents with fewer strings attached, but the tension never entirely dissipated. "When we wanted the human gene," Greider told me, "we just called Bob Weinberg, and he was great. He sent it out to us right away." Just to underscore that last point: Greider, who had been courted by Mike West and Alex Barkas, who had collaborated for years with Geron scientists and served as a founding member of its scientific advisory board, found the Whitehead lab much more cooperative than her former colleagues in California. "Almost anybody in the field who's in the know got it [the telomerase gene] from Weinberg," agreed Counter.

In 1999, Thomas Okarma, Geron's president, scoffed at this notion and almost belittled me for asking about it. He estimated that perhaps five or ten researchers had been unhappy about Geron's telomerase MTA over the years. I heard a starkly different account from Weinberg. "At least a dozen different groups, and maybe twenty, turned to us for the gene after giving up trying to get it out of Geron, and by now my lab has probably given out the gene to six hundred or eight hundred labs," Weinberg told me. "Suffice it to say that we've gotten a lot of requests from people who had previously gotten the clone from Geron and were unhappy with the strings Geron had attached to their using the clone." Weinberg's lab has made a point — a very political point, in the context of scientific freedom — of distributing its telomerase gene for free to any and all academic comers (the lab charges a fee to companies who want to use it). In fact, Weinberg goes so far as to suggest that Geron's stance on intellectual property may have impeded the speed of research on diseases like cancer. "I happen to know," he said, "that a number of companies have shied away from thinking about developing telomerase inhibitors, for fear that they may run afoul of the Geron patent and the very aggressive policy of Geron in warding off anyone who might come within light-years of their terrain."

It has become a cliché of contemporary biology that academia and biotech often work at cross-purposes, one devoted to basic knowledge, the other to the creation of products (with all the economic, proprietary, and publicity concerns to which a business must legitimately attend). But the logistics and legalisms of obtaining reagents like the telomerase gene from Geron seem to have left a particularly sour taste in the mouths of academics. Many basic scientists felt the company had consistently overstated its science while playing to Wall Street and the popular culture. Indeed, while the company increasingly emphasized the role of telomerase inhibitors as a potential cancer treatment, academic researchers pursued a more nuanced view of telomere biology. Greider has advanced the argument that telomere shortening acts as a kind of tumor suppression mechanism; when cells replicate too many times, as they begin to do in cancer, their telomeres would become short, their chromosomes would become unstable, and the cells would die before they turned into full-blown cancers. Ronald de Pinho, a researcher at Harvard Medical School, has discovered that telomerase offers promise as a treatment for cirrhosis because the enzyme seems to rejuvenate liver cells. De Lange showed that a flotilla of proteins gathers around the tips of telomeres, and that the DNA folds back into hairpin loops to protect the very end.

Blackburn, who had been sitting back, watching all this unfold while continuing to publish terrific science on telomeres, reflected a widespread disdain for the way Geron went out of its way to push a simpleminded hypothesis about telomeres and aging. "Geron went over the top," she told me one day in her laboratory. "They exaggerated so far that everyone just said, 'Oh, give me a break!' I mean, they were just hyping stuff with no scientific data . . . And with a company, of course, you know that this is about commerce; it's not about science." Even Vincent Cristofalo, a former member of Geron's SAB, felt Geron regularly oversold the telomere hypothesis. "I feel very strongly about that, and when I was on the SAB, I was critical of the overinterpretation of things. They were pushing the company line, and I was very disappointed in that."

In a similar vein, Weinberg pointed out the "unfortunate semantic implication" that *cellular immortality,* a venerable biological term, had been mischievously allowed to imply something altogether different: a kind of immortalization of the entire organism. "And the fact is, that's an enormous leap," he said. "There has been the implication that cellular immortalization will translate in some way to lengthening — if not greatly extending — the life span of an organism. That is irrational. But that irrationality has been exploited by companies like Geron to yield the impression that they are about to deal successfully with the aging process through the use of things like cellular immortalization." In fact, a vigorous and very interesting debate has developed in the last several years about whether telomerase might actually play a role in *causing* cancer.

William Hahn of Weinberg's lab led a team that published a paper in 1999, showing that turning on the telomerase gene predisposed a cell to become cancerous and might play a role in the formation of tumors. Cancer cells multiply ferociously, and these cellular doublings run roughshod over the Hayflick limit. In 1992, Cal Harley and Chris Counter were among the first to notice that cancer cells use telomerase to keep their chromosome ends well groomed. This prompted Weinberg, who can become perhaps a tad too accusatory when discussing Geron, to suggest that the way the company hailed the life-extending potential of telomerase was "grossly irresponsible, because to my mind, immortalized cells — independent of whether or not they can extend life span — are actually a danger for the organism because they already have moved ahead one step toward becoming malignant." On the other hand, Geron has launched an interesting clinical trial of a cancer vaccine — the company's first test of a potential drug — that specifically targets cells containing telomerase. The "immortalizing" enzyme's contradictory virtues and drawbacks prompted Elizabeth Blackburn to entitle a recent review "Telomerase: Dr. Jekyll or Mr. Hyde?"

Of all the people in the telomere field, perhaps Carol Greider experienced the rockiest relationship with the company. She had endured years of friction: her turf battle with Geron's in-house scientists, her uneasiness about the way the company hyped her own scientific results, and of course the mistaken paper during the telomerase race. But to Greider's credit, she did what scientists seldom do when they make a mistake; she published a follow-up paper in the literature, setting the record straight. Her problems with Geron, she said, came to a head when she published an article in the *Annual Review of Biochemistry* in 1996 questioning the link between telomere length and aging. "It had kind of been inflated in the literature," she said. "It's like a game of telephone. You talk about cellular senescence, and then it gets written up in the lay press, and then it gets written up again, and then suddenly it's telomere length determining life span. And there was never any data for it." So Greider wrote that telomere length "is clearly not directly correlated" with aging or, by inference, longevity. "And I heard a lot from Geron," she recalled, "because they were not happy that I had written this." (This, too, was not an isolated story; at least one other researcher, Titia de Lange, received scolding phone calls from Geron scientists when she planned to publish a scientific commentary that contradicted the company's commercial message.)

When I asked Greider who specifically had chastised her, the question prompted a painful admission. "Cal Harley, actually," she said, referring to her former close collaborator. "Saying that I wasn't supportive enough of the telomere hypothesis. And I said, 'Cal, you don't support a hypothesis.

You *test* a hypothesis!' And you know, it was clear that he was just doing his company thing, and so that was when I wrote the letter saying I wanted to go off the SAB. It was one in a long string of little things that I just had no need for." "Carol felt very strongly that they were continually and perpetually hemming her in," recalled Maroney. "And they wore her down."

Greider made her estrangement sound like a divorce. "I sort of see it as irreconcilable differences, in the sense of intellectual directions," she told me one day over lunch at Johns Hopkins, where she moved her lab in 1998. "I have learned through this whole thing that I am very much in favor of curiosity-driven research and academic freedom, which I didn't know before. Unless it's tested, you don't know how strong your feelings are in one direction or another. But what a company needs to do is to get its name out there, and, I mean, Geron has been hugely successful at getting its name out in the media, as a young biotech company needs to do. But I just don't think that one needs to say things that are really overreaching the facts. It bothers me a lot when people do that." She was not alone; it bothered a lot of people.

&

There was another divorce brewing in the Geron family, one that was partially obscured by the company's continuing string of scientific successes. At one level, cloning the human telomerase gene represented the crowning triumph of Michael West's founding vision. Barely six years after Redwood Shores, when he laid out the vision of aging research, and the cloning of key genes, as the scientific road that would lead to powerful pharmaceutical weapons against aging, Geron had its vaunted immortalizing enzyme in hand. The company immediately launched an ambitious campaign to test its potential application in medicine. Soon, papers about telomerase's remarkable activities were flying out of laboratories.

But at another level, the triumph masked increasing internal tensions in Menlo Park. Michael West, who had a history of clashes with lab supervisors while a student, had become a problem inside his own company. For years he had been inched further and further from the center of the action. In June 1993, Ron Eastman became president of Geron. Bringing in an experienced CEO, Villeponteau recalled, was a condition for the next installment of investor funds, $5 million. "West was in one sense shuttled aside," he said, "and he wasn't too happy about that." This internal exile was apparent even to outsiders passing through. "I was surprised, when I went out to Geron in 1994–95, that West seemed to have been marginalized," Greider said.

"I think Alex Barkas had his hands full dealing with him," one person familiar with the early days of Geron put it. "To a certain extent, Barkas did what you always do in cases like this: you give the guy a lab and four or five people to work with, and kind of get him out of the way." To be sure, there were people at the company, especially scientists, who would have run through a brick wall if West had asked them to. The problem was that West was perceived, even by some of his most fervent admirers, as someone whose gifts as a visionary biologist and entrepreneur exceeded his skills as a pure scientist, and he didn't seem to fit anyone's idea of a well-organized executive. In conceptualizing research initiatives and areas to pursue, he took big risks and frequently saw those risks rewarded, more often than not by a field that only grudgingly caught up with his aspirations. In his own science, however, he was not considered rigorous, even by admirers. Carol Greider and Vincent Cristofalo both recalled a meeting of Geron's scientific advisory board where West was making a scientific presentation and James Watson, a member of the SAB, interrupted him, launching such a blistering critique of West's argument that he couldn't finish. "Jim Watson asked him, 'Is any of this published?'" Cristofalo recalled. "Mike said, 'No.' And so Jim said, 'I don't believe a goddamn word of anything you just said.'"

But there may well have been something else afoot in the deteriorating relations between Michael West and the corporate leadership at Geron. At some point, according to people who worked at the company at the time, West began to doubt that telomere research and telomerase, although biologically interesting, were going to lead to products quickly, and suspected that they might never have an impact on aging, which remained his central focus and passion. Indeed, a number of the company's original scientists, utterly devoted to aging as a focus of research, became disenchanted with Geron's increasing emphasis on cancer research. Even as the company was enjoying its finest hour in public, West continued to push the rival technology of stem cells.

8

HAYFLICK UNLIMITED
(Or, the Immortal Dr. Hayflick)

IN THE SPRING OF 1997 a Canadian film crew preparing a documentary on the biology of aging showed up at the California headquarters of the Geron Corporation. The company offices had by that time expanded to several buildings on a quiet side street in an industrial zone of Menlo Park, on the far side of the Bayshore Freeway and in the midst of other high-tech start-ups. A low-slung sign planted on the front lawn, with its trademark teal green DNA-and-hourglass logo, marked the company's headquarters, and the path to the front door bore about as much media traffic in those days as rush hour on nearby Highway 101. This was the company, as one press account suggested, that was working on a "Zeus juice" that would reverse aging. That was the story everyone wanted to tell; that was the story everyone wanted to hear.

On this particular day, the film crew had come to interview, among other people, Leonard Hayflick. The rise of Geron had coincided with, and perhaps hastened, the second ascendance of Hayflick as a prominent figure in the world of molecular gerontology. He had been associated with the company since its inception, served on its scientific advisory board from day one, and considered Mike West to be, as he often put it, "one of my scientific grandsons." When West got his first big infusion of venture capital in the early 1990s, he had made the pilgrimage to Sea Ranch and talked Hayflick's ear off about his plans for the company, all day and on into the dinner they had at the Sea Ranch Lodge. "He was so excited he didn't even touch his food," Hayflick recalled.

By the mid-1990s, Hayflick had rehabilitated his reputation from the depths into which it had plunged in the 1980s. His lawsuit against the federal government, while not forgotten by Hayflick, had become a dim recollection in the scientific community's general consciousness, an abandoned and unidentifiable wreck on the shoulder of a busy highway increasingly plied by biologists traveling from academia to private industry as fast as they could, with as many genes and growth factors and cells and other proprietary reagents as they could carry, en route to commercial development (and, in many cases, great personal wealth). In 1976, Hayflick had been browbeaten into leaving a tenured position at Stanford, was pilloried in the pages of *Science* as an unprincipled, money-grubbing opportunist, and was likened to a Shakespearean figure doomed by greed and vain self-interest to a tragic end because he stood to make a million dollars selling "his" cells to industry (a sum, incidentally, he never came close to realizing). A mere four years later, on that watershed day in October 1980 when Genentech, the prototype start-up flying the flag of the New Biology, went public on Wall Street, a Stanford postdoc who had been paid in stock for a glorified summer job at the young biotech company saw his paper worth skyrocket to more than $1 million in two hours, and by the 1990s, dozens if not hundreds of respectable biologists boasted paper fortunes well in excess of a million dollars, many of them without having advanced the treatment of human disease one iota, some of them simply rewarded for telling a good promissory story. Say what you will about Hayflick, his work had immeasurably improved the safety of vaccines given to upwards of a billion children worldwide.

Once he got his wheels back on the road, Hayflick had roared to eminence within the community of gerontological researchers as an articulate and persuasive advocate for funding basic molecular research on aging. He came out of the cold, as it were, moving from the University of Florida to the University of California at San Francisco in 1987. In 1995 he published a well-received popular book called *How and Why We Age*. The lifelong, blue-collar provocateur had become a prominent establishment figure in his adopted field, and he clearly relished every opportunity to play elder statesman, to the point of occasionally indulging in interview requests that bordered on stunts. And something like that occurred with the Canadian film crew that day in Geron's offices.

As Hayflick recalls, he was sitting in a room at the company, pontificating on aging research and the role of cell senescence in aging, when one of the Canadians asked how human cell lines were established. "And I told them," Hayflick said, "that I've done that on myself dozens of times. And they said, 'Oh, you have? Would you like to do it on camera?'" Hayflick said he wouldn't mind, but under one stringent condition: the producer had to promise to use the footage, because Hayflick wasn't keen on lopping off a portion of his anatomy and then having the event end up, so to speak, on the cutting room floor. "So we did it," he continued. Mike West, who was also present, went to fetch a sterile scalpel, some alcohol, and a Band-Aid. "And on camera, I lifted a hair from that part of my anatomy" — he pointed to his knee, where a reluctant little pimple of skin rose as he tugged at a single hair — "and lopped off the top of the pyramid of skin that I had raised. And Mike took the specimen, carried it back to his lab" — and here Hayflick gave a little laugh — "and said he was going to culture me. I said, 'Go ahead, be my guest.' And he did."

West took the cells back to his little lab at Geron, where he and his colleagues digested away the connective tissue and spread out the cells from Leonard Hayflick's world-weary hide, just shy of seventy years old, in a cell culture flask, fed them some nutrient broth, and placed them in the incubator. There, as hoped, they took hold and grew. And then, several months later, West decided to take this ad hoc experiment a step further and throw a wrinkle into the exercise — not for the benefit of the film crew, now long gone, but to satisfy his own curiosity and, it must be said, his own brand of scientific impatience. He decided to try to "immortalize" Hayflick's old cells, using the human telomerase gene that had only recently been isolated. "I remember," he said later, "I wrote in my notebook — on the top of the page, to designate that experiment — The Immortalization of Dr. Hayflick. When I wrote that down, I thought, 'It sounds like a mystery novel.'" It did end up being a kind of whodunit. It's just that West, in a sense, turned out to be the victim.

No one knew it at the time, but those very cells — and by extension, what ultimately could be done to them — would lead to one of Geron's greatest scientific triumphs and make immortality (of the cellular sort) even more of a household word. Unfortunately, West's little experiment also catalyzed tensions between himself and the company he had founded five years earlier, to the degree that they parted ways just before the company went on, improbably, to an even greater scientific — and media — coup.

༄

The logo of hourglass and DNA on Geron's front lawn seemed to advertise the company mission, but in fact it symbolized a kind of growing identity crisis for Geron. The hourglass implied that it was an "aging company," yet few executives working inside the building believed Geron was going to cure aging with the magical enzyme telomerase. In fact, it wasn't even clear what telomerase might be good for, commercially speaking. True, it was present in an unusually high percentage of human cancers, and the company talked up its potential use as a cancer diagnostic, but that was unlikely to be a company-transforming product; Geron had licensed the technology to Boehringer Mannheim and stood to collect only royalties. True, telomerase also seemed to be an important elixir stoking the malignancy of cancer cells, and Geron embarked on an ambitious program to find molecules that might inhibit it, as a potential cancer treatment. But there again, the company had farmed out some of the rights, and, in any event, the odds of bringing a cancer therapeutic to market are daunting. The notion that telomerase could be injected into humans as a rejuvenating cellular tonic was a distant prospect, and was not aggressively pursued. And so there was a kind of tongue-in-cheek, media-weary cynicism about the company's public image even among its own scientists; when a public relations official walked me through the labs one day, Scott Weinrich, one of Geron's most respected scientists, whispered in mock horror to my escort as we passed by, "Don't show him the fountain of youth!"

Underneath that self-deprecating corporate humor, however, tensions continued to brew around Michael West. He still had an apostle's faith in molecular biology's ability to attack the process of aging, but a company needs products and profits. It was unclear whether telomere biology could provide that in the short term, at least with regard to aging, so cancer and "other age-related diseases" began to take a more prominent place as the company's commercial focus. The more cancer and "age-related diseases" became the mission, the more disillusionment set in among Geron's first-generation scientists; several left the company in 1997, including Bryant Villeponteau and William Andrews, complaining that aging was no longer a priority in research. Moreover, for a company whose identity was nearly synonymous with telomere biology, Geron scientists had been having a devil of a time, according to West, taking the obvious next step of testing the immortalizing properties of this powerful enzyme. "We were having trouble getting telomerase into the cells," West said, "and so in frustration, I went back and tried it myself."

Again, West's penchant for self-dramatization undoubtedly colors this story. In the fall of 1997, West claims he set out to immortalize the very cells that Hayflick had lopped off his knee for the Canadian television crew. In West's breezy account of what he later called "the experiment of a lifetime," he inserted the telomerase gene into the cells, and, lo and behold, Hayflick's creaky seventy-year-old cells suddenly sported telomeres as long as a hippie's hair. The telomeres grew longer and Hayflick's cells kept replicating, crashing right through the Hayflick limit and continuing for dozens more cell doublings. To use a vernacular explanation that would get plenty of mileage in coming months, West had "immortalized" Hayflick's cells, reset the clock, and rejuvenated them. These two mavericks refused to succumb to proper scientific decorum as they contemplated the mischievous wordplay that might grow out of this stunning demonstration. West said, "I thought we could call the paper 'The Immortal Dr. Hayflick.'" Hayflick had a similarly modest title in mind: "Hayflick Unlimited." The paper that ultimately emerged from this line of research at Geron bore a slightly more impersonal, but hardly less sensational, title when it appeared several months later in *Science:* "Extension of Life-Span by Introduction of Telomerase into Normal Human Cells."

The phrase "extension of life-span" was guaranteed to raise the eyebrows of scientists, journalists, and generalists. But it was the reaction among investors that really raised everyone's consciousness. Several days before the research was published, rumors began to circulate, and Internet bulletin boards crackled with investor queries about the frenetic action in Geron stock, which nearly quadrupled in value in one day. Professional analysts scoffed at the run-up, pointing out that this was the kind of research that, although scientifically impressive, was years away from clinical testing, much less FDA approval. But it demonstrated once again the mesmerizing allure of "immortalization" as a commercial concept — and a popular topic. Following its front-page account of the research, the *New York Times* did no fewer than three follow-up stories in the span of a month.

The research catapulted Geron into an even higher orbit of cultural awe. Not long after, *Fortune* said the company had "the highest buzz-to-equity ratio in biotech history," and an analyst at the Web site Motley Fool.com gushed with wide-eyed wonder at the company's research agenda. "Adding telomerase elongates the telomeres and makes the cells immortal," wrote Tom Jacobs. "Yes, I said *immortal,* and it's worked over and over in Geron's labs." The biology of telomerase even figured in a bestselling science fiction novel, John Darnton's *The Experiment.*

Within the company itself, ironically, the leading prophet of aging research was becoming the protagonist in a story with a sad ending. The data from West's experiments on Hayflick's cells were not included in the *Science* paper, even though West claims to have been the first scientist at Geron to get the experiments to work (a claim disputed by Geron scientists familiar with the research). West later admitted he had done only a single experiment and "simply didn't have the time" to bring his results in line with the fifty or so separate cell strains simultaneously worked out by Geron scientists and their University of Texas collaborators. "Mike is willing to do experiments with enough rigor to convince himself, but not necessarily with enough rigor to convince me," said Woody Wright, senior author on the *Science* paper. "I wish that I'd had the time to do a bigger number of Leonard's cells," West said, "because I really wanted — in fact, I encouraged the labs at both places to use Leonard's cells, because I thought it would be some poetic justice for him." Hayflick, who reviewed the Geron manuscript prior to publication, told me the company's scientists felt West had not documented his research well enough. "They argued that Mike's record keeping was poor, which is why they didn't want to include this in the group that they published," he said, making clear how dubious he was about this reasoning. As West put it, "It ended up being a little graph pasted on my office door, rather than a publication."

The dispute — and in Hayflick's opinion, it was a dispute, not an oversight — helps explain the beginning of the end of Mike West's tenure at the company he had founded. To hear insiders tell it, West had always represented a mixed box of chocolates to the management and investors of Geron. His ability to "pick the hot-button items," as one former colleague put it, and see a navigable path through the thorny middle distance of scientific research was widely respected. "I've always admired his . . . his avant-gardism — perhaps that's the best word to use," said Günter Blobel, the 1999 Nobel laureate at Rockefeller University, who served on Geron's scientific advisory board during the early years. "He's always been way ahead of his time." "He never shied away from the bold moves," said Bryant Villeponteau, "because he sees the big picture. But day-to-day details, he kind of glosses over. Which doesn't make him the best manager, which was one of the problems at Geron." Indeed, he too often seemed a lone wolf within the corporate structure. "Mike definitely has his own ideas," said Alan Walton, "and if people don't want to play along with him, he takes his ball and goes home with it." In some respects, West was viewed as a bit of a dilettante scientist by his peers, despite the fact that he had a Ph.D. When I spoke with him the first time, I admit feeling a little leery when West repeatedly used the word *we* to describe the scientific enterprise, as in "We've been doing histology for a while, and you'd think we would have noticed a lot of things through the microscope." It seemed a little like the enthusiastic camp follower leveraging inclusion through use of the first-person plural.

So it's not difficult to imagine the awkwardness created by those cells that had formerly resided atop Leonard Hayflick's kneecap. More or less simultaneously with West's ad hoc experiments, Andrea Bodnar led a team of colleagues at Geron that accomplished the same feat with far greater precision, more persuasive proof, and many more cell lines. They showed that "old" cells, with shortened telomeres after many cell doublings, could be rejuvenated in a lab dish when biologists introduced the immortalizing enzyme. Indeed, the addition of telomerase essentially reset the clock on the Hayflick limit, so cells could exceed their normal quota of replications. When this blockbuster paper was submitted to *Science* on December 1, 1997, West's name wasn't even in it. Hayflick, at least, was acknowledged in the fine print for reading the manuscript.

Both West and Hayflick were disappointed by that development, but not to the point of divorce. It was custody of West's most cherished children, the stem cells, which led to the final breakup. West claims to have faced considerable skepticism inside Geron about the potential of stem cells. He felt there wasn't enough support within the company for pursuing the research; only two people, he recalled, were assigned full-time to the project at Menlo Park, in addition to the three academic researchers the company was supporting. Those two in-house researchers were Walter Funk and Joe Gold (it is typical of the company's control-oriented public relations policy that when I attempted to interview them and left messages on their voice mail, I got return calls from Geron's corporate communications department, stating that upper management at Geron could answer any questions I might wish to ask Funk or Gold). But it was no secret inside the building that West felt the company was squandering an incredible opportunity by not pursuing human embryonic stem cell biology full steam. "It was unbelievable," said Bill Andrews, an unabashed West fan. "Here's this intelligent, brilliant person coming up with plans for doing stem cell research, and he just couldn't get people interested in it."

"I really wanted to see the embryonic stem cell thing very aggressively pursued," West said, "and I didn't feel that Geron was willing to do that. I obviously can't go into too much detail on that because a lot of it, I suppose, would be considered Geron confidential information. But yeah, I wasn't particularly impressed by the interest." West saw, for example, that the continuing political controversy in Washington created an enormous opportunity for private industry to run with the technology — run all the way to the patent office, to the best journals, to the best labs for further collaborative agreements. "There was clearly a strong difference of opinion in the direction of the company," Hayflick told me. Geron's management wanted to pursue telomerase inhibitors, "but Mike saw greater potential in the stem cell area." West's stubbornness might have worked against him, too. "There were two scientists and a couple of research assistants assigned to the project," said one Geron scientist familiar with the situation. "But they were keeping the project at a modest level. Of course, they didn't have the money to do it. The company's focus was on telomerase, and properly so, because that was where the financing had come in; and from a practical standpoint, that was where the products were going to be developed."

As West became increasingly frustrated, he approached management in 1997 and proposed spinning off stem cell research under the auspices of a separate, Geron-related subsidiary, while Geron pursued the telomerase work. This idea was, at least for internal consumption, considered seriously enough that the company explored separate financing for the spin-off, and scientists inside the company were aware that the plan was under consideration. Bryant Villeponteau admitted being jeeringly skeptical about stem cell research until he heard West make a presentation to the company's scientific advisory board. "He did just a fantastic job," he recalled. "In Mike's view, this was his baby. He brought it to the company. They were going to allow him to spin it off, and then they didn't. They had even lined up financing, and then they pulled the rug out from under him at the last minute. Geron decided it wanted to do it itself and not let it go, and I know Mike wasn't happy with the decision. I think that's one of the reasons he left." It would oversimplify things to suggest that that was the sole divisive issue between West and Geron's upper management. Another former scientist said that company officials were incensed that West went off and founded another stem cell company while still employed at Geron. "At Geron, they were not sufficiently willing to put up with his failure to follow up, and not doing things with rigor, and not being a team player," Wright said of West's tenure. But the stem cell episode was almost certainly the final straw.

There might even have been a long-range plan to ease West out because in November 1997, just months before the final breakup, Geron hired a new vice president, Thomas B. Okarma, a former emergency room physician and graduate of the Stanford University School of Medicine, who had previously founded a company called Applied Immune Sciences. Okarma's new title, "vice president of cell therapies," clearly augured crossed wires with West's stem cell ambitions, but anyone who has met both men can easily imagine a Wagnerian clash of personalities, too. West was dreamy, philosophical, almost adolescently enthusiastic about scientific ideas, an absentminded manager who sometimes missed meetings and appointments and whose corporate behavior could be erratic — "Not a guy who gets things done operationally," in Villeponteau's estimation. Okarma was crisp to the point of razor cuts, businesslike, articulate and well-spoken, about as confident as you would expect an ER doc to be, sometimes aggressive to the point of abrasiveness. This wasn't oil and water but something more volatile, potentially more explosive. According to Hayflick, West's contribution to the telomerase paper became a casualty of this conflict. "That was a result of his falling out with Tom, I think," he said. Okarma and West clashed bitterly, a former company scientist said, over another of West's ideas — one that eerily ratifies his ability to peer into the future. West wanted the company to apply for a Department of Defense grant to fund research into the detection and treatment of anthrax. Okarma opposed the idea.

The life history of start-up companies can be as ruthless and unsentimental as Darwinian evolution, and in Geron's life history, it made sense, after five years of excellent laboratory science, to bring in someone like Okarma, who had served as a vice president at Rhone-Poulenc Rorer, a major pharmaceutical company, before coming to Geron and knew something about converting interesting science into actual products. But his arrival undoubtedly hastened West's departure. "He did have people in the company blocking him," Villeponteau said of West. "Even before the stem cells, he was relegated to the role of a sort of grand ambassador. When Okarma came in, originally hired as vice president of cell therapies, he took over the embryonic stem cell project. Mike was again shoved aside, so if he wanted to do something, he would have to go outside the company. They kind of neutered him."

Less than two years after arriving at Geron, Okarma replaced Ronald Eastman as president and CEO. Geron's corporate repudiation of West has been so cold, so graceless, and so self-servingly revisionist that it's more redolent of Eastern Bloc historicism than West Coast entrepreneurialism. I came to this conclusion when I interviewed Okarma in 1999 for a magazine article. I mentioned a comment West had made about the origins of the ethics advisory board at Geron, which West had helped establish, and Okarma, clearly peeved, insisted that West "had no business" saying anything about Geron.

"Well," I said, "you can't separate Mike West from Geron."

"Sure we can," Okarma snapped. "He's been gone for two years. Come on," he added testily.

"He founded the company," I pointed out.

"So what," said Okarma.

"He made up the name."

"I don't care," Okarma replied, clearly incensed.

Those sentiments may well have been appropriate from a corporate point of view, but to market them in an interview seemed belligerent, bullying, and small-minded, and I couldn't recall another occasion when a corporate executive had so scornfully disowned a company founder in public (ironically, I spoke to several members of Geron's scientific advisory board, including Hayflick, Wright, and Cristofalo, all of whom *volunteered* the view that the company was poorer for West's departure). Okarma seemed to go out of his way to make sure West's vision no longer animated the company, declaring, "We are not an aging company." "That would not have been said when Mike West was there," said Villeponteau. "I saw it coming, and that's why I left."

It's not hard to imagine the anguish Mike West experienced as he himself felt exiled from the company he founded — founded not merely as a way to make money, but as the physical engine to achieve a lifelong, quasi-religious dream. In January 1998, as he was basically clearing out his desk, the Bodnar paper demonstrating that telomerase could "immortalize" old cells came out in *Science,* creating an enormous splash and setting new standards of media frenzy — the fountain of youth had truly gone molecular! Everyone from the *New York Times* to the *Los Angeles Times* sent reporters scurrying out to Menlo Park to do follow-up stories. "I didn't think I'd live long enough to see this," Leonard Hayflick told the *San Francisco Chronicle.*

To add insult to injury, the first hints of success in the Geron-sponsored stem cell program would occur in Wisconsin within a month or two of the telomerase paper. While all this was going on, in February 1998, West — whose candor on the issue is undoubtedly constrained by his confidentiality agreement — left Geron. Company founders, especially dreamers with as grand a vision on as large a canvas as West's, do not typically walk away from a work in progress after applying the primer coat. When I pressed him about the circumstances of his departure, he said, "You know, it was . . . it was with great reluctance. I really liked the company, and I liked the people there. I was sorry to come to the conclusion that I felt I could do more to advance science on my own than I could get done at Geron . . . I just felt that being there was actually slowing me down and not helping me. And, you know, my goal here is to try to do —" He paused to gather himself. "You know, it's a challenging project. Doing something about aging in our lifetime is a big project. And we have to be very productive and get a lot of things done. And what I was interested in doing was working on human therapeutic cloning, and I just felt that it was best not to try to get all that done at Geron. Geron had way too many things to preoccupy its attention."

So Michael West, the ambassador-at-large for human embryonic stem cell research, was suddenly a minister without portfolio. Although the corporate identity crisis at Geron smacks of the everyday growing pains at struggling biotech companies everywhere, the split between West and Geron had a deeper, more personal edge, if for no other reason than West's lifelong obsession with the biology of aging. Like Hayflick a generation before him, he walked away with a considerable (and understandable) chip on his shoulder. You need only appreciate the immensity of West's scientific and commercial ambitions to understand the tremendous amount of resentment his departure from Geron must have caused, and the hellbent tenacity with which he contrived to get back into a game he believed he had invented and then seen snatched away.

West now found himself operating out of a poultry company. During the summer of 1997, West, his wife, Karen Chapman, and some associates founded a company called Origen Therapeutics, based in South San Francisco. Origen was devoted to avian stem cell biology. It was largely West's brainchild and, as usual, he brought some intriguing ideas to the table; one plan was to reengineer chicken antibodies to make them into human antibodies that could be used as a form of medicine ("I actually think it would be a billion-dollar idea," West said). "I just hung out there because I didn't have anywhere to go," he said.

In the spring of 1998, Chapman, the president of Origen, had an appointment with several executives of a large commercial poultry breeding company on the East Coast, but at the last moment she became ill and couldn't make the meeting. West went in her stead, agreeing to meet with several officials of Avian Farms in the offices of a small biotech company they owned in Worcester, Massachusetts. West had barely spent an hour in the place when he realized he might be able to get back into the stem cell game much quicker than he had thought. But, as he would soon learn from the front pages of virtually every newspaper in the country, he also had a lot of catching up to do.

9

"MAMAS, DON'T LET YOUR
BABIES GROW UP TO BE
COWBOYS . . ."

IN 1995, ALTA CHARO, a law professor at the University of Wisconsin, received a call out of the blue from a stranger who happened to be a colleague on the university's science faculty. Jamie Thomson introduced himself and said he was calling because he knew Charo had served on the ill-fated 1994 NIH embryo panel. He wanted to pick her brain. He specifically wanted to talk about embryonic stem cells. "He said he had been working on primates and was considering moving on to human work," Charo recalled. "He invited me to come to his lab and sit through a presentation he was giving. It was a long talk, at least forty minutes or an hour, with slides, and then we just sat and talked about what it would take to move into humans."

The conversation ranged from Thomson's concerns about the ethics of stem cell research to highly pragmatic (and problematic) logistical questions that would bedevil any scientist willing to take the next, obvious step. Thomson wondered whether the federal ban on embryo research, as specified in the Dickey-Wicker amendment, included so-called overhead costs. It was a subtle question, but one that hints at the reign of terror the right-to-life faction in Congress had stirred in the research community. Universities earmark a certain percentage of every NIH grant they receive for what are called overhead costs — janitorial services, general lab supplies, building maintenance, heating, right down to the lightbulbs — so even the most scrupulous researcher accepting funds from private industry might technically still be doing experiments in a lab indirectly subsidized by federal grants — a form of fiscal miscegenation that could cause enormous ethical, political, and legal problems for both researchers and universities. "We talked about whether the federal restriction of embryo research included overhead costs, and would that preclude having privately funded

158

research?" Charo recalled. She immediately made a few calls to officials in Washington, but the issue was murky. She couldn't get a definitive answer.

Thomson, a soft-spoken but determined researcher, decided to play it safe. He set up shop in a small, closetlike facility in the university hospital reserved exclusively for human embryo work and then, with funds provided by Mike West, began trying to isolate human embryonic stem cells. The first steps were bureaucratic, not scientific. He applied to the university's institutional review board, or IRB, shortly after publishing a paper in 1995 reporting the isolation of embryonic stem cells from monkeys. Norman Fost, a well-known bioethicist and a member of the university's IRB, recalled that Thomson took exceptional pains to make sure all the ethical issues — informed consent by couples donating embryos created by IVF procedures principal among them — were addressed before proceeding. There was "quite a gap," Thomson said in an interview, "before we got our first embryo."

Thomson always seemed to err on the side of discretion and restraint. He didn't pave his route with press releases. He declined even to describe his lab's creation of monkey stem cells at meetings until the experiments were published in August 1995 (although he shared the news with several like-minded researchers, including Roger Pedersen). And he rebuffed potential collaborators like Joseph Itskovitz-Eldor, of the Rambam Medical Center in Haifa, Israel, who approached Thomson's lab immediately after the 1995 paper.

But Itskovitz-Eldor was persistent, and he felt a special affinity for the University of Wisconsin. In 1985 he had traveled to Madison to learn micromanipulation techniques from Neal First, an animal scientist at the university who had been a pioneer in cattle cloning; Itskovitz-Eldor returned to Haifa and was among the first in the world to apply the technology to assisted reproductive medicine for humans. Twelve years later, in early 1997, Itskovitz-Eldor organized a scientific meeting in Israel in honor of First and invited Jamie Thomson to speak. It provided another opportunity to lobby Thomson about a possible collaboration. The two scientists spent a couple of days together traveling around Israel, and Thomson finally agreed to work with Itskovitz-Eldor. It was a fateful decision. As a result, the Wisconsin group was able to work with both domestic and overseas sources of human embryonic material: fresh embryos left over from reproductive medical procedures at a small IVF clinic affiliated with the University of Wisconsin, and frozen embryos provided from the IVF center in Israel. As it turned out, the Israeli embryos generated more stem cell lines than the Wisconsin embryos — an outcome not without legal and scientific ramifications.

That Thomson and Itskovitz-Eldor were not proceeding alone became apparent at a meeting in Boston in the fall of 1996. Alan Trounson, an Australian IVF expert and perhaps the leading innovator in the field after Robert Edwards of England, had been asked to organize a session on human embryonic stem cells at the annual meeting of the American Society for Reproductive Medicine. He asked John Gearhart to give a talk, and it was at that meeting that Gearhart revealed that his lab at Johns Hopkins had begun to work with human fetal material culled from therapeutic abortions, in hopes of isolating human pluripotent stem cells. Trounson, too, had assembled a formidable team to attempt to derive human embryonic stem cells. He had recruited a gifted scientist from England named Martin Pera, whom Trounson describes as "a man with incredibly green fingers for growing cells." Although research on embryos was forbidden in Australia, Trounson's group had forged a collaboration with an innovative researcher in Singapore with the memorable name of Ariff Bongso.

Bongso, an expert in fertility research at National University Hospital in Singapore, was the son of a Sri Lankan rugby player; trained as a veterinarian in Sri Lanka and Canada, he had studied the mating habits of water buffalo and other animals for clues to improving fertility techniques for humans. Although his work remains largely unacknowledged in the West, he has pioneered a number of key technologies for the cultivation of stem cells. Indeed, in 1993 he developed a technique for growing a fertilized cell into a five-day-old embryo, in the hope that this would improve the success rate of IVF pregnancies; the following year, well ahead of everyone else, Bongso published a paper describing his attempts to isolate human stem cells from these embryos. He could indeed extract the cells, but then they would either stop growing or spontaneously trip into more specialized fates. At that juncture, he teamed up with Trounson. By 1995, he may well have had more hands-on experience with human embryonic stem cells than anybody in the world. One prominent American stem cell researcher, speaking off the record, told me Bongso deserves as much credit as Thomson and Gearhart for advancing the research. "This fellow Bongso," he said, "I mean, that group, I think, actually predates the Thomson findings, although the American newspapers and magazines give a lot of the credit, give *all* of the credit, to Thomson."

Gearhart took a different approach from Thomson. In the early 1990s, he had been inspired by animal research conducted by Brigid Hogan and Peter Donovan at Vanderbilt University; they had isolated primordial cells from the developing germ-line tissue of fetal mice, and these embryonic germ-line (EG) cells seemed to have many of the same characteristics as embryonic stem (ES) cells. The fact that they could be isolated from fetal tissue, oddly enough, made the research more ethically acceptable. Gearhart's lab, in the basement of a building at Johns Hopkins, occasionally received fresh human fetal material — fetuses from five to nine weeks old — as a result of medical abortions conducted at Hopkins. Gearhart and his principal collaborator, a postdoctoral fellow named Michael Shamblott, never knew in advance exactly when they would receive fetal tissue, but once they had fetched the material and brought it back to the lab, they would delicately excise a small hump of developing tissue known as the gonadal ridge; this is where the germ-line cells — the eggs or sperm of the developing organism — first arise before migrating to the gonads later during fetal development. Shamblott would "disaggregate" the tissue and isolate what they called primordial germ cells; these would be plopped into growth media and incubated.

For the Wisconsin group, the key to the entire set of experiments may not have had anything to do with the embryos per se but rather with the elixir in which the embryos grew. The Thomson lab received embryos at the cleavage stage, a very early phase of development that is reached a few days after fertilization. As this clutch of identical embryonic cells continues to divide, the primordial clump slowly doubles in roughly geometric progression: next 16 cells, then 32 cells, and so on. About five days after fertilization, the embyo contains 200 to 250 cells. At this point, a very primitive segregation of fate begins to unfold, and the first faint architecture begins to emerge in the form of the blastocyst, with about 30 embryonic stem cells. For a fleeting embryonic moment, each of these cells is endowed with a pluripotent genetic cargo: the biological capacity to become any cell type in the organism.

That was the goal — to capture those fleeting, changeable cells.

☙

For a long while, no one could isolate human stem cells. As Bongso was perhaps the first to discover, they were fragile and finicky. They would only grow on a lawn of so-called feeder cells — a carpet of embryonic mouse cells that presumably oozed chemical signals that tickled their growth. It was even harder to coax a fertilized egg to reach the blastocyst stage. Thomson's group discovered that an improved recipe for the growth medium helped the embryos to develop into blastocysts. The efficiency of the process was not great, but as Thomson recalled, "Given it was our first attempt, it wasn't all that bad. I mean, there are two steps. One is growing the cells, the embryos, from one cell to blastocyst. And that's still pretty bad. But the medium has improved dramatically recently, and that's actually why we were successful. But the success rate is still on the low side. Once you have healthy blastocysts, the success rate from that point was fairly good, and we ended up getting five [cell lines] out of fourteen blastocysts. And a few of those blastocysts looked pretty bad. So we were doing fairly good for a first attempt."

Thomson began to see the first signs of success toward the end of 1997 — just about the time his financial angel, Mike West, was getting his wings clipped by Geron. He had succeeded in creating one cell line, later called H-1, from a fresh embryo provided by the Madison IVF clinic. Then, in January 1998, Itskovitz-Eldor dispatched one of his students, Michal Amit, to Madison, carrying with her frozen Israeli embryos, and these ultimately yielded another four lines. But consider the daunting attrition of the research. The Wisconsin group started out with thirty-six frozen or fresh embryos, donated by couples who had undergone in vitro fertilization. Of the thirty-six embryos, they managed to cultivate twenty from "cleavage-stage embryos" (that is, embryos that had divided only several times) to the blastocyst stage; of the twenty blastocysts, fourteen looked robust enough to plunder for their inner cell mass; of the fourteen viable blastocysts, five yielded stem cells vigorous enough to establish cell lines in lab dishware — and, to hear some researchers tell it, of those five cell lines, only one is truly robust. Of the five Wisconsin lines, according to Itskovitz-Eldor, four were derived from Israeli embryos.

In May 1998 the NIH held a symposium on embryonic stem cells in Madison. Unbeknownst to government officials — indeed, to almost everyone — human ES cells had already been isolated. "At that time," recalled Itskovitz-Eldor, who had traveled to Madison for the meeting, "we never mentioned that there were human ES cells actually growing a few blocks from where the meeting was held." Thomson had originally intended for these initial cell lines to meet several "rigorous criteria" before declaring them to be human embryonic stem cells. He wanted to see if they would continue growing for an entire year, while retaining normal chromosomal features and the same developmental potential — the full metamorphic repertoire — as when they started out. After observing the cells in culture for about six months, the Wisconsin group sent a paper to *Science* on August 5, 1998, describing the creation of its five human embryonic stem cell lines.

As the Wisconsin researchers began to explore the properties of these unusual cellular dynamos, the stem cell story, perhaps fittingly, converged with the telomere story. Unlike normal human cells, of the sort first grown in a test tube by Leonard Hayflick, for example, embryonic stem cells expressed high levels of telomerase; in other words, these cells had managed to switch on the gene that tells the cell how to make telomerase, and the presence of this normally rare and suppressed enzyme effectively immortalized embryonic stem cells. This made sense, because one of the most remarkable things about ES cells is that they seemed never to hit the Hayflick limit. They replicate without limit. They are, in a very real sense, immortal, until they receive a signal to specialize.

The landmark *Science* paper produced by the Wisconsin group makes the argument (among other points) that breakthroughs and brevity can go hand in hand: it covered barely three pages in the journal, described the technique of stem cell isolation with admirable economy, and along the way managed to score both scientific and, subtly, political points. In an obvious nod to medical possibility, Thomson and his colleagues acknowledged that the ability to produce a "potentially limitless source" of neurons and cardiac cells, for example, had widespread implications for transplantation therapy. But they also pointed out that basic research on mouse embryos revealed substantial differences between mouse and human biology and that human stem cell research therefore offered important, if not unique, insights into areas of great clinical interest, including birth defects, infertility, and pregnancy loss. In other words, they made the case that despite social unease, human embryos would provide critical medical knowledge unavailable by any other route.

Gearhart and Shamblott, meanwhile, appeared to be slightly behind their Wisconsin rivals getting into print, although they had previously announced — to little attention — the isolation of embryonic germ cells from their fetal tissue at a meeting in July 1997. They had gone on to show that the cells could differentiate into many different tissues, and then fired off a paper at the end of September to the *Proceedings of the National Academy of Sciences*. In terms of submitting publishable data, they were about two months behind the Wisconsin group, but the two papers appeared almost simultaneously. About a year after their paper came out, I visited the Johns Hopkins lab, and Shamblott showed me what their EG cells looked like. He had walked down the hall to a small room where the cells typically reside in an incubator, and he placed a small rectangular plastic container under a microscope for me to look at. After centering the cells in the field of view, he whispered, "It's really nice," as if a normal tone of voice might perturb the magnificent architecture of the cells. I found myself looking at an "embryoid body," an aggregation of embryonic cells. Under low magnification, the EG cells looked like bubbles in spittle, or jellyfish eggs, a small cellular protuberance against a nebulaic gray background. They represented, literally, an uprising of protolife, a heave of biological potential up from a background of plastic and nutrients, and knowing that each one of those cells had the potential to become human brain or blood, liver or lymphocyte, kidney or knee only increased the sense that I was looking at the microscopic equivalent of Chartres, a soaring cathedral of cells wondrous not only for its sheer physical grandeur, but for the agency of boundless good works it was biologically empowered to deliver. As I stared in awe at this translucent clutch of cells, Shamblott gently reached in to pluck out the tissue culture plate. "I'm going to have to pull this off now," he said, apologetic but also protective. The cells, he explained, didn't like to be out in the cold too long.

While the Thomson and Gearhart labs rushed to finalize their publications in the fall of 1998, competitors were hot on their tail. The Australians, as it turns out, may have been much closer than anyone suspected. Trounson said in an interview that his group at Monash University in Melbourne had isolated human embryonic stem cells by the fall of 1998. In September, Benjamin Reubinoff, an Israeli gynecologist working on sabbatical with Trounson, flew from Singapore to Australia with human embryonic stem cells in his shirt pocket. Reubinoff and Bongso had isolated them, and now the Australians would study them. "We actually had embryonic stem cells at the time James Thomson published his work," Trounson said. "My colleague Martin Pera was one of the reviewers of that paper. And so when Martin told me, I was a bit disappointed. But I also recognized that James had been leading the area anyway for some time, so that was the right thing to have happened."

Even Mike West's new company managed to insinuate itself into the race, at least as a dark horse. In the summer of 1998, a young South American scientist named Jose Cibelli traveled to Israel to work in the laboratory of Itskovitz-Eldor. Cibelli worked at Advanced Cell Technology, where West had agreed to serve as president. He was due to arrive in Worcester full-time in October, but several months earlier Cibelli was dispatched to Haifa with the express intent of trying to clone human embryonic stem cells — in the same lab as one of Thomson's collaborators. "Jose worked in my lab independently," Itskovitz-Eldor said, "and with no relation to my project with Thomson. He received no knowledge from me, only technical help and biological matter." Cibelli's research didn't fare well, and neither did Roger Pedersen's. Although he had advanced the American research, at least logistically, by steering Mike West to Thomson and Gearhart, Pedersen never seemed to have gotten his experiments to work. He certainly didn't have a paper in the hopper (per his original condition to be interviewed, he refused to discuss research that had not yet been published).

Soon, the entire world would learn about human embryonic stem cells — in a way that left no doubt that their discovery was a milestone in medical research. Thomson's paper came out in the November 6, 1998, issue of *Science*. When the scientific triumph was publicly revealed the day before, science writers and commentators broke out a whole new thesaurus of superlatives to describe the importance of embryonic stem cells. They were inevitably described as the Holy Grail of biology, although some framed the research in a bigger picture. "The study of aging is undergoing a possibly profound change," wrote Nicholas Wade in the *Times*, "and a handful of biologists, whose hubris has not yet been punished with a thunderbolt from Mount Olympus, are beginning to think about interfering with the mechanisms that make the body mortal." Geron issued a press release modestly describing the work as a "breakthrough." The B word is often tossed around like pocket change by biotech companies desperately seeking attention, but no one disputed its merit on this occasion. And, as *Science* later put it, on one of the few occasions Yogi Berra has been cited in the scientific literature, it was "déjà vu all over again" on Wall Street, where investors reacted to the stem cell announcement by driving up the price of Geron stock 31 percent in a single day. As a *Fox News* commentator put it,

"There's nothing like the promise of eternal life to give a stock price a boost." But as *Science* also noted, Geron "has been operating in the red to the tune of $40 million since 1994 and is still years away from profitability . . ."

Within a month, Geron president Tom Okarma was testifying before Congress, pleading with senators to allow academic researchers to experiment with the stem cells. As well he might. Geron's commercial license with the University of Wisconsin essentially meant that many academic researchers would be doing research for Geron. It took a while for the lightbulb to go on, but Douglas Melton of Harvard expressed it best: "That would make me an unpaid employee of Geron!"

⌇

The real comic relief came a week later, when the world could sit back and watch Mike West, the architect of Geron's stem cell program, attempt to elbow his way back into the public spotlight. In the interest of corporate transparency, he managed to steal Geron's thunder, invoke the wrath of Bill Clinton, alienate countless scientists, and, in another of his unannounced appearances, exasperate members of the president's National Bioethics Advisory Commission.

In the spring of 1998, exiled from Geron and divorced from the stem cell race he had personally organized, Michael West was wandering the world like an entrepreneurial nomad, searching for a way to become a player again in the stem cell field. Around March, he traveled to Scotland and met with Ian Wilmut, the main architect of Dolly, the cloned sheep. West's trip was motivated by a typically interesting and provocative thought: theoretically, you could use the technique of cloning, in which an ordinary adult cell was inserted into a nucleus-free egg cell, to create a customized, short-lived embryo, and from this embryo you could theoretically isolate stem cells. This would solve two problems at once, in West's view. It would overcome any problems of immunological rejection if a patient's own cells were used in this procedure, which West liked to call therapeutic cloning, and it would provide a new, patentable route to stem cells. West traveled to Scotland specifically to ask Wilmut, based at the Roslin Institute near Edinburgh, if he would like to collaborate in an effort to create human stem cells through cloning. Wilmut, a quiet and wry agricultural biologist, rebuffed these entreaties by making the entirely reasonable point that West was asking the wrong person.

"I don't know why you're talking to me," he said, according to West. "I do not have any human technology." "His point was, he had never developed any technology for application in human therapeutic cloning," West continued. "He had only developed technology for animals." Wilmut claimed he knew nothing about human embryology, and West returned to the States empty-handed. What Wilmut didn't reveal at the time was that his laboratory was in the process of becoming a company, Roslin Biomed, which would attempt to commercialize the research that led to Dolly. Within about a year, in May 1999, the Scottish biotech company would be acquired by Geron. Here was further evidence that, at least in the minds of biotech entrepreneurs, cloning and stem cells revolved around each other.

West continued to talk to anyone who would listen. He had broached the subject of possible collaboration with the Australians, Alan Trounson and Martin Pera. He stalked animal cloners. He became infatuated with "nuclear transfer," or cloning, as a way to obtain stem cells. Then, around April 1998, even as human embryonic stem cells began to tick along in Jamie Thomson's incubator back in Wisconsin, West found himself taking a fateful meeting about chickens. That's when he ended up speaking with officials of Avian Farms, a leading breeder of commercial poultry based in Waterville, Maine, to discuss Origen's plan to cultivate stem cells from chickens. "Rob Saglio, the president, said, 'Well, why don't you fly into Boston? I've got this little biotech company down in Worcester called Advanced Cell Technology,'" West recalled. "So we met there." It was a small company devoted to the highly profitable area of livestock breeding, where tinkering with bovine embryos is big business, and animal cloning promised huge returns from developing clones of prize-winning beef cattle.

On the day West met with the Avian Farms delegation in ACT's small conference room, several company scientists dropped in to give brief seminars on their work. One of them was Jose Cibelli, who had worked in the field of animal reproduction all his life. He had done graduate studies at the University of Massachusetts in the laboratory of James Robl, one of ACT's founders, and eventually joined the company. As Cibelli spoke, West's jaw dropped, because the Argentine scientist described experiments he had done in Robl's lab in which he had created a hybrid embryo, albeit short-lived, through the union of a cow egg and an adult human cell.

When Cibelli showed a slide of this cow-human embryo, West couldn't believe his eyes. "I was just flabbergasted," he said. "I mean, he showed me human embryos that had been made by cloning. And I had no idea — no one in the world had any idea that it'd been done. And of course, ACT and Jose had filed patents. They were the first to do it and had filed patents on all this stuff, and I thought, 'Oh my gosh, this is *exactly* what I want to be doing for the next ten or twenty years of my life.'" West realized that cloning — or "somatic cell nuclear transfer," the same technique the 1994 embryo panel had so presciently included in its report — provided another potential route to obtaining embryonic stem cells, and that a cow-human hybrid would be a . . . a *thing,* but not necessarily an embryo. What West probably didn't know is that the 1994 embryo panel specifically mentioned interspecies cloning as one of those Rubicons of experimentation that should *not* be crossed. An embryo that was half-cow and half-human was wholly unusual, and probably wholly unviable, but it gave West what he desperately needed: a chance, so tenuous that it bordered on illusion, to get back in the game.

James Robl had stepped in to stop Cibelli's original line of research at U Mass, but West has always given his scientific colleagues very free rein. West joined ACT as president in the fall of 1998 and immediately instructed Cibelli to get the cow-human work going again. Meanwhile, West decided — in a gesture suggestive of public responsibility but also corporate self-promotion — that he needed to disclose the company's intention to pursue this line of experimentation, given the burgeoning controversy surrounding cloning and embryo research (he did not feel similarly compelled to mention Cibelli's efforts in Israel just a few months earlier). And, as he has often done, he went shopping for a friendly media outlet to retail his story.

"We had decided to announce the human ES cells via nuclear transfer to clear the table on this issue," he explained in a 1999 interview. "I didn't want to be accused of doing this in secret, and I wanted to be open and honest about it. I knew Nick Wade from his previous work — he had written about the early days of recombinant DNA, of mixing human and animal or bacterial DNA, and I thought he had handled it really well in his book *The Ultimate Experiment.* So I called him up and I said, 'You know, Nick, I want to get this out there, that this has been done. Could you help me and just write up a nice story that explains what these technologies are about, and what we've done?'" What's ironic about this call, aside from the way it reveals how a tiny, privately held company thinks it can go to the *New York Times* for "help," is that Wade is the same writer who "outed" West's hero, Leonard Hayflick, twenty-five years earlier as a reporter for *Science*, kicking off the events that led to Hayflick's celebrated (and ultimately successful) lawsuit against the U.S. government.

West even presumed to tell Wade how to do his job. "There are two ways you could tell the story," West recalled saying. "One is mermaids and minotaurs — you know, scientists create half-human, half-cow by cloning. And the other is, there's a silver lining to this dark cloud of cloning. We can actually show we can make these totipotent cells from a patient's cells." In point of fact, West and his colleagues at ACT could demonstrate no such thing; the lone, original blastocyst created by Cibelli had long since died, and no stem cells had been preserved from the experiment. Indeed, the "blastocyst," if it could be called such, lasted barely long enough for Cibelli and Robl to file a patent application. The cells were not characterized, the work had not been reproduced, even by ACT's own scientists, and the research results had never been subjected to peer review, much less published in the literature.

Like many West anecdotes, this one had a false bottom. The more he talked about this episode, the more it became clear that the motivation of his pitch to the *Times* was something more than a gesture of outstanding civic virtue. It turns out that ACT was anticipating negative coverage from a television newsmagazine show and hoped to preempt its impact by giving the sensational cow-human cloning story to another high-profile outlet. Several months earlier, according to West, he had been contacted by a producer at CBS's *48 Hours*, asking what ACT was working on in connection with aging. "Look, I want to be open and tell them what we're doing," he recalled. "I mean, what am I going to say? I'm working on cloning, but I'm not going to talk about anything that we're doing? So we just showed them the nuclear transfer [work] and we talked about it. But after I filmed with them, I came back into the lab, and I had used a hard-boiled egg as a prop, just to talk about nuclear transfer to an egg. I came back into the lab, and there's shattered egg all over the floor. And I asked the technicians, 'What's this?' And they said, 'Oh, they asked for some liquid nitrogen, and they dumped the egg in the liquid nitrogen and then dropped it on the floor and showed it exploding.' I told our PR representative, and we were a little bit concerned. We're not sure what this program is going to end up turning out like. And that's what triggered my thought, 'Well, look, I know this guy Nick Wade . . .'"

The beauty of the anecdote is not only West's shrewd, seemingly guileless and consistently successful shopping of his company story to selected members of the media, but the way he played the media's panting interest in aging, stem cells, and then cloning as if it were a Steinway (I mention this fully cognizant that I, too, was one of those shamelessly panting reporters eager to do one of those stories). More important, the incident suggests West's willingness to market controversy, and even court public censure, as part of a larger, long-term strategy of preparing society for the science to come. By dumping the issue of human cloning and embryo research on the table, as messily as that egg dropped on the laboratory floor, he managed to change the parameters of the conversation, changed the inconceivable into something that had to be discussed — and not in the abstract, but with the urgency that attends a development that might be imminent. In any event, West said the *Times* agreed to hold the story until the CBS show aired, which was scheduled for November 12, 1998. By coincidence, that turned out to be just days after Geron's announcements on the breakthroughs that had occurred in Jamie Thomson's and John Gearhart's laboratories. Roger Pedersen had an inkling that something was up, because Nicholas Wade had called him, described the cow-human embryo experiment, and asked for his reaction — without revealing the identity of the researchers, according to Pedersen. "It's hard to say this is a total sham," declared Pedersen, not knowing West was involved, "but I smell a sham here." That devastating remark was part of the story describing ACT's cow-human experiments that appeared on the front page of the *Times* on November 12. When I later asked Pedersen about his harsh assessment of a former benefactor, he still seemed mortified by the comments attributed to him, since they had clearly strained relations with the person who had organized and subsidized the research effort that had long been the dream of his entire field. "It certainly led to a distance between myself and Mike which had never existed before," Pedersen admitted quietly.

On November 12, when the *Times* story appeared, Michael West did not merely inherit the wind — he had to button up his collar against a tornado of negative reaction, from the White House on down. More than the successful use of the immortalizing enzyme, more than the reports on human embryonic stem cells, perhaps as much as the initial reports of Dolly and her ramifications for the possibility of human cloning, this line of experimentation set off all kinds of moral and political trip wires — the creation of embryos solely for research, the integrity of species boundaries, the ethics of publicizing unverified work outside the scientific literature, and the unregulated environment in which private companies could pursue such controversial work, with no apparent social or scientific accountability. West's story had something to offend almost everyone. The mere notion of a cloning experiment using a cow egg and a human cell seemed guaranteed to violate every commonsense scruple about the natural propriety of scientific inquiry. Religious leaders railed against West and ACT. Bill Clinton denounced the research as "deeply troubling"; in fact, the president ordered his National Bioethics Advisory Commission to investigate it immediately. Scores of newspaper editorials attacked West. Scientists dismissed the research as doomed to failure, the announcement as shameless grandstanding.

And yet, to some, the experiment did not seem entirely frivolous. Indeed, it sounded like exactly the kind of imaginative, daring, and desperately resourceful detour scientists willingly, if reluctantly, undertake in order to maneuver around political constraints placed on experimentation — in this case, all the social and ethical constraints evoked by the word *embryo,* especially as it was defined by religious conservatives, right-to-life activists, and natural philosophers, who argued that embryos were a form of human life due all the rights and respect accorded to other, more fully formed members of the species. While it was true, as West's scientific critics claimed, that a cow-human clone was unlikely to yield viable embryos, and therefore perhaps viable stem cells, for very good biological reasons, it is also true that experiments are done precisely to separate conjecture from reproducible fact; given the occlusive politics of embryo research, this was a conjecture worth pursuing. "I was intrigued," admitted Alta Charo, who was at the time a member of the Clinton bioethics panel. "If it were the case that you could take an enucleated cow egg and a human cell, and create an entity that functioned like a human embryo early on, so that you could get the stem cells, and then it decompensated later on, you would have evaded the problems of human embryo research, because you would not have destroyed a viable human embryo."

This was exactly the kind of contorted, "mousetrap" scenario scientists found themselves concocting when nonscientists were allowed to define life and its beginnings. West was pilloried for announcing that ACT was conducting the experiments, but in fact a number of leaders in the field would, to varying degrees, end up pursuing variations on the same general theme: create a "thing" that couldn't technically be called a human embryo but could be used to harvest immunologically compatible stem cells. Geron began trying to figure out how oocytes could reprogram adult cells, and there were rumors that the company was sponsoring the creation of cloned human embryos for research purposes. John Gearhart was working on a strategy to merge the nucleus of a human somatic cell with the "jelly," or cytoplasm, of egg cells to create what he called karyoplasts. Other researchers, in England and Israel and Singapore, where the political constraints on research were not driven by the same religious concerns, were trying to create embryos in IVF clinics and harvest the cells for research. And everyone was trying to figure out what magic resided in the egg cell — what spells it could cast, in the form of molecules or messenger RNA or other biochemical inducements — that allowed it to reprogram an adult cell. To paraphrase Henry Adams, any path that arrived at stem cells was the right one. And according to a former Geron scientist, Mike West was willing to push the issue, regardless of the scorn and public censure it brought him. "I think Mike is sacrificing himself for our sins," he told me.

Less than a week after the furor triggered by the *Times* story, the National Bioethics Advisory Commission, responding to Clinton's urgent request, addressed the issue of cow-human cloning at a meeting in Miami. When West got wind of the fact that NBAC intended to discuss cloning, he called up, invited himself, flew down to Florida, and appeared on November 17, 1998, to testify. Harold Shapiro, chairman of the panel, reflected an almost national state of confusion when he announced, "I do want to tell the commissioners that Michael West is here. He performed this experiment we're all talking about, or referring to, and thinking about — and have limited knowledge about."

The exchanges with panel members ranged from the scientifically skeptical to the ethically adversarial. Commissioner David Cox, a geneticist at Stanford University, later described West's testimony as "just an infomercial for the company." But perhaps West's most revealing remarks came in response to a question about the ethics of the research his company was undertaking. "As we inevitably, in the coming years, have more and more sophisticated technologies, which raise more and more red flags, I think it's absolutely critical to keep public trust and to have open and honest discussions," he told the commission. "Forgive me for telling a personal story, but I read a newspaper editorial by an individual who wrote that science should stop so that ethics can catch up. And my personal ambition in biomedical research is to communicate to people in public policy and in biomedical ethics, so that, simply, ethics can walk hand in hand with science, and science does not have to stop, because there is so much to be done in so little time." West seemed to be saying that ethics were fine, as long as they kept pace with the science and didn't slow it down. He was all for "open and honest discussions," but as both his past and future would confirm, he never stopped to ask for permission. He simply did what he felt he had to do to make his long-standing dream come true.

All in all, it had been a brutal couple of years for Mike West. After clashing with the management of Geron over the stem cell program, he got shoved out the door of the company he had founded, then watched his successors take credit for research he had single-handedly organized, supported, and championed, despite all the in-house skeptics. He had tried to be open with the media about his new company's work in therapeutic cloning, but one television show implied he was shattering long-standing moral boundaries on biological research, and Nicholas Wade had written in the *Times* that his tiny company was "venturing deep into uncharted realms of ethics and medicine."

Like all apostles, however, West was born with a bullet-proof vest; he just let the public scorn bounce off him. As if to show that he maintained a sense of humor about this personally turbulent period in his scientific and business life, or perhaps to show just how idiosyncratic his judgment could sometimes be, he wandered among the members of the president's bioethics commission during a break in the Miami meeting and, according to at least one member of the panel, his old nemesis Carol Greider, tried out the same joke on a number of panel members. "This gives a whole new meaning," he was heard to say, "to 'Mamas, don't let your babies grow up to be cowboys . . .'"

10

DEAD IN THE
WATER

ALI HEMMATI BRIVANLOU stood at the window of his office and squinted into the distance. His office is on the seventh floor of a building at Rockefeller University in New York, and what had captured his attention, on the morning of November 10, 2000, was an unusual beehive of police activity on the East River near the Fifty-ninth Street Bridge, about five blocks south from where he stood watching. Under a gray autumn sky, flashing red lights blinked from police vehicles gathered on the bridge, and police boats cruised back and forth in the water below. Brivanlou, a small and energetic man with a high waist and slightly hunched shoulders, who tends to rock on the balls of his feet, idly remarked that something bad must have happened.

As he stood there watching the scene, I was struck by how unusual an office this was for a scientist — several large plants flourished in pots, and, even more remarkable, the usual utilitarian linoleum had been paved over with a luscious terra-cotta tile, exactly the kind of idiosyncratic perk that makes a tenured professorship at Rockefeller one of the most coveted positions in the world of international science. The floor and the greenery and the stark white walls smuggled a tiny bit of the Mediterranean onto the Upper East Side, which was even more of a trick given that Brivanlou's building sits directly above the fumes and automotive hubbub of Manhattan's FDR Drive.

I had come to see Brivanlou that morning not because of the scientific work he was doing, but rather because of the scientific work he was not allowed to do. But before we got to that, I innocently asked about his background, which was an invitation for him to tell a remarkable personal story. About twenty years earlier, he had been an accidental immigrant in the

United States, working as a substitute teacher by day and busing tables at a Los Angeles restaurant by night, struggling to learn English, getting fleeced by immigration lawyers in a desperate bid to obtain a green card, and offering to do anything, including working for free, to get a letter of recommendation to graduate school. In some respects, he would turn out to be a more credentialed version of Michael West: he too wandered into university labs, volunteering to work for nothing, was at heart a provocateur, and was impatient to move forward. His personal story amply suggests why he is such a determined scientist.

As his name implies, Brivanlou is an exotic hybrid of old-world Europe and the Middle East. Born in Tehran in 1959, raised partly in Paris (where his father, a physician and researcher, worked in the lab of Nobel laureate Jacques Monod at the Pasteur Institute), reared as a Shiite Muslim (praying five times a day) yet educated at French-language schools in Iran and then at a university of science and technology in Montpellier, France, Brivanlou possesses dark Middle Eastern features but speaks in a rapid, Franco-frenetic tumble, with an accent so thick and slithery that you sometimes struggle to understand everything he says, even while enjoying its lubricious zest. He pronounces his calling in life with a certain old-world charm: "om-brreee-ohl-o-geest." He studies embryos. Indeed, he is in awe of them, having spent many years staring into microscopes, watching the miracle of life unfold repeatedly before his dark green eyes. In describing his history, Brivanlou vividly recalled his first days in graduate school, when he was ordered by his adviser to do nothing but sit and stare at a fertilized frog egg as it convulsed and cleaved and reshaped itself into a living organism. And so, when Brivanlou declares himself to be an embryologist, there's a kind of subtle provocation embedded in the very job description. Just as the word *embryo* has become a red flag in political debate, Brivanlou believes that avoiding its use represents a form of social and cultural denial. Embryologists devote their lives to and invest all their intellectual capital in studying embryos, which is to say everything — genes, growth factors, cells, and cell fates — that involves the unfurling of life from a fertilized egg to a living organism, of whatever species. Including humans. *Especially* humans, Brivanlou would say.

"There is a lot of police activity," he said, nodding again toward the bridge. Flashing red lights continued to blink on the span, and police boats continued to circle in the water below. At one point, he thought he might have seen a body bag. "Something bad has happened," he said again. With a shrug of his hunched shoulders, he returned to his chair.

Brivanlou was especially impatient during the week I visited his lab. The results of the 2000 presidential election were still very much in doubt, a week after the voting, and the outcome was of more than idle political interest to scientists like Brivanlou. Not once during the entire presidential campaign, not even in the customarily science-centric preelection candidate interviews conducted by *Science* magazine, was either candidate asked in a significant public forum about his views of embryonic stem cell research, or research cloning, or the larger issue of federal oversight for a host of emerging and potentially problematic areas of biological research with far-reaching implications for human health and human society. Albert Gore, the self-styled omnivorous technophile, was known to support stem cell research but never brought it up; on one occasion, George W. Bush responded in writing to a questionnaire from a Catholic group with the statement, "Taxpayer funds should not underwrite research that involves the destruction of live human embryos." That was it. Aside from an op-ed piece by actor Michael J. Fox, hardly any commentators noted that the 2000 presidential election had enormous ramifications for medical research. And until that unprecedented election was resolved, scientists like Brivanlou — and, more to the point, the institutions where they worked — were terrified of moving ahead on research even grazing these controversial areas.

But Brivanlou, as his life history suggests, is not easily deterred. That very week, he was preparing a proposal to show to both a prominent West Coast venture capitalist and a venture capital group associated with Rockefeller University. He had already made arrangements with Zev Rosenwaks, director of the Center for Reproductive Medicine and Infertility at New York Weill Cornell, to collaborate. "He has already given me eight embryos," Brivanlou told me, "and I still haven't touched those eight. But I think I'm going to do it immediately, especially seeing what's going on in the election scene these days . . . So I'm looking at the elections with a completely different view than other people are looking at it. I'm continuously thinking of my embryos when I listen to the results, because I know that that's going to be night and day."

Brivanlou did his graduate work at the University of California at Berkeley, but he made his reputation as a postdoctoral fellow in the Harvard lab of Douglas Melton. In research on the earliest moments of embryological development in a laboratory animal known as the African clawed frog (*Xenopus laevis*), Brivanlou and his Harvard colleagues discovered the chemical signaling and genetic activity that control the development of the vertebrate nervous system — in other words, the way in which our brains begin to form (and, by extension, what happens when they don't form properly). It was a fundamental biological question that had remained unanswered since early embryological organization had first been described in 1924 by Hans Spemann.

As Brivanlou (or almost any other molecular embryologist) will tell you, the fertilized egg is a wondrous theater of activity. The mother leaves love notes for her eventual offspring in the form of RNA "messages" that swim around in the plasm of the unfertilized egg, bearing important biochemical instructions that tell the zygote how to make the proteins that kick off the development of life. Soon, newly activated genes produce new proteins, and those proteins drift through the egg in a gradient, heavier concentrations demarking the basic geographies of head and tail, front and back, right and left; cells divide, genes turn off, proteins that control the basic blueprint of shape and form ("morphogens") mass in certain regions of the developing embryo, just as continents mass on the surface of a planet, giving rise to three primordial "germ layers" — mesoderm, endoderm, ectoderm — that in turn begin to specialize into recognizable tissues (brain, stomach, lung, kidney, heart, liver, muscle, bone, skin, and so on). In subsequent research, Brivanlou and others in the field began to piece together the step-by-step, moment-to-moment sequence of biological events — genes turning on, proteins wheeling into action, stem cells peeling off their immortal cycle and hurtling down the differentiation exit ramp toward one physiological fate or another, structures taking shape — that add up to the development of a mature amphibian.

But that was in frogs, and Brivanlou wanted to do better than that. He wanted to map the early developmental events in human embryos, too. And he had the embryos, sitting in a freezer in his lab, waiting to be thawed. In the summer of 2000, he had reached a tentative collaborative arrangement with Zev Rosenwaks, an Israeli-born expert in reproductive medicine, who had agreed to supply Brivanlou's lab with surplus human embryos left over from in vitro fertilization efforts across the street at Weill Cornell's IVF center. You could almost see lines of impatience on Brivanlou's expressive face as he agitated to get going. Critics of modern biology like to interpret this impatience as the body language of the "research imperative" — an insistence on doing experiments *now*, without moral pause or philosophical reflection, regardless of social consequences. But to a scientist who has lived his or her entire intellectual life in the shadow of a mystery as monumental and compelling as this one, it is simply the body language of consuming curiosity. In purely pragmatic terms, the earliest moments of human development held the key to tissue regeneration and cellular repair, but in some ways, the less pragmatic reasons for the research were equally compelling, more fundamentally satisfying, of greater grandeur. Brivanlou, for example, wanted to take a time-lapse genetic and biochemical sequence of snapshots of human embryonic stem cells as they developed, moment by moment, analogous to the earliest hours and days of the embryo; he wanted to catalogue the choreography of some 40,000 human genes, hoping to identify the ones that caused premature termination of pregnancy or led to congenital maladies. And he wanted to start working on it quickly, because researchers in England had already embarked on the same kind of research, with the approval and oversight of the United Kingdom's government-run Human Fertilisation and Embryology Authority.

"As you know," Brivanlou said that day, "nobody can stop scientists from doing the experiments they want to do. Nobody." The fierceness of the remark made me wonder how long those embryos would remain untouched in the freezer.

༄

Embryo research, although always controversial, never had much to do with the pursuit of regenerative medicine until the late 1990s, when embryonic stem cell research and cloning for biomedical research brought the two streams together. To its critics, embryo research has often been portrayed as cold-blooded murder conducted by emotionally desensitized and morally inert scientists. William May, the conservative bioethicist, has described this dilemma as a tension between "molding and beholding," between tinkering and admiring. In the context of stem cells, religious and political conservatives have characterized the research as a form of "embryo farming" and as an "industry of death," where artificially created embryos would be "harvested" or "strip-mined" for spare body parts and then tossed in the trash — in short, as a kind of *Brave New World* nightmare, with shelves of embryos piled up like auto supplies at a Wal-Mart, nascent beings "instrumentalized" and "dismembered" for the benefit (and this was the cultural component of their moral contempt) of baby boomers who wanted to live forever. But you rarely hear the purveyors of this arch and extreme language use it in the presence of the scientists who actually do this research, and I think the reason is that it is difficult to demonize or trivialize the intentions of people whose interest in relieving human suffering is indisputable. The harsh language of homicide and dismemberment willfully caricatures the scientific enterprise, and while there were hundreds of biologists in the fall of 2000 who were eager to do research on stem cells, few of them rebutted that mischievous caricature better than Ali Brivanlou and his mentor, Doug Melton.

Brivanlou reflects a more expansive, European sensibility than many of his American colleagues. His journey to Rockefeller began, oddly enough, with the Iranian Revolution of 1979. "At the same time the Persian revolution was happening in the streets of Tehran, there was an internal revolution going on in my family," he said, explaining that his parents decided to get a divorce. Brivanlou's father was granted custody of his eight-year-old son, Iman. "And the issue became: What's going to happen to my brother in view of the outside revolution and the inside revolution?" Brivanlou said. It was decided that the youngster should be sent away from the turmoil of Tehran to live with his older siblings in France; at the time, Ali, then in his early twenties, was pursuing a Ph.D. in biochemistry and living with his sister, two years younger, who attended the medical school at Montpellier. "So my sister became Mommy, and I became Daddy, and we all lived in a nice little apartment. I was in charge of external affairs, like school and insurance and immunization, and she was in charge of internal affairs."

In 1982, Brivanlou's mother, who had moved to California, desperately wanted to see her youngest child. With deep reservations, and having assured his father that he would return with his brother after a short visit, Ali Brivanlou obtained two-week visas from the American embassy in Paris and then flew with his young brother to Los Angeles. Two calamities immediately ensued. First, immigration officials at the airport, incensed that visas had been issued to Iranians in the first place, reduced the time of the stay from two weeks to forty-eight hours. Second, Brivanlou's mother refused to let young Iman return to France, and Brivanlou refused to abandon his brother. In an instant, he had violated his father's trust, abandoned his doctoral work in France, and become an involuntary immigrant. "We came in with a forty-eight-hour visa, and we never went back," he admitted. "So this is how I ended up in the United States. Not only in my private life was I going against the wishes of my father, who had entrusted me with my brother, but also my entire professional life was in ruin."

Brivanlou couldn't speak English, couldn't do science, was under constant threat of deportation, and held only odd jobs. He worked as a substitute teacher of physics and general science at the French-language high school in Los Angeles. Then, from 5 P.M. to 2 A.M., he worked at "a kind of restaurant where you go serve yourself and then sit outside. I was a busboy there for about three years. That was $3.50 an hour. That was really harsh." Brivanlou sought political asylum because of the revolution in Iran, but his request was turned down. And then one of his lawyers asked, "Are you a scientist, and can you do something that other scientists can*not* do?"

At this point, Brivanlou made a list of all the cellular and molecular biologists at local universities, and then started going door-to-door. "I got rejected by everybody at USC first, and then at Caltech," Brivanlou recalled. "And then one day, I came across a professor, a junior faculty member at the time, at the Molecular Biology Institute at UCLA. His name was George Fareed. Now, Fareed is a Persian name, a Persian last name. And so I thought, Well, that's interesting. Could this guy be Persian? Might I have a chance of actually landing with somebody who speaks my language and offering my services for free, in exchange for a letter of recommendation if he liked me?" Fareed did not speak Persian and felt quite removed from his Iranian ancestors. But he did speak French, and as soon as Brivanlou heard the sound of his adopted language, "it was like the doors of heaven opened up." While teaching at the French high school and busing tables at the restaurant, Brivanlou began to work four hours a day in Fareed's lab. Ultimately, Fareed and his colleagues started a biotech company called Ingene, offered Brivanlou a job, and, most important, sponsored him for a green card.

Legal at last, Brivanlou immediately began applying to graduate schools and was accepted at the University of California at Berkeley in 1985. Students there routinely do short stints, or "rotations," in several labs before deciding on a graduate adviser with whom to work. Brivanlou was quickly invited to join the lab of Robert Tjian, a prominent molecular biologist; but to satisfy the requirement to do all three rotations, he dutifully showed up at the lab of a newly arrived embryologist named Richard Harland. Since Harland was just setting up his lab at Berkeley, Brivanlou did little more than help him unload boxes of books. The first week Brivanlou was there, however, the frogs arrived. "Richard squeezed the female frog [to release its eggs] and fertilized it with sperm," Brivanlou remembered, "and then he put it under the microscope. And he said, 'So here's your first experiment. I want you to just look at this embryo.' I said, 'So should I, you know, record something, or draw, or take a picture of it?' 'No, just look at it. Just look at it for an afternoon.' And that was amazing! He put that one cell in there, and in an hour and a half, there were two. And then after that, every twenty minutes, they would double. You would pick up a phone call, you would come back, you would look down in the microscope: there were twice as many cells. And these are *huge* embryos, okay? I was absolutely fascinated by the intrinsic ability that that one cell had to generate an entire tadpole in a day. You would go from one cell to forty million cells in something like twenty-four or thirty-six hours." In the time it took to make a tadpole, Brivanlou became an embryologist. And it was not just a job description; it was a philosophy about the value of both fascination and curiosity in achieving basic knowledge as well as medical good. "If you look at the beginning, and how things are being set up and made," Brivanlou said, "you have a better chance to understand how they work at the end than if you just look at them when they're done."

Brivanlou worked three years in Harland's lab without producing a single publication. He also suffered a devastating spinal injury while bodysurfing during a vacation in Hawaii, broke his back, and was in a full-body cast for five months. But after years without experimental success, he switched to another project and began to make progress. In 1991, on the basis of this work, he applied and was accepted for a postdoctoral fellowship in the Harvard laboratory of molecular embryologist Doug Melton.

Lineage is important to the way scientists think about problems, and to the aggressiveness and imagination with which they search for solutions. Melton had done research in Cambridge, England, in the 1980s, and had sat next to Richard Harland in the same laboratory. Melton studied in the lab of John Gurdon, who is among the two or three scientists who, in the 1950s and 1960s, most explicitly paved the way to Ian Wilmut and Dolly. In landmark experiments published in the early 1960s, Gurdon showed that by inserting a mature, differentiated frog cell into a frog egg stripped of its nucleus, you could create an embryo that would develop into a genetic copy of the same animal — could, in fact, create clones of frogs. And Gurdon's philosophy about cloning and reprogramming adult cells informs both Melton's and Brivanlou's interest in nuclear transfer. They never saw it as a way to create copies of animals, or humans for that matter. Rather, they saw it as an immensely powerful biological tool that provided a unique way to tease out the secrets of normal development, disease-related mutations, and other aspects of basic biology.

Following his studies in England, Melton returned to Harvard (he is currently the head of Harvard's Department of Molecular and Cellular Biology, ensconced in a new building on Divinity Avenue in Cambridge). With researchers like Brivanlou, Melton began to catalogue the proteins that control the development of organs, especially in the nervous system. They learned that certain of these growth factors, or "morphogens," could basically generate nearly any cellular fate, depending on their concentration. Brivanlou developed a clever technical trick to show what would happen in a developing embryo if a particular gene was blocked. Knocking out the action of a growth factor known as activin right after a frog egg had been fertilized, for example, produced a startling result. The embryo became stuck. It couldn't tell head from tail, front from back; moreover, it failed to develop a mesoderm — the layer of embryonic tissue that ultimately gives rise to bone, muscle, and connective tissue. But perhaps the most unexpected and stunning observation came when Brivanlou went back and examined the cells in these failed frog embryos more closely. He was astonished to find that when you blocked activin, virtually all the cells in this aberrant embryo became brain cells. In other words, by *blocking* a chemical signal, all the cells in the embryo were consigned to a kind of default fate, and that fate was neural. Slowly, the vertebrate embryo was beginning to yield some of its secrets. On the basis of that work, Brivanlou was hired as a lab chief at Rockefeller.

In November of 1991, meanwhile, shortly after Brivanlou arrived at Harvard, Douglas Melton underwent a dramatic and involuntary conversion of his own scientific interests, the kind scientists rarely experience. His son Sam, only six months old, nearly died in the hospital. Soon after, he was diagnosed with juvenile diabetes, the potentially fatal disease that occurs when the body cannot make insulin to metabolize blood sugar. "I was there when that happened," Brivanlou recalled, "and we went through a very scary period." Melton didn't show up in the lab for weeks, and it was unclear at first if his son would survive. "And Doug — I remember vividly — came back one day after his long absence," Brivanlou recalled, "and talked to the group in a very emotional way. He said that he had been questioning some of the things he had been doing, and that he wanted to reassure us that as far as he was concerned, things were going to remain stable at Harvard, but he was now thinking about starting something independent of Harvard, to address perhaps a more direct application of our biology to modern medicine." Venture capitalists began to show up in the lab, and Melton eventually founded a company called Ontogeny, with the aim of developing molecular medicines for diseases like diabetes (in the promiscuous and savage world of biological commerce, Ontogeny merged with two other biotech companies in 2000 to become Curis). Back in his Harvard lab, Melton began to work with mouse embryonic stem cells in a bid to understand how the cells that make insulin are created. But, literally betting on his son's life, he didn't put all his eggs in that basket; he has continued to search for an adult progenitor cell in the pancreas that might accomplish the same thing.

Almost inevitably, Melton received a visit from Michael West, just around the time I dropped in on Brivanlou. "He came to see me to ask me to be on the scientific board of ACT," Melton recalled. "And I just — well, I declined, but I just found something uncomfortable about him. My grandfather always said, 'You know, life is short, and you should be careful who you associate with.' There's actually something unusual about him. I can't quite put my finger on it." In any event, Melton had a fruitful partnership with the Howard Hughes Medical Institute, an elite private foundation that lavishly funds many of the top biologists in the country. So he had compelling personal reasons to push stem cell research as fast as he could, as well as a very wealthy patron. Brivanlou, meanwhile, had prepared an inch-thick document, "A Proposal for Molecular Human Embryology," in which he outlined an ambitious plan to use human embryos to create a complete genetic map of early embryonic stem cell development, funded either by the private sector or by a privatized arm of Rockefeller. The experiments, Brivanlou argued, "will change human embryology from a traditional descriptive science to state-of-the-art molecular embryology." They had the embryos, the robotics, the DNA microarrays, and the expertise to forge ahead, and yet everything, politically, was still up in the air, and had been for the better part of two years.

৯৯

It is probably safe to say that in the recent history of biology, few if any discoveries comparable in importance to human stem cells have inspired so little follow-up experimentation, even though *Science* declared their isolation to be the "breakthrough of the year" in 1999. The research became mired in a bog of talk, an endless cycle of ethical and political consideration, followed by recommendations or proposed legislation, followed by inaction. In the summer of 1999 the National Bioethics Advisory Commission recommended that the federal government fund the derivation of human ES cells from frozen leftover IVF embryos, as well as vigorous research on them; as the *Washington Post* put it, "President Clinton thanked the commission and never mentioned the report again." In December 1999, after a long delay, the NIH issued its preliminary guidelines for the research, but the timing was entirely political; a source close to the process said the NIH was "terrified of jeopardizing the fiscal year 2000 budget, which was being debated all fall" in Congress, by releasing the guidelines sooner. Members of Congress held serial hearings and huffed and puffed about legislation that would liberalize stem cell research; in the spring of 2000, Arlen Specter promised a "knock-down, drag-out battle" in the Senate, but the issue never came up for debate, much less a vote. In August 2000, after nineteen excruciating months of deliberation, the NIH issued final guidelines for researchers who wished to use federal funds for human ES cell work, but the rules arrived as a dead letter in the midst of the presidential campaign. With virtually no chance for the plan to be implemented on his watch, Bill Clinton finally found his voice on the issue, celebrating the "potentially staggering benefits of this research" and adding, "I think we cannot walk away from the potential to save lives and improve lives, to help people literally get up and walk, to do all kinds of things we could never have imagined, as long as we meet rigorous, ethical standards."

It would be misleading to describe the stalled progress as purely political, however. More than a hundred scientists had requested human embryonic stem cells from both University of Wisconsin and Johns Hopkins researchers, but only a handful succeeded in getting them, and it was more than the uncertainties about federal funding that accounted for the delay. Here, the problem was as much legal as political. Geron controlled all uses of the cell line developed by John Gearhart at Hopkins and kept a tight rein on distribution. The University of Wisconsin had negotiated more leeway in its dealings with Geron. The university had patented Jamie Thomson's primate work, of course, and turned administration of the patent over to the Wisconsin Alumni Research Fund. WARF was a nonprofit organization that had been established in 1925 to oversee the university's intellectual property, beginning with synthetic vitamin D supplements licensed to breakfast cereal makers and later including the rat poison Warfarin and the trademarked "Wisconsin Stem Cells." In February 2000, WARF set up a separate organization, the WiCell Research Institute, to distribute Thomson's ES cell lines (as a historical aside, WiCell had a separate facility, a staff of twelve, and a multi-million-dollar government grant to do essentially what the NIH had asked Leonard Hayflick to do by himself a generation earlier). The University of Wisconsin worked hard to make the cells widely available while attempting to monitor their ethical use. But frustrations were mounting. "What I'm hoping," said one stem cell researcher at the time, "is that I can get hooked up with people who are totally out of the NIH system, and sort of lead a double life."

Even a double life wouldn't have solved many of the problems. Just as in the case of telomerase a few years earlier, the stem cells came with so many legal strings attached that many scientists found the terms unacceptable. One of the most vocal was Doug Melton. "I was surprised at that time by the absurd, in my view, restrictions, which WARF has now dissolved and wants to forget that they ever had," said Melton, who didn't need to wait for the NIH guidelines because of private funding from the Howard Hughes Medical Institute. The original Wisconsin MTA for embryonic stem cells had "quite unpleasant and idiotic" language, in Melton's estimation, including threats of court injunctions and demands of confidentiality of Wisconsin-provided information and materials. "There were restrictions that I found so onerous as to be laughable. But beyond that, neither Howard Hughes nor Harvard would allow me to accept anything under those conditions . . . They [the Wisconsin MTA] said that at any point they could give ninety days' notice and require the return of all information and material. So I said to them, 'So you're expecting me and my students to begin research projects, when you could at any moment just want everything back?' And they said, 'Well, our lawyers tell us we have to have those restrictions . . .' And this is above and beyond any concerns about intellectual property." "We had some problems with it at the Whitehead as well," said George Daley, a stem cell researcher at the MIT-affiliated institute. "Basically, there was a clause that said that you had to destroy the cells [under certain circumstances], so your research is fundamentally beholden to what they want you to do. That's sort of a Damocles' sword that hangs over your work."

Complicating the matter was the University of Wisconsin's agreement with Geron. Having funded Thomson's work, Geron had received exclusive commercial rights to medical applications of embryonic stem cells in six key therapeutic areas: heart, nervous system, liver, pancreas, hematopoietic system, and bone formation. Put simply, if Doug Melton figured out a way to cure diabetes using the Wisconsin cells, Geron could lay claim to the commercialization of his work, and if Melton took his research to another pharmaceutical company, Geron could sue that company. Geron had already established its swaggering reputation regarding intellectual property with telomerase, and it's not difficult to surmise Geron's attitude about the company's control of embryonic stem cell research from the unambiguous vocabulary of CEO Thomas Okarma, who told the *New York Times*, "This is where the center of the universe is in stem cells . . . We were working to derive these cells before anyone else thought about it. I'm not apologetic for our intellectual property. We paid for it, we earned it, and we deserve it."

By the fall of 2000, Gearhart's lab had received 150 requests for the Hopkins cell line, and Wisconsin had received more than 100, but according to a report in *Science*, WiCell had agreed to send its cells to only a half dozen researchers. "I don't think anybody will get cells for another two years," predicted NIH scientist Ron McKay. Melton, too, saw the handwriting on the wall at a stem cell meeting in Utah in the spring of 2000. "It became completely clear to me that no one was getting cells out of WARF and Geron, and the whole field wasn't really going anywhere, even though there was great interest in it, because there was just not enough material for anyone to work with," he said. And so, early in 2000, long before the presidential election, Melton approached the Howard Hughes Medical Institute and asked them to support experimentation with — and, more controversially, the derivation of — human embryonic stem cells, as a way around the federal, legislative, and legal roadblocks.

៚

As the public and political arguments traveled in circles, private companies and privately sponsored researchers moved ahead in a straight line, pushing the science further along and patenting each new development, with neither public oversight nor social vetting. And the uncertain outcome of the presidential election merely added an extra dollop of confusion to the picture. Would the White House ultimately be occupied by a friend or foe of stem cell research?

Real life does not stand still, even in the midst of political stalemates, and later it grew grimmer for Douglas Melton. In November 2001, he told me he was "reeling" from the news that his fourteen-year-old daughter, Emma, had just been diagnosed with juvenile onset diabetes, too. For some researchers, stem cells were not an indifferent and bloodless exercise in molding rather than beholding. The research was — as it was for all patients and all their loved ones — a matter of life and death. "I'm outraged by the implication that I want to do this out of some ego-control problem," Melton said, "where I want to control nature or make financial gain." But those were the terms in which the ethical battle was being framed.

Brivanlou, too, felt deep frustration. "As a vertebrate embryologist, I have to tell you I'm . . . I am ashamed to say that at the beginning of the twenty-first century, we know *much more* about how a worm embryo, a frog embryo, or even a mouse embryo develops than we know about ourselves," he told me. "And every single one of us justifies in our grants that the reason we are using these animal systems is to ultimately know about ourselves. And then somehow, when it gets to that point, when you have to do the specific experiment, you're not allowed to do it. Not because of scientific reasons, but because of legislative reasons. My wife, who went to law school at Harvard and quit science after getting her Ph.D., keeps reminding me that progress in science will not happen in this country at the bench. Progress in science in this country will happen at the legislative level." Or, he implied, will *not* happen because of legislation.

Brivanlou and Melton, each in his own way, would spend much of the next two years exploring, poking, pushing to find ways around the legislative and political impasse over stem cells. It was not the only activity in the field of regenerative medicine, but it became the critical battleground, not least because the eventual winner of the 2000 presidential election turned to a philosopher named Leon Kass to be his moral consigliere on issues biomedical. Kass not only opposed human cloning (that was the immediate battle); as far back as 1983, in an essay called "Mortality and Morality," Kass had argued forcefully against any biomedical attempts to extend the human life span. "Finitude" — Kass's preferred euphemism for a death unforestalled — was, on balance, a good thing. It enriched our sense of what it meant to be human and strengthened the bonds of family. Indeed, the desire to live longer (to say nothing of "practical immortality") was yet another form of cultural decay. "It is probably no accident," Kass wrote, "that it is a generation whose intelligentsia proclaim the meaninglessness of life that embarks on its indefinite prolongation and that seeks to cure the emptiness of life by extending it."

Of all the uncertainties on that overcast day in November, only one minor mystery lent itself to a swift, if grisly, resolution. The New York tabloids reported the next day that the police had fished a body out of the East River in the vicinity of the Fifty-ninth Street Bridge. It turned out, according to subsequent press accounts, to be the body of a fifty-five-year-old Hong Kong immigrant, the victim of his seventeen-year-old daughter and her boyfriend, who later confessed to strangling both the man and his wife and then dumping their bodies in the river. So much for finitude and family values. Brivanlou's fascination with the crime scene was serendipitous but apt. Stem cell research, for all but a chosen few, was dead in the water.

11

ELIXIR

CYNTHIA KENYON ARRIVED A LITTLE LATE on the day I went to speak with her, and as I sat in her office at the University of California at San Francisco awaiting her arrival, I noticed several books on her bookshelf. Lewis Carroll's *Alice's Adventures in Wonderland* and *Tales of Mystery and Imagination* sat alongside James D. Watson's *Molecular Biology of the Gene*, a combination that hinted not only at the breadth of Kenyon's knowledge, but the way her interests straddle popular culture and hardcore science. What really caught my eye, however, was a green poster taped to her office door. The poster announced a lecture Kenyon had given several months earlier at the National Institutes of Health about the area of research that has consumed her interest for the past decade or so: the search for genes that influence the life span of tiny nematodes and, by biological inference, perhaps our own life span as well.

In addition to the time and date of the lecture, the poster featured a picture showing a pool of water in a sylvan glade, with caricatures of strange frolicking creatures. The longer I waited, the more my eye kept returning to the quirky illustration. I noticed that the "pool" in the picture was actually a fountain, and that the cartoonlike creature in the foreground was actually a wizened nematode, hobbling along with the help of a cane, while half a dozen infant worms, in diapers, splashed in the spray of the fountain. Finally, I realized that the pool was meant to represent the fountain of youth. Indeed, the title of the lecture — one of numerous talks Kenyon has delivered on the same theme in recent years — was "Genes from the Fountain of Youth." Nowhere else, in all my travels, was the connection between molecular biology and life extension made more graphic, more explicit, and frankly more whimsical. And the notion of longer life

189

through genetic manipulation could hardly have a more appealing spokesperson than Kenyon.

When she finally arrived, about a half-hour late (she'd returned the night before from a trip to Taiwan), Kenyon — tall, strawberry blond, looking at least ten years younger than her forty-six years, dressed in a white turtleneck, jeans, and black lace-up boots — appeared justifiably jetlagged. Her subdued demeanor surprised me a little, and I later realized this was because I had come with expectations shaped by her numerous appearances on television. Both because of her excellent science and, I suspect, her photogenicity, Kenyon has been a fixture on the lecture and *Nova*-type science television circuit as perhaps the most articulate exemplar of molecular biology's assault on aging. To cite but one of many instances, she riffed with Alan Alda on a PBS–*Scientific American* documentary about aging. I conversed once with a network medical correspondent at a social gathering who practically drooled at the mention of Kenyon's name — in sheer gratitude to have as a source a hip, well-spoken biologist who communicated both optimism and excitement about the complicated genetics of aging, and who had the credentials to back up her enthusiasm. Consider the clarity of this statement: "The process of aging influences our poetry, our art, our lifestyle, and our happiness, yet we know surprisingly little about it. Genetics has taught us a great deal about gene regulation, development, and the cell cycle. Can it teach us how we age?" That wasn't from our interview; that was from one of Kenyon's scientific papers in *Cell*, one of molecular biology's most turgid, data-thick, jargon-strewn journals.

In person, she came across much the same: inordinately bright, gently aggressive, thoughtful in a big-picture way, and utterly undeterred as a scientist by potential hurdles. She spoke in a rapid, breathless whisper that conveyed both intimacy and urgency, seemingly allowing you to see the gears of her mind turning as she processed a thought. And the thought that brought all of us — TV crews, reporters, book writers — to this small office was simple, yet audacious: there were genes that controlled life span, at least in the animals biologists routinely study, and Kenyon saw no reason why similar genes shouldn't be found in humans, too. In fact, she was about to put other people's money — that is, venture capital — where her mouth was, by starting a biotechnology company in 2000 devoted to commercializing precisely those ideas.

It is a hard argument to ignore, for several reasons. Kenyon had an excellent pedigree: she trained with several top-notch biologists, including Sydney Brenner, who was arguably the greatest twentieth-century biologist never to receive a Nobel Prize until he finally got one in 2002. She also came of age as a developmental biologist during the 1980s and learned a lesson near and dear to the hearts of all molecular embryologists: when it comes to genes, evolution keeps going back to the same well. The genes that control the development of nematodes and frogs and mice turn out to be the same genes that operate in humans (albeit in slightly more complicated ways). And when Kenyon began to pursue aging research in earnest in the early 1990s, she quickly discovered that aging in nematodes was related to a hormonal signaling system surprisingly similar to insulin signaling in humans. She wasn't shy about the implications, either. In the early 1990s, at meetings, she began to talk about the "fountain of youth gene."

∽

The marriage of hard-core molecular genetics, our most precise and vaunted life science, with perhaps our oldest and most alluring human fantasy — eternal youth in a bottle — is one of the most intriguing (not to say amusing) aspects of recent gerontological research. If you consider the moral embedded in the misfortune of Tithonus — he is the poor sadsack of Greek mythology, who mistakenly asked the gods for eternal life instead of eternal youth and wound up in an endless purgatory of decrepitude — the fountain of youth myth has been part of the written record of human longing for at least two millennia. Indeed, the story of Tithonus has an appropriate echo in one of the truisms of modern scientific research: If you don't frame the question (or wish) the right way, you are doomed to get the wrong answer.

If Tithonus gave the myth a precautionary quality, ancient cultures, notably the Egyptians and Chinese, gave it an aura of quackishness (at least to our modern eyes) by seeking immortality through alchemy. Nearly 2,500 years ago, Chinese alchemists known as "thaumaturgists" devoted considerable energy to creating "drinkable gold" as a means to prolong life. "The king of the state of Chu was presented with an 'elixir of deathlessness' by thaumaturgical technicians," notes a Chinese text dating to around 400 B.C. The historian Gerald J. Gruman, who has surveyed the long-standing human fascination with life extension, writes that, "Interest in the fountain of youth had reached an apex in the fourteenth and fifteenth centuries, and the discovery of America gave a new impetus to the tradition in the early years of the sixteenth century."

Juan Ponce de León gave this myth an enduring ethos of futility. His benighted search for the mythical fountain of youth is a standard chapter in human folly. Perhaps it is appropriate that the Age of Discovery, in which Ponce de León had more than a walk-on role (he sailed on Columbus's second voyage to the West Indies), included as part of its mission this hoped-for deliverance from short and mortal life. Around 1513, the year in which Ponce de León discovered Florida, the average life expectancy of men and women was probably no more than three decades. Famine, plague, infant mortality, war — to the sixteenth-century mind, there were plenty of reasons to husband and nurture this myth, like a small candle in the big wind of Nature's indifference.

Born in 1460 of a noble family, Ponce de León lived to the relatively ripe old age of sixty-one. His nonmythical exploits were considerable: he fought the Moors, made his way early to the West Indies, and, under Spanish sponsorship, extensively explored Hispaniola, where he later served as governor. Legend has it that while serving as governor of what is now Puerto Rico, Ponce de León first heard tales from the local population about a fountain or spring on an island to the north called Bimini that reputedly rejuvenated anyone who drank from it. Writing to Pope Leo X in 1514, the Italian geographer Peter Martyr, who resided in Spain, reported, "Among the islands of the north side of Hispaniola, there is one about 325 leagues distant, as they say who have searched the same, in which is a continual spring of running water, of such marvelous virtue that the water thereof being drunk, perhaps with some diet, maketh old men young again." After conceding his own doubts, Martyr added, in words that still resonate half a millennium later, "they have so spread this rumor for a truth through all the court, that not only all the people, but also many of them whom wisdom or fortune hath divided from the common sort, think it to be true."

Intrigued perhaps more by reports of gold than rejuvenating waters, Ponce de León mounted an expedition. Here the tale is instructive on two counts. First, this voyage of discovery, launched in March 1513, nearly five hundred years ago, was privately sponsored; then as now, business rather than government took the lead in underwriting the search for the fountain of youth. Second, although Ponce de León didn't find what he was looking for, he found something that came to be strongly associated with aging. He discovered Florida.

Since the early sixteenth century, the fountain of youth story has been passed down almost exclusively as a precautionary fable about impossible human desires, a fable where social wish fulfillment overcomes biological inevitability. Since the wish has existed from the beginning of recorded time, however, it has been a myth with legs — many of which have belonged to the shamans, quacks, and patent medicine salesmen who have, from time immemorial, peddled nostrums and potions to increase life span. The tale took a notably pseudoscientific turn in 1889, when the respected French scientist Charles Edouard Brown-Sequard claimed he could rejuvenate old men with a tonic containing the crushed testicles of dogs. There has always been a market for the promise, and that is probably why public attention, through its proxies in the media, became so focused when modern biology began to hint at a different ending to the same old story.

It is both the gift and bane of every generation to believe it views age-old problems in completely new ways, but you can understand the special swagger that molecular biologists brought to the notion of biological rejuvenation. Cynthia Kenyon, for example, came from the High Church of can-do reductionism. "The thing about aging," she told me that day, "was that everybody thought, 'Well, you know, the animal just breaks down, and what's to study really? You know, what is there? It's not going to be very interesting.' But when I looked at it, I could see that different animals have different life spans. And my first slide for every talk I give shows a picture of a mouse and a canary and a bat. The mouse lives two years and the bat lives fifty years and the canary lives about thirteen or so years. They're all small animals. They're warm-blooded. And they're not that different, really, in such a fundamental way from each other, and yet they have enormously different life spans . . . So it seemed to me, for some reason, animal life span was under [natural] selection, probably coupled to reproduction in some way."

Born in Chicago in 1954 and raised mostly in the Northeast (she is a descendant of William Bradford, who came to America on the *Mayflower*), Kenyon attended the University of Georgia. "I was wandering from major to major," she recalled, "and my mother worked in the physics department there, at the University of Georgia, and she found this book." Kenyon nodded up at the bookshelf, at Watson's *Molecular Biology of the Gene*. "She

brought it home to me when I was in college, and I thought, 'This looks really interesting.' I learned about switching on and off genes in bacteria, and
I just thought it was fabulous. So that's really what got me into it." She
graduated from Georgia as class valedictorian in 1976 with a degree in biochemistry.

Kenyon moved on to MIT to get her Ph.D. but also spent a year in the
Harvard laboratory of Mark Ptashne, a biologist who had plowed a lot of
the early ground in explaining how genes are regulated by studying a
simple virus that infects bacteria. In 1981, Kenyon went to the "other"
Cambridge, in England, and spent five years in the laboratory of Sydney
Brenner, studying the way genes control early development in the nematode *Caenorhabditis elegans.* Brenner had tirelessly championed experimental work on this organism, with its 19,000 genes and its easily observed
959 cells. Soon Kenyon became part of an army of researchers who,
through genetic and bioluminescent tricks, could show polarities, axes, and
even stripes forming in very early worm and fruit fly embryos when genes
became activated. She learned not only the exquisitely sensitive genetic
choreography of developing organisms but that these master-control genes
kept cropping up in different organisms, up and down the phylogenetic
ladder. Not with 100 percent similarity, to be sure, but if you found a gene
that controlled the front-and-back (dorsal-ventral) development in fruit
flies, for example, you were likely to find something similar in yeast, frogs,
nematodes, mice, and, if you were lucky, in humans, too. "So that made me
extremely interested in evolution," she said.

That was the intellectual incitement that Kenyon brought with her
when she came to the University of California at San Francisco in 1986.
That, and a maturing sense of ambition about tackling aging, with the help
of her little worms. "Basically, I saw aging as being —," she began, and then
stopped herself in midthought, as she often does in conversation, and made
a 90-degree turn. "See, the other thing is, I have really been struck in my life
by how unimaginative scientists tend to be. How pessimistic they tend to
be. In a certain way. They tend to think something is boring and will be too
complicated to solve, unless it's really forced on them. There are many people who thought we couldn't use genetics to solve the problem of development, for example. There'd be raging debates about this in the literature . . .
And then, when development was really understood, it was quite clear that
there were a whole set of genes that were just *dedicated* to making patterns

of development. Which makes a lot of sense, because that's what evolution does." And Kenyon knew that Tom Johnson, a biologist at the University of Colorado, had been studying a mutant nematode that lived longer than usual. It was called *age-1*. "So it seemed to me that aging was just exactly perfect to study. Nothing had been cloned. So that told me that it would be possible to study aging genetically — it might be — because it seemed like there was a long-lived mutant. But no one was interested in the field. By the time I started studying it, I was very well established, so I could study it. But it was hard to get people in my lab to study it. People thought if they did, they would just disappear. You know, scientifically." In other words, as recently as 1990, the study of aging was considered a form of career suicide for a molecular biologist.

"So I saw aging as potentially being like development, in a sense. Something that was controlled by genes and that was subject to perturbation during evolution that would lead, in one fell swoop, to animals with different life spans. So I thought that there was going to be something really interesting there. Something kind of beautiful and elegant. And that we could study it by looking for mutations that affect aging. And everybody would — at the end, they would look at it and they would go, 'Wow! Look! It's really neat!' You know? So that was my very strong bias going into it. And then we found a long-lived mutant."

Kenyon wasn't wandering through virgin scientific territory. Michael Rose, an evolutionary biologist at the University of California at Irvine, had been breeding fruit flies to get increased life spans since the early 1980s; he didn't know exactly what genes were responsible for life extension in his flies, but he knew it was possible to, as he put it, "postpone senescence" genetically. Seymour Benzer of Caltech, one of the grand old men of fruit fly genetics, was exploring the possibility, too; he would later isolate a gene that influenced life span, which he called *methuselah*.

In the early 1990s, Kenyon finally managed to persuade several graduate students in her lab to work on aging, and they began to set up what are called life-span screens. They created mutations in the nematodes and then looked for mutant worms that seemed to live longer on average than their mates. Worms provided a quick and dirty test strip for life-span studies; whereas mice live on average two years and fruit flies about forty days, nematodes typically live less than three weeks. "The first one we found, we kind of bumped into it setting up our screen," Kenyon recalled. The longevity gene they discovered was already very well known. It was called *daf-2*, and had been studied by many labs for another reason: it conferred a kind of suspended animation, called the "dauer" phase, on larval worms that found themselves in a stressful environment, either without enough food or in too dense a population. That work was published in 1993.

Kenyon's nematodes, as big and exciting as lint, kept throwing off scintillating clues about the genetics of aging. As the UCSF group teased apart the biology of these long-lived mutants over the years, they learned that two genes implicated in life span, *daf-2* and the related *daf-16*, seemed to sit at a crucial junction of cellular activity connecting life span, stress, metabolism, and sexual reproduction. In fact, hormones seemed to cue the system to age more rapidly, while sterility or infertility seemed to promote a longer life. Worms in which the gonads had been excised by a laser beam lived the longest of all — a tactic, alas, with limited commercial potential.

In the early 1990s, shortly after the first papers came out, Kenyon received the obligatory call from Michael West. He had followed her work, of course, and invited her down to Geron to give a talk on life-extension genes. It was a bit of a reach for Geron to get involved in the genetics of aging at that point, but West's interest merely ratified the notion that Kenyon's research fell within the same larger orbit.

༜

Other people were gravitating to the study of aging. One of Kenyon's old buddies from the Ptashne lab at Harvard was a quiet, meticulous, and somewhat reticent molecular biologist named Leonard Guarente. A researcher at the Massachusetts Institute of Technology, Guarente has studied important, but rather esoteric, aspects of gene regulation for twenty years. He had attended MIT as an undergraduate, obtained his Ph.D. at Harvard in the laboratory of Jonathan Beckwith, and then stayed on at Harvard to do a postdoctoral fellowship with Ptashne before moving to MIT to start his own lab in 1981. Like Kenyon, Guarente had large ambitions. Despite his informal attire (extremely worn blue jeans, peach-colored sports shirt, and old tennis shoes on the day I visited his lab) and his seeming discomfort at being interviewed (he spent nearly an entire hour gripping his elbows and dispensing short, clipped answers), he thought big and had been feeling a need to work on a larger scientific canvas. "The primary thing was not an interest in aging per se," he told me, "but an interest in something else, something different from what we had been doing, which might be viewed as high-risk, high-reward. I spent a number of years mulling over possibilities, and we settled on aging. And I say 'we' because it was really two incoming graduate students and I, as a joint decision, because if they weren't going to do it, it wasn't going to happen." Around 1991, "with some trepidation," the three researchers — Guarente, Brian Kennedy, and Nick Austriaco — began to tackle the biology of aging.

Guarente's research in a sense recapitulated the boom-and-bust cycle of the aging field for the past half-century. It had been known since the 1950s that yeast cells — specifically, the matriarchs of the strains, known as mother cells — eventually grew lumpy and showed their age; in short, they senesced. Guarente, like Kenyon, had known about Leonard Hayflick's work, and its limitations, early in his training. "I was aware of the Hayflick limit from college on," he recalled. "And the issue always has been, What does the Hayflick limit mean? And now that we understand that the Hayflick limit is due to short telomeres, the question is, What do short telomeres mean in real aging? That's still to be resolved. But at least in that system, the Hayflick system, it did suggest that you could bring the tools of molecular biology to study a problem in the realm of cell senescence."

Guarente had made his reputation studying gene regulation in yeast, and it was on this single-celled creature that his group initially focused. The first breakthrough arose through serendipity. The lab had created "a bunch of different strains" of yeast, Guarente recalled, and the two postdocs plucked out two strains with significantly different life spans. They mated, or crossed, them — a classic strategy to tease out genes of interest. Some of the progeny were long-lived; some were short-lived. Then that invaluable but rarely acknowledged laboratory technique known as negligence provided the next crucial clue. As Guarente explained, "What happened was, these strains from the cross were sitting on a petri plate in the fridge. *For a while*, okay? A few months. And my student just went to grow them all up one day. And when he inoculated them from this old plate, some grew and some didn't. It seemed peculiar. He followed that up and noticed that the ones that grew were the ones that happened to have the longest life span. And he was able to show that the reason they grew is that they were generally just more stress resistant, and survived longer in the fridge. So these two qualities, stress resistance and longevity, seemed to be related, at least in this particular set of strains."

That accidental insight gave them a way to screen many mutants quickly. They just dumped all manner of stress on their yeast cells and then looked for strains that had the right stuff. To make a long story short, they managed to identify one long-lived, stress-resistant strain of yeast that also happened to be sterile. The genes that cause sterility in yeast are very well known and have been catalogued; moreover, there was a history linking sterility to longevity. Guarente knew, for example, that calorie-restricted animals tend to be both long-lived and relatively infertile. The same little nexus of biological activity that Kenyon had seen — metabolism, stress, reproduction, and longevity — floated up out of the data. Quickly, Guarente's lab identified and cloned a gene that seemed to confer longevity, stress resistance, and sterility. It turned out to be a gene known as *sir-4*, one of a class of so-called silencing genes — a group that encodes proteins that in turn throw a blanket over long stretches of DNA, smothering and therefore silencing large arrays of genes. The MIT group published this result in *Cell* in 1995.

There is an instructive moral attached to this "fountain of youth gene," especially because of the dead end to which it led. When Guarente and his colleagues tried to figure out what the mutant gene actually did, they stumbled upon the fact that it might have something to do with telomeres. Telomeres? The bells began to ring with that observation. "Yeah," Guarente admitted, "we thought this must mean telomeres are important. That was our initial thought. Because if we mutate this complex so that it's no longer in telomeres, the cells live longer." This was not merely an attractive working hypothesis in general; following the publication of the *Cell* paper, Guarente had been invited to join the scientific advisory board of Geron, where the telomere hypothesis was gospel. All that remained to merge telomere research with longevity genes was a few straightforward confirming experiments. Exactly the kind that tell you what you least expect. "So what we did was a simple control experiment," Guarente continued. "The experiment did not extend the life span, but instead told us that telomeres were not what was important." It turned out to be something else entirely.

Experiments often capsize in the minimal swells of these little details, but it's worth lingering on this point. The telomere hypothesis, and its link to the Hayflick limit and also to aging, had been the fuel that lit Mike West's meteoric rise in the world of biotech venture capital. It had become the

Holy Grail of researchers during the early days at Geron. In the public imagination, the equation linking telomere length to longevity might as well have been carved in Carrara marble, for all the reverence and awe it inspired. But outside the self-serving precincts of private enterprise, there remained considerable skepticism about the role of telomere shortening in the aging process — a skepticism all the more sobering when you hear it from a onetime member of Geron's own SAB. "There was no reason to think that telomere shortening was playing any role here," Guarente told me. "Now of course, you should know that the only system where telomere shortening really seems to happen is in humans. The data are a little bit less compelling in vivo, but there's a general trend toward telomere shortening with age in people. But, for example, not in mice. So, if one thinks dispassionately about it, there's no real reason to think that telomeres are doing anything [in aging], and our experiments said that they weren't." That last remark is inferentially damning, suggesting as it does that some scientists in fact have not been able to think dispassionately about the telomere hypothesis. It's hard to imagine someone as mild-mannered as Guarente deliberately insulting anyone, but it's equally hard to imagine a more knowledgeable, albeit indirect, indictment of the telomere fanatics.

Once they'd overcome their disappointment that telomeres weren't involved in the life span of yeast, Guarente's group started to take a closer look at the entire family of *sir* genes. It was a long grind, marked by one spectacular wild-goose chase in which they kept searching for microscopic circles in cells as if they were the signature of UFOs. But by focusing on the silencing genes, the group discovered that another member of the family, a gene called *sir-2*, strongly influenced the life span of yeast. When the MIT biologists genetically engineered yeast to have an extra copy of the *sir-2* gene, the yeast cells lived longer. "That pushed us to do the biochemistry of the gene," said Guarente, "which really gave us what I think is the most important result of all, and that is its exact biochemical activity, what it does." The actual purpose of the gene — making an enzyme that removes an acetyl group from histone, an important component of DNA in the nucleus — is not nearly so important to lay readers as what the gene needs to do this. It needs a chemical called nicotinamide adenine dinucleotide, or NAD for short. And the most interesting aspect of this discovery is that cells, and indeed organisms, produce an abundance of NAD only at certain times — during reduced caloric intake and reduced metabolism. The bells went off again, and this time it wasn't a false alarm. The *sir-2* gene and its biochemical pathway linked the high-tech molecular manipulations of life extension in yeast to that low-tech life-extension strategy that had been known for the better part of the twentieth century in rats: a starvation diet. "We were," Guarente said, "acutely aware of that link."

To understand Guarente's excitement, you have to go back to the 1930s, to a bunch of scrawny rodents that have danced like sugarplums through the dreams of countless researchers who have studied aging. In a classic set of experiments conducted by Clive McCay at Cornell University in 1935, rats were fed about 30 to 40 percent less than their normal allotment of daily calories. These underfed animals then displayed a remarkable quality. On average, they lived about 20 to 40 percent longer than rats raised on a normal amount of food; some lived 75 percent longer. More than half a century later, this remains the only proven strategy of life-span extension in mammals. And ever since, the intersection of diet and longevity has intrigued researchers — and life-extension fanatics. A former UCLA scientist, Roy Walford, has undertaken a much-publicized experiment on himself — he is eating a reduced diet in hopes of extending his own life. And according to a recent *Wall Street Journal* article, hundreds of Americans are restricting themselves to a diet as low as 1,500 calories a day in hopes of living longer. Guarente was the first biologist to trace that intersection to a humble molecular cog in a universal metabolic wheel that can be found in any higher organism, from yeast to humans.

A subsequent, tantalizing convergence of yeast and worm longevity occurred only recently. Guarente's group showed that by inserting a single extra copy of the *sir-2* gene into *C. elegans*, the worms live longer, too. Moreover, the researchers discovered that the mechanism that accounted for life extension in nematodes used the same metabolic pathway as the dauer formation in nematodes, one that involves insulin signaling. That made sense to Guarente, because dauer is a kind of dormant survival strategy adopted by larval worms that are stressed by a scarcity of food. Again, evolution seemed to be telling them that once it found a useful and efficient way of protecting survival, it preserved it in other organisms. That evolutionary view has gained even greater currency recently. Linda Partridge and her group at University College London have shown that similar hormonal genes control the life span of fruit flies. Toward the end of 2002, a group in France had extended the findings to mice.

The appearance of hard-core molecular biologists like Guarente and Kenyon in the aging field about a decade ago signaled an important transition, even though neither knew of the other's interest at the time. "There are a lot of very smart people in the aging field," Kenyon told me. "But it's interesting in biology to look at the cultures of different fields. My field — molecular biology and molecular genetics and gene expression and developmental biology and stuff like that — the people in those fields did not have such a high regard, I think, for the majority of [people in] the gerontology field." If you recall Leonard Hayflick's comments from the prologue, you'd have to conclude that the feeling on the part of gerontologists was mutual.

When I first spoke to Guarente, in the fall of 2000, he mentioned that his group's experiments showing the intersection of the *sir-2* gene and dauer formation had not yet been published. I wondered aloud if he had told Cynthia Kenyon about these results, since they so obviously complemented her research, and he quickly set me straight.

"Yes," he replied, a small smile spreading on his face. "She and I are starting a company together. We talk all the time."

༄

Around 1996, Cynthia Kenyon began showing her "movie" to venture capitalists. The film clip shows the difference between geezer nematodes and worms that have been genetically rejuvenated; it lasts less than five minutes but communicates more of the promise of this line of research than fifteen peer-reviewed papers in *Nature*, especially to nonscientists. "I show them the movie, of worms that are normal that are about to die after two weeks, and the altered worms that are still moving," Kenyon said. "They *see* it with their eyes, you know? And I can tell them how many things [genes] are conserved between worms and other animals, and that it seems to be happening in flies as well, and yeast. And that's really all you have to do. It speaks for itself." While it would be fair to say that the venture capital crowd did not rain cash on Kenyon as they did on Mike West in 1991, in part because biotech investment entered a very rough patch in the late 1990s, the interest was there, and the money eventually followed.

Kenyon had been approached for years by venture capitalists, in part because of the continuing press reports, and by December 2000, she and Guarente had incorporated, prepared a term sheet "for our investor friends," as Guarente put it, and received a first round of financing of $8.5 million from several venture capital outfits, including ARCH Venture Partners, Oxford Bioscience Partners, and Tredegar Investments. The company even had a name, although Guarente sounded unsure how long it would remain the name. It was Elixir Pharmaceuticals. I took it as a sign of Guarente's ambivalence that he even asked me what I thought of it. When I replied that it had "a certain connotation," he laughed and added, "Good and bad, right? I would really like a name that would connote youth, so I think Elixir does that. But I also think the word *elixir* in the past has had a connotation of, you know, a traveling salesman and a tonic. So I don't know." Kenyon seemed even less enthusiastic when I spoke with her. "I don't think it should be Elixir," she said flatly, "for several reasons. Apparently there was a drug that was developed in the 1930s that killed a lot of children that was called Elixir . . ." The full name was actually Elixir Sulfanilamide, and more than one hundred children indeed died horribly protracted and painful deaths after taking a liquid form of the antibacterial drug specifically formulated for children. The discovery that the drug's manufacturer, S. E. Massengill of Bristol, Tennessee, had used a solvent similar to antifreeze to create the product sparked a national scandal that led to dramatic changes in food and drug law, including the requirement that companies must prove their products are safe before obtaining FDA approval. "And I think," Kenyon continued, "it can have slightly . . . I don't know, in some ways it's good, but it can sound like pseudoscience also, you know? So I'm hoping that we change it."

In any event, Kenyon and Guarente joined forces with Cindy Bayley, formerly a partner at ARCH, a Chicago firm, to create a company with the long-term but ultimate hope of selling a product — ideally, a small molecule that could be packaged as a pill — that will extend the human life span by tinkering with the *sir-2* pathway. By the end of 2001, Elixir — they stuck with that name — was up and running in temporary quarters in Cambridge, Massachusetts, with about forty employees, the requisite scientific advisory board, and funding from, among other places, Oxford Bioscience. Alan Walton, Mike West's original angel, was once again backing a company that hoped to treat aging. Elixir quickly obtained exclusive licenses from MIT (to work with Guarente's now-patented *sir-2* genes) and from the University of Connecticut (to work with a gene, related to longevity in fruit flies, with the fanciful name of INDY, an acronym for I'm Not Dead Yet).

There is good scientific reason to believe that genes may indeed play a role in establishing life span. More than fifty mutations affecting life span have been identified in animals. A number of scientists in recent years have studied identical twins who have lived to one hundred and beyond, in the belief that they provide a window onto the genetics of human life span. A group headed by Thomas Perls at Harvard Medical School, for example, has investigated a large cohort of centenarians in New England, and in the summer of 2001 his team identified a large chunk of DNA on chromosome 4 as being a likely spot for genes related to life span (this report got considerable media attention, including a front-page story in the *Wall Street Journal*, but it is always worth pointing out that well over a decade elapsed between the time a tentative neighborhood for the Huntington's disease gene was identified and the time the gene was actually found; in the search for genes, as in urban warfare, the endgame almost always involves incredibly time-consuming, house-to-house combat along the chromosome). Similarly, a group in Italy that has followed some Sardinian centenarians believes that genes related to either stress or lipoprotein metabolism may influence life span in its cohort. "When single genes are changed, animals that should be old stay young," Guarente and Kenyon wrote in a *Nature* article. "In humans, these mutants would be analogous to a ninety-year-old who looks and feels forty-five. On this basis we begin to think of aging as a disease that can be cured, or at least postponed." In reviewing the genetic extension of life span in animals, an early Elixir press release noted that such a doubling would be "equivalent to extending the human life span to approximately 170 years."

Nonetheless, the biological likelihood of extending life span through genetic manipulation is by no means either inevitable or even possible. Leonard Hayflick has long argued that these experiments conflate life-span extension with developmental biology, in the sense that the genetic manipulations necessary to extend life must be present in the developing embryonic stages of the organism in order to work; he doubts that dramatic increases in life span will occur any time soon, if at all. S. Jay Olshanky told a meeting of actuaries in 2002, "There are no death or aging genes — period." Judith Campisi, a researcher on aging at Lawrence Livermore laboratory in California, expresses similar doubts. She has argued that senescence evolved to protect cells from the type of runaway growth typical of cancer, a mechanism that protects a young organism but may sabotage an older one, regardless of other genetic manipulations. "God gave us this wonderful response against potentially oncogenic influences," she told a recent meeting, "and it works pretty well for about half of our lifetime. But senescent cells may *promote* tumorogenesis, and this is going to be a hard wall in terms of extending the life span of multicellular organisms. I don't think we're going to be able to change it by going into the genome and starting to muck around."

But life-span extension has now become a respectable (if modest) goal of the drug discovery process, and the search is not limited to longevity genes. William Andrews, who left Geron when the company shifted its focus from aging to cancer research, started a new company called Sierra Sciences in Reno, Nevada, and has been trying to harness the power of telomerase to rejuvenate cells, and bodies. The company claims to have discovered molecules that might transiently allow the telomerase gene to be turned on, so that people could periodically take a pill that would lengthen their telomeres, theoretically resetting the clock of cellular aging. While highly speculative and preliminary, this suggests that there may be a number of paths to the molecular fountain of youth.

With the mere hint of such possibilities, molecular biologists are now squeezing themselves ever more snugly between a rock of their own scientific confidence and the hard place of public expectation. When you ask about the creeping use of the term "practical immortality" and the I word in scientific conversation, Kenyon properly recoils from the notion, but does not entirely retreat. "We don't have anything!" she insisted. "These worms aren't immortal. They live twice as long. Or they live four times as long. In fact, now we can get 'em to live six times as long. So we found that the germ line controls aging. And if we take *daf-2* mutants and kill their germ cells, or their whole gonad, now they'll live four times as long as normal. But that's not immortal." Pause. "That's not to say that you *couldn't*. But we haven't." And therein may lie the difference between molecular biologists and philosophers: the latter often inflect their pronouncements with a palpable, if unspoken, *but*. The unspoken word in the biologist's rhetoric is always *yet*.

There is one final — and by now perhaps predictable — episode to relate in the *sir-2* story, and it of course involves a visit from Mike West. The *sir-2* gene creates a protein that interacts with what's known as the chromatin structure of a cell nucleus; this basically refers to the way DNA (and genes) get bound up and wrapped and organized. It is precisely that organization, that virgin state of DNA, that has become the Holy Grail of cloners, because it could help explain how an egg cell "reprograms" an adult cell back to the very first moments of developmental time. This may just be another way of saying that all roads in biology lead back to the egg, but it is yet another opportunity to point out that Michael West always seems to be scurrying along those highways. In early 2000, West and his colleagues at ACT drove from Worcester to MIT and made a presentation to Guarente's lab group about their cloning work. In fact, they came with hopes of striking up a collaboration.

"You know, the question is, How is the oocyte reprogramming the nucleus?" Guarente said. "And very likely, it's by reestablishing the proper chromatin structure and boundaries. So to that extent, we are interested in looking at the same question. But it wasn't quite yet to the point where we could see specific experiments that we would do, that we should do." So the collaboration never materialized. But West and his fellow scientists were very much in the thick of things — the thick of chromatin, the thick of cloning, and the thick of a social debate that seemed to grow, in decibels and derogation, with each passing month, in part because of developments at a small company in Worcester.

12

Unk!

Almost every morning during the fall of 1999 and the spring of 2000, usually before 10:30 A.M., an overnight delivery truck would roar up Research Avenue on the outskirts of Worcester, Massachusetts, hang a right on Innovation Drive, and come to a halt in front of a nondescript brick building called Biotech Three. It was one of those ubiquitous pods of architecture housing today's version of entrepreneurial biology, a low brick-and-tinted-glass affair that had taken root, like a windblown seed, across the road from the University of Massachusetts Medical Center. The courier would disappear into the building with one or two large gray containers that looked like toolboxes.

When I first saw those gray boxes, sitting on the floor in the reception area of Advanced Cell Technology during a visit in December 1999, I thought that a maintenance worker had left his tools behind. It turns out that I was right, but not in the way I had imagined. Those boxes were used to convey the starting material for an audacious series of biological experiments almost guaranteed to offend anyone who heard about them. The gray cases contained egg cells. Cow eggs, actually, technically known as oocytes. Hundreds of them.

They had been harvested the day before in Iowa from the ovaries of slaughtered cattle. The eggs had then been placed in small plastic tubes, roughly the size of the cap of a pen, and dosed with a marinade of enzymes that primed them for fertilization; the tubes were then plunged into a dense mass of small steel beads that filled a thermoslike container embedded in the insulated body of the gray box. A large battery pack warmed the steel beads to a steady temperature of 38.5 degrees centigrade (roughly our body temperature) so that, biologically speaking, the eggs were coddled in the warm, inciting liquors of imminent fertilization, even as they made the overnight, express-mail journey from Iowa to Massachusetts. By the time they arrived in Worcester, they were primed for the experiment I had come to see. These cow eggs would be deliberately fused with adult human cells, in the hope that they might form an embryo. A short-lived embryo, that is, from which human embryonic stem cells might be plucked.

On the dreary, drizzling mid-December day in 1999 when I visited, Jose Cibelli, the lead scientist at ACT, had forgotten about the appointment and was still at home, working on a scientific paper; the oversight reinforced the kind of overworked, understaffed, hurried, and ad hoc air that surrounds the company, a tiny enterprise that at the time didn't occupy much more square footage than a ranch house. While I waited in a conference room for Cibelli to arrive, Mike West popped his head in the door. It seemed like a completely spontaneous encounter, and it helps explain why so many scientists feel galvanized (not to say enabled) by West's enthusiasm and imagination. The man just loves to talk, and the conversation roams all over the place. He plopped down in a chair and within ten or fifteen minutes had spoken about ACT's interest in using cloning (or, to use the term he preferred, "somatic cell nuclear transfer") to create offspring of endangered species, dashed off to photocopy a wire service article about the cloning of deceased pets, and recounted the times when, while working late at the office, he found himself taking calls from desperately ill people who wanted to know when cures from stem cells might be available for their particular ailment. "We just don't have the people power to do all these experiments," he lamented. It occurred to me that one of the reasons West has so successfully courted the press is that he always makes time for reporters, and always seems to hold the door open to them (you just needed to be aware that other doors, perhaps less obviously apparent, were kept firmly shut). West knew that the cow-human experiments were controversial and perhaps even biologically doomed to fail, but in his headlong interest to rewrite the history of disease treatment and human life span, he said, "We're pushing it as aggressively as we can, as a small company."

ॐ

Nothing undermines the requisite awe — or, depending on your point of view, amplifies the inherent revulsion — provoked by a cloning experiment involving human DNA than to see its repetitive, almost mindless piecework procedures firsthand. Experimentalism in general often resembles assembly line work, and that is especially true of cloning, but not because it's already viewed as a routine, industrial procedure; to the contrary, the failure rate is so extraordinarily high that lots of starting material has to be processed for a chance at even one success. These experiments at ACT took place in a long, dark, narrow room in the middle of the lab area, where the lights are kept low to allow technicians to view the cells through microscopes while poking, turning, and prodding them with micromanipulation devices. A young woman named K. C. Cunniff, blond hair spilling out over a white lab coat, began by vacuuming all the bovine DNA, all of the cow's genetic material, from the egg cells, one after another, for more than an hour. "The oocytes are horrible today," she remarked to a colleague, with vivid — if nonscientific — disdain. "Really immature. Really hairy."

As I watched a magnified view of the proceedings on a television monitor, it was hard not to be impressed by the nimble micromanipulations of the technicians, and by the silence. The techs refrained from the usual "slam your head against the wall" rock music from the boom box near the door, K. C. told me, in deference to the visitor. Nor did I hear the massive organ chord you keep thinking you should manufacture inside your head while witnessing the genetic trespass of one species into another. K. C. deftly maneuvered a pipette to hold a round, plump egg cell, which had previously been stained with a fluorescing dye that lights up DNA, and then moved it around until she found the telltale blue-green glow of the nucleus. Using her other hand, she nudged a sharp, hollow needle up against the surface of the egg and, with a swift precise thrust, entered the egg, sucked out the DNA, and withdrew the needle in a matter of seconds. "Very quick," she said. In, suction, out. In, suction, out. "E-nukes," they called the eggs, because they were removing the nucleus from, or "enucleating," them.

"It's like playing a video game," said Nancy Sawyer, an ACT technician who was watching along with me.

"The same video game over and over and over again," Cunniff added.

As Cunniff moved another cow egg into position, Sawyer gazed at the TV monitor and made a sound like "Unk!" as the needle punctured the egg. "That's so cool," she said of the image. "So pretty!" On the monitor, the image indeed possessed a strange, abstract beauty, like a grainy and somewhat murky photograph of a passing asteroid, or the kind of dark and mysterious landscape Arthur Dove might have painted if he did scientific illustrations, with one isolated lunar mare on the large, moonlike oocyte glowing an eerie greenish blue, betraying the location of the DNA that would soon be suctioned away.

At a nearby microscope, Sawyer, dressed more for a woodland hike than for a cloning experiment, in jeans and a green flannel shirt, performed the next step. Human skin cells, known as fibroblasts, had been provided by an anonymous donor; they dotted the screen like lumpy, transparent potatoes, much smaller than the egg cells. Sawyer loaded up a hollow needle with the human cells ("like BBs," she said), held each cow egg immobile with suction, and, applying gentle pressure, inserted the needle tip between the rind of the egg, known as the zona pellucida, and the cytoplasm. She likened the task to "putting a golf ball on a beach ball." With a barely perceptible squeeze of one hand, she deposited a single adult skin cell, with its unique payload of human DNA, into the egg of a creature that moos. To my surprise, Sawyer even managed to maneuver her needle tip to literally tuck the hole closed. Tuck, tuck. "Yeah," she said, acknowledging the gesture, "it's important that they have a nice home." A tiny figurine of Winnie-the-Pooh, clutching a blue umbrella, hung suspended from the top of her microscope.

As I watched on the monitor, I confess to experiencing a visceral reaction to the microscopic violence of the procedure; it is hard to watch the needle tip press in and then finally puncture (unk!) the outer membrane of the egg, for it has all the dynamics — force, resistance, breakthrough — of an unnatural violation, an uninvited intrusion. It might have been an example of what Arthur Caplan, the Penn bioethicist, has termed the "yuck factor," and what Leon Kass says leads to a "wisdom of repugnance" — that reflexive, emotional recoil from something that appears unnatural and inspires our revulsion. But as I've thought about it, and I've thought about it a lot since that day, I've come to the conclusion that the "yuck factor" should only mark the beginning of our reaction to procedures and technologies that give pause, not the sum total of our thinking. It is also hard to watch, as I have, an incision being made in human flesh at the start of open-heart surgery, a similarly unnatural, even barbaric violation of flesh and norms that just happens to restore health and prolong lives. At both the real-life and microscopic levels, much of the controversy about research cloning — indeed, its very future — may depend on how that visceral reaction gets processed, and whether we stop at the egg or look beyond it to the posthumous gifts it might confer.

After stuffing every cow egg with its little spud of human DNA, Nancy Sawyer prepared to give them a zap of electricity. She lined the cells up in an electrical field and pushed a button on an Electro Cell Manipulator, giving them a quick, 15-microsecond jolt of 120 volts. The jolt effectively fused these two odd biological bedfellows, part human and part cow, into one cell, which, after a final step, believed it had been fertilized. "So that's about it," Sawyer said.

To Jose Cibelli, who had been conducting similar nuclear transfer experiments for almost a year at the time of my visit, these fused cells were nothing more than "pre-embryos." Indeed, the odds were extremely long that any of the 130 or so biological mergers performed that day would yield a blastocyst, that cozy little bubble in which the sought-after stem cells form. "If I'm lucky," Cibelli told me later, "I'll get one blastocyst. That's how low the efficiency is. But I'm trying a new technique."

The purpose of the research was to use the cow egg as an incubator of sorts, one that could take the human DNA so precisely injected into it and, as Cibelli put it, "rejuvenate" it — reset the clock of the human adult cell back to zero and give it a second chance to develop from scratch into an embryo. If any of the fused cells reached the blastocyst stage, Cibelli might be able to harvest embryonic stem cells, which was the real object of the exercise. And the reason for the cow eggs is something any businessman can appreciate: they cost only about $1 apiece, whereas human egg donations could easily run about $2,000 to $4,000, to say nothing of the ethical issues *that* would raise. So at ACT, West and company made it sound like it was purely a business decision. And that, buried amid all the minutiae, was the larger point: the first tentative steps toward the creation of cloned human embryos and the production of stem cells had become, precisely because of right-to-life politics, the exclusive province of business. Or, to put a slightly finer point on it, the exclusive province of whatever Michael West wanted to do.

In West's grand vision, these cow-human fusions were a necessary, if controversial, first step in obtaining embryonic stem cells, which he believed would revolutionize medicine. Anticipating ethical objections, West and several colleagues had written in an issue of the journal *Nature Medicine,* "Does a blastocyst warrant the same rights and reverence as that accorded a living soul — a parent, a child, or a partner — who might die because we failed to move the moral line?" And as West said to me, "I think a lot of the problem we have in trying to develop these new technologies for medicine is people's knee-jerk reaction to words like *fetal* and *embryo*. You know, they use the word *fetal*, and people just go completely irrational. I feel like we're in the Dark Ages sometimes."

There were a number of experts who found this line of research not only willfully controversial, but biologically nonsensical. John Gearhart fairly snorted in derision when I asked him about West's *Nature Medicine* commentary. "Ridiculous," he said dismissively. James Thomson, the stem cell biologist at the University of Wisconsin, said, "Even if the cow nuclear transfer physically can't work because of biology, it'll create a good debate." Was he being facetious? "No," he said. "I mean, nobody knows if that will work. There are biological reasons to think it may not work, but until somebody actually did it, you wouldn't know . . . But doing it straight with human stuff in the absence of a debate is maybe inappropriate." I asked Jose Cibelli when he would know if any blastocysts had formed from the cells I'd observed being fused. He said he wouldn't expect to see anything before nine or ten days. I realized I'd have to get in touch either on Christmas Eve or Christmas Day.

A truly surprising development did come out of those experiments, however. By chance, I happened to meet with Leonard Hayflick shortly before witnessing the cow-human cloning experiments. He had come to New York to attend a meeting of the American Foundation for Aging Research, on whose board he served. I asked his opinion about the experiments West was conducting up in Massachusetts.

"Did Mike talk to you about the nuclear transfer work?" Hayflick said with a conspiratorial little smile. I couldn't help noticing that he started rolling up one leg of his pants. "The human cells he's using for the cow work came from here," he said, pointing to a patch of skin just above his right knee, where the site of a skin biopsy was still slightly detectable. "Unfortunately, my scar is six months old, and it's not very visible, but that's the remains of it. To the best of my knowledge, the human material they are working with is my cells."

Here was a worthy sequel to the Immortal Dr. Hayflick. The man had agreed to have himself cloned! (Practically, though, the chances were negligible that a viable embryo would result from the use of his cells, and reproduction wasn't the aim.) To my knowledge, neither Hayflick nor West ever discussed or confirmed this in public (although in conversation with me West confirmed Hayflick's assertion), but the story makes clear that while society was just beginning to wrestle with the implications of human cloning, West and Hayflick were once again two cowboys back in the saddle together, rushing ahead to stake out the territory and terrifying the natives as they sped by. Miller Quarles, one of West's earliest investors, later told me he, too, had donated cells for ACT's cloning experiments. It was beginning to sound like the DNA version of the Price Club.

༄

Public fascination with — not to say dread of — the technology of cloning erupted with full fury in February 1997, when a team of Scottish scientists led by Ian Wilmut published a paper in *Nature* describing the creation of the cloned sheep Dolly. The reaction to this news drew as sharp a line between what C. P. Snow called "the two cultures" as the San Andreas fault. Nonscientists reacted with horror, immediately fearing the creation of cloned human beings — a fear actively promoted by renegade fertility experts who vowed to do the very experiments that elicited overflow amounts of the "yuck factor." On March 4, 1997, reflecting this unease, Bill Clinton urgently requested "a moratorium on the cloning of human beings until our Bioethics Advisory Commission and our entire nation have had a real chance to understand and debate the profound ethical implications of the latest advances." Most scientists, however, were stunned by the development for an entirely different reason: reprogramming. The successful cloning of Dolly, despite the technique's difficulty and low rate of success, meant that Wilmut and his colleagues had managed a profoundly powerful trick. They had taken an adult sheep cell and restored its DNA to the kind of virginal, pristine condition it possessed at the moment of fertilization.

To explain why this so stunned scientists, consider this analogy: as a single fertilized egg develops, its DNA is constantly being modified and sculpted, by both design and use. Some genes turn on very early in development, for example; once they've done their job, they become permanently muffled, as if duct tape had been wrapped around the gene. As the embryo, then fetus, then neonate continues to develop, its DNA continues to accumulate these silencing overlays, until, in an adult cell, the DNA may look like a raggedy pipe covered with layer upon layer of duct tape at various intervals. The dogma of biology had been that once taped over, embryonic and fetal genes would never express themselves again during the life of the organism. But cloning took adult DNA and stripped away its genetic wrap and developmental duct tape — put the DNA, as Mike West liked to say, in a little time machine that carried it back to the Edenic genetic state of fertilization. Now that DNA, in the right environment, could make a blastocyst, stem cells, possibly an *entire* organism all over again. What's more, the power to "reprogram" that adult cell clearly resided in the egg. The egg cell possessed factors, chemicals, an embryonic magic that could restore DNA to its virginal, prezygotic state. As Günter Blobel once said of cloning, the oocyte "overwhelmed, overpowered" the somatic cell. "It all comes back to the egg," he said.

And the power wasn't limited to a *human* egg. This became clear in a little-known and never-reported experiment carried out around 1990. Although the conventional history of embryonic stem cell research typically traces its origins to basic science in the academic communities of England and the United States, a parallel history unfolded in a more pragmatic discipline that typically receives much less attention (and, frankly, respect): animal science. This oversight in part accounts for the social shock that Dolly caused. Veterinary scientists had for years been fiddling with embryonic stem cells, animal reproduction, and selective breeding, for obvious commercial agricultural reasons. One laboratory involved in such research, at the University of Massachusetts, was the incubator of what would eventually become Advanced Cell Technology, and that research took a decidedly bold and controversial turn in 1990. Unbeknownst to almost everyone, a semihuman embryo was created, with rabbit eggs.

The laboratory was run by James Robl, a genial and well-respected animal science researcher then on the faculty of the university. In 1990 a graduate student in his lab named Philippe Collas (now at the University of Oslo) embarked on a series of experiments attempting to improve basic techniques of cloning, using the rabbit as an experimental species. At some point, as something of a lark, Collas decided to remove some cells from his own cheek and fuse them with the egg cells of rabbits. "We were thinking pretty far out in those days," Collas said. "I did some pretty wild stuff, including cloning myself and also Jim's technician. I was making some beautiful embryos, and they all went to the 32- or 64-cell stage." "They didn't tell me until after they'd done it," Robl recalled. "They were laughing about it, not taking it seriously. So we started watching them [the eggs], and they started cleaving, or dividing. And the more they started cleaving, the more nervous I got." To the best of Robl's memory, the "accidental" rabbit-human embryos went through roughly five cycles of division, over the course of about four or five days. When it became clear that the embryos might continue to develop, Robl had second thoughts. "They were doing it just for the heck of it, but it was touchy, so I just said, 'Stop.'" Robl's group never even tried to publish the results, and given the casual origins of the work, it would not be surprising if similar interspecies experiments were conducted in other labs at the time. Robl's group went on to other things, but the message of Collas's ad hoc experiment made a lasting impression on anyone who saw or heard about that little ball of cells. (This wasn't a fluke: when I interviewed Cibelli in 1999, a paper sat on his desk written by Chinese scientists reporting on efforts to clone giant panda cells in rabbit oocytes, and the *Wall Street Journal* carried a front-page story in March 2002 reporting that Chinese scientists claimed to have inserted an adult human cell into a rabbit egg, coaxed it to reach blastocyst stage, and then harvested human embryonic stem cells from it.)

Over the next few years, Robl's lab group at U Mass focused on trying to clone several barnyard animals — cows, pigs, and chickens — for agricultural purposes. "We felt that it was time for us to move to the cow, because the cow would be something of commercial interest," Robl said. But cows were expensive animals to work with on federal research grants, so Robl and one of his students, Steven Stice, started to think about forming a private company and raising money to subsidize the research. Although Robl had no declared interest in aging research, he ended up founding a company whose corporate mission has since become inextricably associated with both human cloning and age-related therapies — and a company that had a preposterously auspicious godfather attending its birth.

"The idea kind of floated around a little bit," as Robl tells it, "and didn't go anywhere until I had lunch with another colleague of mine in the Department of Veterinary Science." The colleague, a Peruvian by origin, was a molecular veterinarian named F. Abel Ponce de León; he was in fact a distant relative of the sixteenth-century explorer. This latter-day Ponce de León was exploring a typical twentieth-century financial landscape: he consulted on the side for an agricultural company based in Maine called Avian Farms, which was constantly on the lookout for new applications of biotechnology in agriculture. Robl described his interest in cow cloning, and Ponce de León responded enthusiastically because he was, at the same time, interested in cloning chickens. "We're not completely ignorant fellows," Robl said, "and we put the two ideas together, and he said he would contact Avian Farms about this idea and see what we could do." It didn't hurt that both Robl and Ponce de León had been on the dissertation committee of Milton Boyle, the vice president of research and development at Avian Farms. In 1994, Avian Farms agreed to fund the new company. Called Advanced Cell Technology, it was basically run out of Robl's lab at U Mass. Ponce de León viewed the enterprise with a unique sense of history — "I have not changed the family's line of work," he said with a laugh.

Advanced Cell never set out, of course, to create cow-human embryos. Indeed, a series of failures and setbacks in the company's research perfectly illustrates the zigzag trajectory by which cutting-edge research unfolds. The company first set out to clone barnyard animals, especially cows. Robl wanted to obtain bovine embryonic stem cells for a simple biological reason: theoretically, you could pluck these cells out of an embryo, genetically modify them (insert, for example, the gene for a pharmacologically desirable human protein like albumin), and reinsert them into another cow embryo, to create (if you were lucky) a pharmaceutical factory on hoofs. "We got off, unfortunately, to a *very* slow start," Robl said, "because everything we tried to do didn't work." While struggling to improve the technology, however, they performed a crucial experiment in which they compared the success rate of cloning using stem cells with that of a control group using adult cells known as fibroblasts from a mature cow. These mature body, or somatic, cells had been included in the experiment with the expectation that they would yield poorer results; instead, they "worked a whole lot better" than embryonic stem cells. All of a sudden, Robl's group had stumbled upon the very technique that the 1994 NIH embryo panel, a year earlier, had mentioned as a theoretical route to the creation of human embryos: "somatic cell nuclear transfer," or cloning.

At this point, ACT stopped being an agricultural company and suddenly found itself pursuing human medicine. Here was a way, possibly, to get fetal animal cells for clinical use; the initial interest was the creation and harvest of dopaminergic neurons, the kind of brain cells that steadily disappear in Parkinson's disease. It was Robl's belief — a naive one, he now admits — that genetically engineering animal cells through cloning would be far less controversial than trying to obtain neural cells from human fetuses. So a major effort in the lab was directed at creating cow embryos using somatic cells in order to harvest fetal cells for human therapy. "We thought, Gee, if we could get fetal cells, that would be a great application for this," Robl said. "And again, there was never a thought," he added with a chuckle, "that you could get offspring from these things. So the next step was just [cow] fetuses."

☙

Around that time, a razor-thin, dark-haired veterinarian from Argentina with bright eyes and a lively smile arrived in Robl's lab to pursue a master's degree. Jose Cibelli had received his veterinary degree at the University of La Plata and had worked several years in the Argentine cattle breeding industry before coming to Amherst in January 1994. He almost didn't last more than a month. Dismayed by the rigors of a New England winter, and by his wife's unhappiness with the weather, he was ready to return home to Argentina until he was talked out of it by a fellow South American on the U Mass faculty — namely, Abel Ponce de León. Robl quickly put Cibelli to work trying to create cow embryos through cloning. It is a testament to his patience and determination (and also to his ambition) that, by his own estimation, Cibelli attempted no fewer than ten thousand microinjections of adult cow cells into cow eggs in his effort to make the technology work.

Eventually, it did. Robl's lab sent the cloned embryos to the Midwest to be implanted in cows; the plan was to remove the developing fetuses at around forty days of gestation and harvest genetically engineered fetal brain cells for possible human therapy. One of ACT's collaborators on this project was Curt Freed, the University of Colorado neurosurgeon who has pioneered fetal cell transplants into Parkinson's patients in the United States. Robl recalled the day when the first cloned cow fetuses arrived back in the lab. "We were looking at them," he said, "and they were absolutely perfect fetuses! And so it was at that time that we thought, Gee, if we can make these nice little fetuses, then we better start thinking about taking them to term."

Somewhat similar thoughts were occurring to scientists in Scotland and elsewhere. In the mid-1990s, Jim Robl, Jamie Thomson, Ian Wilmut, Neal First, and a few other scientists straddling the worlds of animal science and developmental biology were all circling around the same general scientific area: the creation (or use) of embryos, through cloning or more natural means of fertilization, to obtain embryonic stem cells in a variety of animal species, and possibly even to clone live animals, for pharmaceutical and medical purposes. Robl's lab at U Mass proved to be an incubator for many of the leading animal cloners: not just Cibelli, but Steve Stice (the first scientist hired by ACT and now head of Prolinea, a rival company in Georgia) and Paul Golueke (who went on to do animal cloning and human embryonic stem cell research at Infigen, another of ACT's biotech rivals, in Deforest, Wisconsin).

All three of those prodigies found themselves in an animated conversation one day in the spring of 1996 that would have legal, scientific, and social implications in this story. They were driving west along the Massachusetts Turnpike, returning to Amherst from Cambridge after attending a meeting of the Massachusetts Biotechnology Association at MIT. One of the hottest sessions at the meeting had focused on the enormous potential of embryonic stem cells as a form of human medicine. "We were talking about cells and organ transplants," Stice recalled, "and saying, Wouldn't it be great if there were a source of human tissue for transplants? That's sort of where it started." The three scientists — Cibelli, Stice, and Golueke — knew they couldn't obtain human embryonic stem cells because of the federal ban on funds for research involving human embryos. But they bandied about the idea of an audacious, somewhat transgressive experiment: Why not fuse the contents of an adult human cell with a cow egg? If the resulting embryo survived for just a week or two, Stice said, "you might be able to get something to develop, and get that to form an embryonic cell line of some sort. That was a blue-sky idea."

Like a lot of blue-sky conversations in science, this one dissipated as swiftly as the exhaust of their car. And indeed, for a number of months, the idea went nowhere. It's not clear whether anyone in the car was aware that the precedent had already been set by Philippe Collas's similar experiments with rabbit eggs in 1990, although Collas told me, "Jose knew perfectly well what had gone on in Jim's lab. How could he not know?" Robl recalled mentioning it at some point, too. "We always thought about doing it, and we never actually did it," Cibelli recalled in an interview, "so I finally said, 'Well, I'll give it a try.'"

This time Robl's students didn't keep him in the dark. Cibelli approached him with the idea of trying to create a human embryo in cow eggs to get stem cells. Robl went to the dean of the university to get approval, and although in both Robl's and Cibelli's recollection it was a back-burner project, Cibelli began working on the experiment during the summer of 1996. Thus, at least half a year before the world learned of the existence of Dolly, another experiment, pushing the technology even closer to human cloning, began to unfold in western Massachusetts.

This first attempt at human cloning began with a mouthwash. Cibelli swished his mouth with 70 percent alcohol, which was enough to dislodge and rinse out some large cells from the inside of his cheeks. Known as epithelial mucosal cells, these provided the adult cells he would use for nuclear transfer. Using the same techniques honed by years of cattle cloning, he vacuumed out the DNA from the cow eggs and then, using a tiny needle, inserted his own cheek cells, one per egg. "And after that, if you don't do anything, the egg's going to sit there, and it's just going to . . . *die* after a few days," Cibelli explained. "It's waiting for the sperm. So what you do is, you fool the egg and pretend that it is fertilized." This bit of biochemical chicanery is achieved by dousing the cow egg with a chemical that makes it behave as though it has been impregnated by sperm. Cibelli deposited dozens of these fused and primed hybrid cells in little plastic wells and then placed them in an incubator. "He did a bunch of them," Robl remembered. "It's pretty easy to get them to go to the 16- or 32-cell stage, but very difficult to get a blastocyst." You can surmise from his travel plans that Cibelli didn't hold great hopes for these experiments: during a time when fertilized eggs need to be scrutinized on almost a daily basis, Cibelli arranged to take his family up to Maine for a few days.

"One day, Jose was about to go on vacation," Robl recalled. "And he was about to throw out the dish. I don't generally look at these things, but I did that day, and there was a blastocyst." In fact, to Robl's seasoned eye, it looked pretty damn good. In other words, at least one of the cow-human eggs had begun dividing and had then moved beyond mere cell division to the stage of development where the cells begin to organize themselves into a first, rough architecture — a hollow ball, inside of which, within hours or days, fluid accumulates and stem cells begin to form. Following a routine procedure in the Robl lab, Cibelli carefully teased open the outer part of the blastocyst with glass needles and then placed the inner cell mass — that is, the putative stem cells — on a bed of fetal mouse cells, which typically nourish further development. And then he left for vacation. "We followed it," Robl said. "And it seemed to grow. And it grew for — I don't know — maybe a couple of weeks. It then started not to look so good." They tried to transfer the blastocyst to another dish, but it didn't survive the move. It had, however, produced an inner cell mass of embryonic stem cells. But what kind of stem cells?

"It looked, to my eye, not like a cow blastocyst," Robl said. "The morphology, the shape of the cells, was different. In all respects it was interesting, in that it was a colony that looked, in my recollection, different from a cow embryonic stem cell colony, more like a mouse embryonic stem cell colony, which is much more like what a human embryonic stem cell colony would look like."

When all was said and done, Cibelli had managed to create several short-term cow-human embryos during about two months of experimentation. Several reached the 8-cell stage, one reached 16 cells, and one not only reached blastocyst stage but also sported an inner cell mass apparently of embryonic stem cells. There is a photograph of this . . . this *thing* that people at ACT still pass around like a baby picture. The collection of cells is murky and indistinct, just like the scientific result it purported to achieve, but the photo formed part of the international patent application prepared by the University of Massachusetts, which claimed priority as of August 19, 1996. The biologists never characterized the cells in their hybrid embryo, and never even bothered to submit a scientific paper reporting it (which alone tells you something about the relative standards of a patent application versus a scientific publication). "We never considered a publication," Robl said, "because there was not nearly sufficient data." Indeed, after several more attempts, ACT abandoned the project altogether — not because it was controversial, but because it was so difficult compared with the company's many other projects. "We had eight or ten people pursuing efforts in chicken, pig, and cow," Robl said. "The somatic cell cloning work in the cow was coming along. And it [the cow-human work] was just considered way too risky, and just not very high up on the priority list."

At the top of that list, ironically, was a daring project to create a cloned cow. One such animal, created by Cibelli in ACT's laboratory at U Mass, was gestating toward birth in January 1997, when Robl traveled to Nice, France, for a meeting sponsored by the International Embryo Transfer Society. It was there that he heard for the first time about Dolly, a month before the news was publicly disclosed. With that, everything changed.

༄

Although no one can say with any biological certainty what exactly Jose Cibelli had created in the union of his cheek cell and a cow egg, that translucent blob of cells may ultimately have more value in terms of corporate sociology than scientific or medical utility. The mere image of it, projected on a conference room screen, persuaded Michael West to cast his lot with ACT, setting in motion all the controversy he would subsequently stir over three years of biotechnological provocation.

West wandered into ACT at a particularly propitious moment — so propitious that the 1998 meeting in Worcester might have been a not-so-subtle form of recruitment. As ACT founder Robl recalled, the company had gone through "a whole series of ups and downs" and was once again in flux. "ACT was having some difficulties with a person it had brought in to be the president and CEO of the company," Robl said, "and things were looking pretty bad at that point." This previous CEO had left just about the time West showed up. So he wandered into both a scientific and corporate vacuum and was only too happy to step into the breach and be running the show again.

By the fall of 1998, James Robl's involvement with Advanced Cell Technology had virtually ceased (he remains a consultant but says he is never consulted, and is otherwise unaffiliated with the company he founded). He concentrated most of his efforts on two companies spun off from ACT: Cyagra, which is devoted to agricultural cloning, and Hematech, which is devoted to engineering human antibodies in cows. Indeed, Hematech has scored several impressive scientific coups in areas that ACT originally pursued. The company has created human antibodies in cattle through cloning techniques and, in collaboration with Philippe Collas, showed that extracts from egg cells can partially reprogram adult cells and change their fate. But Robl retains a founder's pride in ACT and qualifies as a more-than-casual West watcher. "You know, people have described him as visionary," Robl told me. "He is certainly somebody that does have a vision, and in fact sometimes when I've had conversations with him, you think he's really on another planet!" Laughing, Robl continued, "He is somebody that I think looks at the world very differently than most other people. He comes across as being somebody that is very sincere and very committed to what he's going to do — and he *will* do it, come hell or high water. But I think beneath all of that there's somebody who's fairly shrewd about manipulating public opinion and, well, even the opinion of investors and colleagues."

In terms of public opinion, Michael West knew ACT would be playing to a much larger, and somewhat more hostile, audience in the post-Dolly era. Nonetheless, one of the first decisions he made on arriving at ACT was to press ahead with the cow-human embryo work. West arrived in Worcester around September 1998, and without much hesitation, he launched Cibelli again on the cow-human hybridizations. I later asked West if, before making that momentous corporate decision, he had consulted with his board of directors or with ACT's scientific advisory board. It was a reasonable question, given the magnitude of the experiments and their potential to inflame public opinion. West told me he'd basically made the decision on his own. Robl, still a consultant, was not consulted, even though he had overseen the original work in his lab at U Mass. "If it had been me," he said, "I probably would have suggested that they *not* continue with it." But, he added, "Jose was *very* enthusiastic about this project, and I think Mike must have picked up on that."

Cibelli, the lead scientist on the project, was indeed eager to resume the cow-human cloning experiments. "Jose is somebody that does tie onto something and sticks with it," said Robl. "Like with the cross-species nuclear transfer, sticks with it probably a fair bit longer than I would. And I think that may be the situation with human therapeutic cloning." To the ACT scientists, "therapeutic cloning" had all the appeal of a customized, multipotent, futuristic form of medicine. You would create a blastocyst using an adult cell from the body of a patient, for the express purpose of obtaining immunologically compatible stem cells that might later be given back to the patient as a form of medicine. As tantalizing as the idea sounded on paper, it never had much traction in the financial community; the production of customized forms of medicine, especially cellular forms of medicine, does not offer the small-molecule, one-size-fits-all economy of scale that pharmaceutical companies (and investors) want in a simple pill. As one biotech executive put it, therapeutic cloning is "a real tough business model."

But the scientific hierarchy at Advanced Cell was as much interested in scientific missionary work among the lay public as in convincing their financial backers. West, Cibelli, and Robert Lanza — a brisk, youthful-looking tissue engineering expert trained at MIT, who joined the company in 1999 — argued frequently and loudly for scientists to be allowed to proceed with therapeutic cloning. They wrote commentaries for *Nature Medicine* and *Nature Biotechnology.* They managed to round up eighty Nobel laureates as signatories to a letter that appeared in *Science* calling for federal support for embryonic stem cell research and therapeutic cloning. For a businessman constantly on the move, West seemed always to have time for ethical debates, radio interview shows, and television interviews, and he gave his home phone number to reporters — anything to get the message out. But then, that was the first job of an apostle. You could argue — and many did, at least privately — that this particular apostle was not the most august or the most credible, or the most respected, member of the scientific community to be making the case for stem cells, to say nothing of the way he wanted to push the public debate, pedal to metal, on human cloning. But it was also true that no one else in the scientific community cared to venture so far out on a limb, nor to argue with as much passion and persistence, especially after Harold Varmus left the NIH in December 1999 to become president of Memorial Sloan-Kettering Cancer Center in New York City. The combination of West's zeal and the absence of a single dominant voice from the scientific community goes a long way toward explaining why a small company in Worcester has driven so much of the public debate on stem cells and cloning.

Cibelli and his colleagues at ACT doggedly pursued their interspecies experiments for months without success, and never published a word on what must have been an unremitting stream of negative results. There were substantial scientific reasons to think that the cow-human cloning experiments were doomed to failure. Neal First, an expert in animal cloning at the University of Wisconsin, says there appears to be a mismatch between the DNA of the donor cell and the DNA in small organelles called mitochondria in the egg cells; First's group has attempted to insert the DNA of five different nonhuman species into cow oocytes, and he noted that "there was a very big mismatch somewhere down the line," suggesting that the interspecies mix of DNA was incompatible. When I asked Shirley Tilghman, a molecular biologist at Princeton (and, later, the university's president), about cloning experiments, she mentioned another hurdle. During normal sexual reproduction, both egg and sperm bring to the fertilized egg different and antagonistic versions of the same genes, and there is a battle between them to assert control on aspects of subsequent development. This phenomenon is called imprinting, and cloning — even using eggs and cells from the same species — would likely strip all the architecture of imprinting from the genes, resulting in flawed or compromised development. Indeed, these would become prominent criticisms a year later when the public debate on human cloning heated up. But Michael West brushed the criticisms aside when I asked about this. "I just don't see the mitochondria as being a problem," he said with a laugh.

For all their self-congratulatory openness with the public, however, the scientific group at ACT practiced a selective form of candor. In December 1999, a week or so after witnessing technicians insert human cells into cow eggs, I sent Jose Cibelli an e-mail, asking about the results. He was a little cagey, stating in reply that he had checked the nuclear transfer units and "they have developed at the 'predictable rate.'" As soon as new protocols were in place and running consistently, he added, "we will report the results using the appropriate channels (peer-reviewed)." In August 2001, when Cibelli was asked at a public forum if the use of cow-human embryos was a realistic approach to research cloning, he replied, "Not at all. It is fascinating to watch, but eventually the cells stop growing. We don't know why." That wasn't the message West suggested on the day of my visit, December 15, 1999. "We could publish now, the data we have," he assured me. "But we're trying to generate a real killer paper here. We're going to do a paper we're proud of." As of this writing, nearly three years after those remarks, ACT's scientists have not published a single word about the cow-human experiments in the scientific literature — although they have spoken exhaustively about them in the popular press.

At the time of my visit in November 1999, West and I, along with a representative of ACT's public relations firm, went to lunch at a hotel restaurant a short walk from ACT's labs. West barely touched the chicken on his plate as he recounted his odyssey in science and his fervent belief in therapeutic cloning as a way to obtain human embryonic stem cells and cure disease. There was, as always, a bit of evangelical fervor in his voice as he described the limitless promise of these protean cells. But it was also clear, even then, that cow-human embryos were less than ideal as a source of human stem cells. The best source, obviously, would be an intact human embryo, an embryo that could be created by merging an adult cell (from a patient with Parkinson's disease, for example) with a human oocyte — in other words, by human cloning.

So I asked West if human oocytes were under consideration at ACT as a way of creating human embryonic stem cells. This wasn't just an interview question about biological reagents; it was a question of keen public interest because of concerns about human cloning, and the irony was that private companies, at least in this area, were under no compulsion to provide society with an answer at all. Before replying, West leaned over to whisper with the PR representative, and after consulting with her for an inordinate amount of time about a straightforward and unambiguous question, he turned back to me — and changed the subject. In retrospect, I understand that Michael West the shrewd businessman had no intention of revealing ACT's plans. But in his highly structured, if idiosyncratic, moral universe, West prides himself on being both truth seeker and truth teller and didn't want to simply blurt out an untruth. And so, in moral contortions of evasion that would be the envy of Olympic gymnasts, he simply wriggled and squeezed and danced around the question, confessing at length his concerns about the dangers of harvesting oocytes from women and how he would hate for a market to develop around that.

He did everything except answer yes or no. As the rest of the world would learn in good time, human oocytes were indeed under consideration at ACT. Like the situational ethics of the baby boomers he sought to serve, West had just exhibited his unique brand of situational candor.

13
STREET-FIGHTIN'
MAN

THE MORNING BEGAN — truly began, with that first gamy whiff of what lay in store — shortly after 9 A.M., when Bradley Martin, his assistant Jin-Quang Kuang, and a researcher named Ellen Flynn marched along a dreary, dimly lit, institutional corridor high in a building at the Johns Hopkins Medical Center in Baltimore. After pausing to take a deep breath, they pushed through a green door and entered a small room where several robust Yorkshire pigs greeted them with braying squeals and frothing curiosity. Flynn wheeled an echocardiogram machine into the narrow aisle between rows of cages, and then Martin, a flimsy yellow surgical gown covering his blue jeans and sports shirt, stepped gingerly into one of the cages and gently wrapped an arm around the huge porker, a gesture that wavered between a hug and a headlock. "All those years of graduate school," he grunted over his shoulder, "are finally paying off."

Wrestling a 400-pound pig into submission and then holding it steady while a technician gingerly rubs a jelly-coated probe over the animal's chest in search of a good echocardiogram signal, set against deafening squalls of porcine protest and the in-your-face odorama of big animals kept in small rooms — that's not how most people imagine the elegant world of cell biology, or the future of regenerative medicine. But the road that connects a promising biological idea, like stem cells, to its ultimate application in the clinic must pass through rooms like this one, where the noise, the odor, and the sheer mess echoes the complex commotion of human biology. And on that lovely morning in May 2001, while politicians on both sides of the embryonic stem cell and cloning controversies were maneuvering for a wild summer of hearings, debates, and ethical hullabaloo, Brad Martin's foray into the animal room represented the quiet pursuit of a kind of biological wild card in the controversy. He had become, by training and employment, an expert in adult stem cells, and he was close to finishing what he hoped would be the penultimate step of preclinical testing for a highly touted and futuristic form of coronary medicine, one that might ultimately be tested in human heart attack victims sometime in 2003.

Martin, a sandy-haired, good-humored senior researcher at Osiris Therapeutics, a Baltimore-based biotechnology company, had been paying weekly visits to the Hopkins animal room for six months. It was a cardiac ward of sorts, since all the pigs in the room had suffered surgically induced heart attacks. Several of these pigs had been treated with stem cells — not embryonic stem cells, but an adult form of these versatile progenitor cells isolated from the bone marrow of pigs (the technical name, a real mouthful, was "adult mesenchymal stem cells"). The reason these — and other kinds of stem cells — are so laden with medical possibility is that they seem to possess an uncanny ability not only to find areas of physiological damage in the body but also to organize the process of healing and repair at precisely those sites.

Even as Martin tiptoed around his study subjects that sunny morning, the stem cell debate was producing much more of a ruckus about fifty miles down the parkway in the nation's capital, where a seemingly endless series of congressional hearings, scientific symposia, and theological lobbying had been going on for months. Embedded in that larger debate over the wisdom of allowing human embryonic stem cell research was a smaller, somewhat more technical question that hinged on precisely the research that Osiris, several other companies, and a number of academic researchers were pursuing: Why even bother with the ethical vexations of human embryonic stem cells, which require the sacrifice of an early human embryo, when you could use adult stem cells? Opponents of embryonic cells ceaselessly argued that adult stem cells were just as good, perhaps better, and didn't involve human embryros at all. "It seems like there is a really good alternative here that is noncontroversial," biologist David Prentice told the *Washington Post* in the spring of 2000. "We really ought to be investing in this more thoroughly, and people need to be more aware of this alternative." Indeed, while embryonic stem cells had received the lion's share of attention (and public adulation) in the three years since Jamie Thomson's publication in *Science,* their supposedly less potent and seemingly less glamorous biological cousins had quietly been writing a fascinating story of their own — a story that in many ways was more advanced, clinically and commercially, than the ES cell story. Several new forms of adult stem cell therapy had moved into clinical testing as early as 1999, and more high-profile experimental treatments were soon to be tested.

During congressional hearings in the spring and summer of 2001, as George W. Bush struggled with his difficult policy decision, conservative politicians — and one biologist with a right-to-life background, the afore-mentioned David Prentice — sang one long aria about the virtues of adult stem cells. The message these lobbyists conveyed was always the same, and always essentially scientific at its core: adult stem cells could do everything that embryonic stem cells could. When it turned out that Prentice was al-most alone among his scientific peers in saying this — and that this suspect view would be forcefully repudiated by a special report issued by the NIH over the summer — it proved to have been an important detour from stan-dard right-to-life arguments, and perhaps a disastrous one. "They found the one scientist they could find who would say that scientifically, embry-onic stem cells were not necessary and you could do it with adult stem cells," said one biologist who requested anonymity. "As soon as the pro-lif-ers moved off their territory and into science, they were in trouble."

In any event, converting the theoretical promise of adult stem cells into medical practice had a very long way to go. Occasionally, it could even get downright dangerous. At one point during his visit to Hopkins, Brad Martin found himself knocked halfway across the pen when one of the two pigs undergoing echocardiograms that morning casually turned its head. But the reward for these early-morning exertions could be seen in the im-age on the monitor of one of the Osiris pigs that had suffered a massive heart attack and then received adult stem cells. This pig's heart pumped in firm, resolute beats on the screen. "That's a cross section of the left ventricle squeezing," Martin shouted over the din, "and it's squeezing pretty well!" Indeed, Osiris researchers had accumulated and reported data at meetings showing that despite significant damage to the main pumping chamber of the heart following a heart attack, many adult stem cells, when injected di-rectly into the coronary circulation, zeroed in on the site of injury, took up residence in and around the dead tissue, and literally remodeled the archi-tecture of a damaged heart. They seemed, in fact, to interrupt the typical progression toward a lopsided architecture in damaged cardiac tissue that typically results in heart failure — and, sometimes, the need for an organ transplant — in heart attack patients.

The new cells, Martin quickly added, lack several critical features typical of cardiac muscle, so Osiris did not claim — at least not yet — that their adult stem cells had matured into functional, bona fide heart muscle cells. And in any event, the fact that the control pig in the adjoining cage, boisterous and feisty despite recovering from a heart attack, *didn't* receive stem cells, was a high-decibel reminder that the cardiac stem cell story is far from settled. But these preliminary results, and several other reports that surfaced in the spring of 2001 from academic labs, suggested that with the help of adult stem cells the heart has remarkable potential to heal itself, and helped make the cardiac application of stem cells, either adult or embryonic, one of the likeliest candidates to reach clinical testing in the near future. Martin told me that if all lingering regulatory and safety concerns with the FDA could be satisfactorily resolved, Osiris hoped to launch a phase I safety study of adult stem cells in humans with heart attacks by the middle of 2003. If all went as planned, the company's collaborators in this trial would be its own form of endorsement; the stem cells would be tested at the Clinical Center on the campus of the National Institutes of Health.

⚬◞

Adult stem cell research has made impressive strides in the past few years, with private companies and academic scientists publishing one striking finding after another. In the gush of scientific comment and press coverage, embryonic stem (ES) cells have been the A+ students of regenerative medicine, but adult stem cells have been, by a long shot, the first to graduate from school and enter the real world of medicine.

Adult stem cells have important limitations. They are merely "multipotent" — that is, they typically can form only a narrow range of tissues, compared with the trademark versatility of "pluripotent" embryonic stem cells. Some scientists complain that they are hard to grow. But they also come without the political and ethical baggage of ES cells. Geron and Advanced Cell Technology have monopolized public attention, but a number of lesser-known companies, many devoted to adult stem cell therapies, have quietly laid the groundwork for clinical testing and in some cases have already taken stem cells into the clinic. Osiris, for example, was formed as a stem cell company in 1993, years before Geron even committed itself to stem cell research. StemCells, Inc., of Sunnyvale, California, first got off the ground in 1995, inspired by the groundbreaking research on blood-forming stem cells by Irving Weissman's group at Stanford University. Several other biotech companies, including NeuralStem Biopharmaceuticals of College Park, Maryland, and Layton Bioscience of Atherton, California, have mounted serious efforts with fetal cells.

"Adult stem cells have been shown to do more things than we suspected even two or three years ago," Daniel Marshak of Cambrex Corporation told me in the spring of 2001 at a stem cell meeting he co-organized at Cold Spring Harbor Laboratory. Even an impassioned proponent of embryonic cells like Ron McKay of the NIH has admitted, with important qualifications, that the interest in adult stem cells has been warranted. "For certain diseases, adult cells appear very promising — for hepatic and cardiac diseases in particular," he said. "However, if you're asking for a solution to Parkinson's disease or diabetes, I would say the cells that offer the best way are fetal and embryonic." That fine teasing apart of potential, that segregation of greater and lesser promise, case by nuanced case, addresses precisely the point that so much of the political debate strives to bulldoze and obscure: both approaches have their merits, and given the early stage of the research, with tens of millions of life histories (and diseases) hanging in the balance, it seems foolish to foreclose the scientific road to potential breakthroughs by choosing one avenue of research to the exclusion of the other. Still, in the unforgiving crucible of clinical studies, where medical potential meets the fickle realities of the human body, adult stem cells have clearly advanced to the stage of critical testing, while preliminary tests of ES cells in humans may not get under way until 2004 or 2005 at the earliest.

As a body part, the bone marrow has never inspired as much rapturous Shakespearean poesy as, say, the heart, liver, brain, or even spleen; for the better part of history, it's been of greater value in a soup pot than in the clinic. But this spongy matrix of tissue, encased in a sanctuary of bone, is increasingly being recognized as the body's safe-deposit box, a secure physiological repository for some of the body's most precious regenerative jewels — namely, cells that can differentiate into many other tissues. In fact, adult stem cells from the marrow have actually been a proven, prominent, and respectable feature of medicine for about four decades. It's just that for most of that time, no one referred to them as stem cells.

Bone marrow transplants, first attempted in the early 1960s, achieved routine success by the 1970s. That success occurred, scientists realize now, because patients received, in the slurry of donor marrow infused into their bodies, hematopoietic (or blood-forming) stem cells. These are progenitor cells of the blood system and possess the ability to "differentiate" (or specialize) into all the cells of a healthy and whole blood system. In this case, one mother hen of a blood cell gives rise to red blood cells, lymphocytes, macrophages, granulocytes, eosinophils — all the foot soldiers of our immune system and the hod carriers of our circulatory and metabolic functions.

The existence of that marvelously prolific cell was first proposed in the early 1960s and grew out of medical experiments related to nuclear warfare and radiation exposure. Two Canadian researchers, J. E. Till and E. A. McCulloch, proved the point with a simple yet brilliantly provocative series of experiments published in the 1960s. They subjected mice to lethal doses of radiation that destroyed the rodents' entire blood-making apparatus in the marrow; then they infused small amounts of blood from matched donor mice. In some cases, the transplantation of a *single blood cell* was sufficient to rebuild the animal's entire blood and immunological systems. "That was the initial experiment that defined the stem cell for us," said Irving Weissman. Although that blood-forming cell, now called the hematopoietic stem cell, was known to exist for decades, the human version wasn't actually discovered until the early 1990s by C. M. Baum, Ann Tsukamoto, and their colleagues, working at a now defunct company started by Weissman known as SyStemix. Here is a lovely example of a medical therapy providing breathtaking cures decades before the science underlying the miracle could be fully and convincingly explained.

In defining the characteristics of the hematopoietic stem cell, however, the scientists who studied it also attempted to establish ground rules for stem cell research in general. Weissman in particular has taken on the unofficial and somewhat controversial role of scientific referee, insisting that any researchers claiming to have found a stem cell must demonstrate a certain set of unique surface fingerprints (or "markers") and show that a single cell can do everything it is supposed to do. And of course, other researchers had to be able to reproduce those results. These criteria have created a classic sociological schism in the stem cell field between hard-core scientists who systematically catalogue reliable markers for each cell, and scientists who are more interested in observing and describing the big-picture capacities of these cells. Some of that tension has spilled, without much annotation or background, into the political debate about adult stem cells versus embryonic stem cells. Nowhere has it been more apparent than in the science surrounding an odd, seemingly modest cell isolated from the marrow — the same type of cell that Brad Martin had injected into his heartsick pigs. The restorative potential of that cell became clear during the 1980s, largely through the work of a former street gang member who found redemption as a biochemist. He was a professor at Case Western Reserve University named Arnold Caplan.

৩৯

At his office in Cleveland, Caplan appeared the very model of an avuncular, buttoned-down midwestern geek, complete with an unmistakable Chicago accent. He wore a white shirt and tie, gray slacks, and black sweater vest, his dark hair creamed and combed back, his black horn-rimmed glasses perched on an amply ethnic nose. This turned out to be the middle-aged look of an erstwhile juvenile delinquent — "somebody," in Caplan's own estimation, "who's genetically aggressive, who keeps trying to make things work and stumbles, actually, into the concept of mesenchymal stem cells." There is a surpassing irony to the fact that such a bare-knuckles personality as Caplan has been pummeled for much of his adult scientific life with the label "phenomenologist." That's a polite term for someone who does fuzzy, undefined biology. Or, as Caplan put it, "That's a word that's synonymous with *bullshit*, okay?" It's also the word that's hung around the neck of many researchers who study adult stem cells.

Caplan is a loud, tough reminder of the bustling marketplace of science. He grew up in Chicago's Wicker Park neighborhood, on the city's North Side, the son of second-generation Russian immigrants who never finished high school; his father drove a delivery truck for the *Chicago Tribune*. This self-described "city rat" no more had the pedigree to become a serious biochemist (which he ultimately did after training at the Illinois Institute of Technology, Johns Hopkins, and Brandeis) than Oliver Twist was destined to be a brain surgeon. Caplan politely dodged any discussion of the particulars of his youthful indiscretions, except to admit in clipped terms, "I ran in a gang. There were guns. There were drugs. There was violence." He told me that once, when he was a senior at Roosevelt High School in Chicago, during a period when he "sort of got 'middle-classed,'" he went on a date with a girl who wanted to see "this play, this thing, this musical." Sitting in the nosebleed section of a downtown theater, Caplan grew increasingly agitated as the action onstage chronicled gang warfare, violence, the sudden firing of a gun, and a killing. "I was, like, gripping the chair as I was viewing this thing, because it was an instant replay of my life," he said. This play, this thing, this musical, was *West Side Story.*

Caplan has clearly outgrown, but never quite lost, this edge; profanity, insult, and an exuberant adult humor flow from his mouth as from a broken hydrant, and you can still feel the restless, almost physical impatience of a swaggering young turk beating beneath the Uncle Arnold wardrobe. Indeed, he came within a hairsbreadth of deep-sixing his scientific career over a trivial ethnic taunt while at Hopkins; at a fraternity mixer, Caplan recalled, he got drunk and was challenged by a teenager who said, "Oh, you Jews are all alike." As Caplan recalled, "I grabbed the kid's head and was propelling it toward an old-style Coke machine, which has a glass panel with fluorescent lights behind it and the cup drops out and the liquid goes into the cup. But it's a glass panel, and what I was physically going to do was, in the old neighborhood style, bring his head through the glass, *and down.* And I caught myself at the last moment, and was so upset with the fact that this . . . history was still programmed into me."

Fortunately, Caplan benefited from a series of prominent mentors, including the legendary Hopkins biochemist Albert Lehninger and the Brandeis embryologist Edgar Zwilling, who helped him channel his energy and work ethic, his voracious curiosity, and his useful disdain for authority and dogma into a productive, unexpectedly intuitive kind of science. Through a circuitous route, Caplan became fascinated with the development of muscle, bone, and cartilage in chicken embryos. And just to show how small the world of biology can be, a formal portrait photograph hung on his office wall of A. J. Friedenstein, the Russian biologist who pioneered studies on the formation of bone — the same Friedenstein who lectured in Moscow about the Hayflick limit, inspiring Alexey Olovnikov to imagine telomerase as a subway train.

The formation of bone and cartilage is controlled by a single progenitor cell that nestles in the spongy tissue, otherwise known as stroma, lining the inside of bone. After proposing the existence of this cell, Caplan, his colleague Victor Goldberg, and his postdoc at the time, Stephen Haynesworth, ultimately succeeded in isolating the prize: a surprisingly multipotent stem cell from the bone marrow. They called it a mesenchymal stem cell (or MSC) because it controlled the development of tissues and organs associated with a region of early development known as the mesenchyme, and this cell possessed the remarkable ability to form not only bone and cartilage but also muscle, tendon, fat, and stroma. There isn't scientific unanimity on how convincing Caplan's evidence was for the existence of this cell, and when he began to call it a stem cell, "I got nailed," he recalled. "And the people who are the worst, the people who are still having trouble with this, are the people who study hematopoietic stem cells . . . And so what I say at the end of lectures now is that I prefer that you call these Arnie Caplan cells, because the term *stem cell* causes me so much heartburn when I'm giving lectures and writing grants."

In 1993, Caplan and Goldberg joined forces with a venture capitalist named James Burns to form a biotechnology company in Cleveland based on their adult stem cell research. I couldn't help laughing when Caplan recited the story of how the company got its name, for it was a far earthier and less romanticized version of the Osiris myth than the one Mike West had described in hushed tones the first time I met him. Burns, the venture capitalist, called Caplan at home one night and suggested the name Osiris, after the Egyptian god of immortality and regeneration. Caplan says he hated the name for two reasons — one, Osiris was Egyptian, and two, he could also be considered the god of resurrection. Burns prevailed. "Then a very prominent and highly regarded orthopedic surgeon who was associated with the company called me up one day," Caplan continued. "And he said, 'My wife and I were just in Egypt, and we finally heard the story of Osiris straight, without it being embellished by you or Burns." In this version, the surgeon explained, Osiris was dismembered, and his body parts were dispersed all over Egypt. Osiris's wife, Isis, systematically retrieved all of his body parts and reassembled them. "Save one," Caplan said. "Which was thrown into the Nile, and was consumed by the bottom fish, the scavenger fish. And that body part happened to have been his penis."

Despite that grim augury, the company set out to develop adult mesenchymal stem cells as a therapy.

✑

Osiris (the company) still lives near the water, now located in a renovated fish cannery in the Fell's Point section of Baltimore, its low-slung redbrick lines abutting the busy harbor. Like many biotech companies, it has the usual tearoom traffic of underdressed scientists (Bob Deans, head of research at the time I visited, sported a handsome ponytail that reached to his waist). And like many young biotech companies, it has had a high turnover of personnel and constant money worries; Deans left the company in 2001, and neither Caplan nor Burns is still affiliated with it. Unlike any other company in the stem cell field, however, Osiris has reduced the harvesting and expansion of stem cells to practice and has been shipping IV bags of the cells to more than a dozen hospitals, where its several clinical trials are under way.

The harvesting process, detailed in a *Science* article in 1999, basically works like this: a small sample of bone marrow, about 25 milliliters, is obtained, typically from the pelvic bone of a donor. The cells are not exactly plentiful: roughly 1 in 10 million marrow cells is, by Osiris's estimates, the desired stem cell, but these cells can be plucked out by a combination of centrifugation and proprietary cell-sorting technology. Once isolated, they are expanded in cell culture flasks to about 500 million stem cells per IV dose and then frozen in liquid nitrogen (the company says frozen cells have a shelf life of up to one year). By altering the culture conditions, Osiris scientists learned that these stem cells could be nudged toward various fates, whether muscle, cartilage, or bone. Interestingly, the cells not only respond to biochemical cues but choose their fate based on physical cues as well, including their three-dimensional environment and even mechanical forces resembling the extension and flexion of joints during walking. Stem cells, whether embryonic or adult, are very attuned to their environment.

When Osiris first began clinical tests in 1995, the treatments were autologous — that is, patients donated their own marrow, and company scientists would then isolate stem cells and expand them for about eight weeks before giving them back to the patient. After several years of that arduous procedure, Osiris scientists realized that stem cell populations harvested from unrelated donors might work in all patients. In the course of assessing the cells in preclinical trials, the scientists had stumbled upon a totally unexpected phenomenon. According to Deans, these mesenchymal stem cells are conspicuously denuded of typical immunological markings. "The cells are MHC Class II negative," Deans told a scientific meeting at Cold Spring Harbor Laboratory in 2001, meaning that they can evade a basic form of scrutiny by immune sentinels known as T cells; they also lack a marker known as B-7, which typically revs up an immune response. Deans added that these stem cells may even secrete a factor that actively inhibits a normal immune response. Jay Shake, a resident at Johns Hopkins who has collaborated with the Osiris group, told a meeting of thoracic surgeons recently that the cells "seem to be totally immunoprivileged and we are currently doing a study where we are injecting cells from one animal into another with no immunosuppression, and we are seeing no rejection." The cells, in other words, seem to deploy a biological stealth technology to remain immunologically invisible. This obscure biological trait has enormous economic implications. If the preliminary observation holds up, this quality could obviate the need for customized (or autologous) treatments, which would open the door to the use of universal donor cells, even in emergency medicine. "It changes the whole ball game in terms of commercialization," said Brad Martin.

When this immunological invisibility was first observed, Osiris scientists were stunned. "In the last year, this all came to light, and we are all just amazed," said Martin in May 2001. "We were *flabbergasted*," added senior scientist Frank Barry, who heads cartilage research. "We still are." Many other scientists remain unconvinced the phenomenon is real — one prominent stem cell researcher, who asked to remain anonymous, said, "I think all of that is hugely exaggerated" — but a clinician who has used the cells in humans told me, also off the record, "It appears to be true." The observation holds considerable importance for the commercial development of two areas of stem cell medicine with potentially large markets: cells that repair hearts, and cells that repair joints worn down with osteoarthritis.

Heart disease is the leading killer in the Western world; more than a million heart attacks a year occur in the United States alone. As a result, heart disease has been one of the most intense — and most impressive — areas of stem cell research in the past year. In the spring of 2001, two separate academic groups published studies showing that the damage caused by heart attacks either in rats (Silviu Itescu and colleagues at Columbia) or in mice (Piero Anversa at New York Medical College in Valhalla, New York, working with Donald Orlic and Ron McKay at the NIH) could be repaired by injecting stem cells in or near the injury. In Itescu's work, for example, a primitive stem cell he calls an angioblast was isolated from human bone marrow and, once injected into a rat's bloodstream, migrated to the scene of cardiac injury and built new blood vessels around the damaged tissue. In Anversa's study, primitive stem cells isolated from mouse bone marrow formed new heart muscle in areas of damage and also helped form new blood vessels. "With cardiac regeneration," McKay explained, "we take the bone marrow of the mouse, sort out the hematopoietic cells, get on a plane, and put them into a mouse with a heart attack. Bingo! And the stunning part is, the heart regenerates."

Osiris began a cardiac program in rodents several years ago, Martin said, but quickly moved into pigs because porcine anatomy is more comparable to that of humans. In the first round of experiments, veterinary surgeons at Johns Hopkins performed open heart surgery on the animals and for one hour tied off the left anterior descending coronary artery, which carries blood to the muscle of the heart's main pumping chamber, triggering a massive heart attack. After waiting two weeks, Osiris researchers injected about 50 million autologous mesenchymal stem cells directly into the hearts of five test animals. The cells were genetically tagged with a marker so their location could be traced in the body. These pigs, as well as half a dozen control animals, were closely followed for up to six months.

All the pigs that did not receive stem cells died within a month or two of the heart attack. Autopsies showed that the damaged hearts in these pigs looked much like those of human heart attack victims: extensively scarred, excessively large (or hypertrophic), and lopsided, in compensation for diminished pumping capacity. Eventually, the wall of the heart thinned and heart failure ensued. In animals that received stem cells, however, the cells found their way to areas of cardiac damage, filling in scar tissue and remodeling the shape of the heart; echocardiograms showed that in these pigs the heart retains a more normal architecture. "Wherever you have necrosis [dead tissue], you see the mesenchymal stem cells," said Martin. "They really fill in these scars." Like all early animal research, these results came with many caveats, but they quickly inspired a follow-up study.

Osiris initiated a second round of experiments in pigs — the same pigs Martin visited the morning I accompanied him — and the results appeared to confirm the initial tests. This second preclinical trial used universal donor cells and delivered them immediately after the heart attack. Echocardiograms, including the ones Martin gathered that morning, have shown a statistically significant improvement in the pumping capacity of the heart as a result and overall improvement in heart wall motion. The company began to explore the possibility of delivering adult stem cells to precisely the right spot in a damaged heart with the use of a catheter, similar to the type used in angiograms or angioplasties, but the researchers soon discovered another "amazing" property of these cells, Martin told me.

When injected into the body intravenously — in a vein in the arm, for example — enough cells find their way to the site of injury in the heart to congregate there and initiate repair. As a result, in the clinical study Osiris hopes to conduct with NIH researchers, the protocol calls for heart attack victims in the Washington, D.C., area to be stabilized at local hospitals and then transferred to the NIH's Clinical Center, where they will receive stem cells by way of an ordinary IV needle in the arm. The ultimate aim, Martin explained, is to manufacture "a universal cell, cryopreserved, which could be in the emergency room of every hospital in the country and used in emergent situations with heart attack patients."

At a farm north of Baltimore, Osiris scientists have also been testing the regenerative power of mesenchymal stem cells in goats that have sustained severe knee damage. Frank Barry, who heads the cartilage repair program at Osiris, brought out a model of the knee joint to illustrate this unusual experiment, which re-creates a common joint ailment. Veterinary surgeons sever the anterior cruciate ligament and remove the medial (or inner) half of the meniscus, a resilient patch of cartilage that forms a cushioning pad between the thighbone — the femur — and the tibia, the larger of the two bones that form the lower leg. The goats then spend several weeks walking on an exercise treadmill using this wobbly, unstable joint — a regimen that chafes the remaining cartilage off the ends of the long bones. This activity creates a harrowingly accurate model of osteoarthritis, a painful condition that afflicts at least one joint in half of all Americans over the age of fifty. Using an ordinary syringe, Osiris researchers then inject approximately 5 to 10 million adult mesenchymal stem cells into the synovium, a little purse of tissue inside the knee. Although the stem cells have been tested in only about one hundred animals, company scientists have presented preliminary data at meetings showing that the injection of stem cells not only restored the surgically removed meniscus, but within twelve weeks recarpeted the eroded surface of the thigh and calf bones with new cartilage.

ᝍ

Osiris is hardly the only company pushing adult stem cells toward the clinic. Across the continent, in a nondescript building near Silicon Valley, I sat in a darkened conference room one afternoon while Ann Tsukamoto showed me a picture of cells glowing with fluorescent green exuberance on a projection screen. What was remarkable about the picture was not merely the seemingly single-file parade of cells marching across the image; it was the physiological terrain traversed by that cellular pack train. The whorled landscape, with its folds and ravines, belonged to the brain of a mouse, and the cells migrating in such orderly fashion across that cerebral wilderness were stem cells. Human fetal stem cells, to be exact. "If you step back and look, we've injected the cells just into this area," explained Tsukamoto, the chief scientist at StemCells, pointing out a red-stained region of the brain known as the subventricular zone, "and the cells have now gone throughout the brain. They've migrated everywhere. There's no sign of tumors, as far as we can tell. And they've incorporated themselves into these active neurogenic areas, where they themselves are continuously self-renewing." This was a picture, in other words, of a self-renewing brain.

StemCells has had a complicated corporate history, but it grew out of the research on blood-forming stem cells in Irving Weissman's laboratory at Stanford University. Although it has not yet launched any clinical trials, the company may have a more impressive history of publications than Osiris and has made steady progress toward testing human stem cells, of either adult or fetal origin, to regenerate brain, liver, and pancreatic tissues.

For decades, the conventional wisdom about the brain has been that we're born with our fixed lot of neurons, and that's it. The brain could not regenerate tissue, according to this view, and every insult to neural anatomy over the course of a lifetime — stroke, seizure, alcohol, Alzheimer's, Parkinson's disease, or just plain old age — irrevocably depleted the allotted population of brain cells. But in the late 1990s, work at the Salk Institute, in the laboratory of Fred Gage, began to turn this orthodoxy on its head, so to speak. Gage, another of StemCells' cofounders, and his colleagues electrified the field of neuroscience when they reported the existence of adult neural progenitor cells in the mammalian brain. Gage deliberately shies away from the term *stem cell* to describe these adult cells, because he's still not sure what they are. But they can be recovered from the brain tissue of rats, for example, and possess the capacity to differentiate into either of the two main neural cells, neurons and glial cells. Gage and his colleagues discovered that new neurons formed in the subventricular zone, a vestige of embryonic tissue near the center of the brain; from this spot the progenitor cells migrate, like ants at a picnic, to several other critically important neighborhoods in the brain, including the olfactory bulb (the neural seat of smell) and the hippocampus (one of the centers of memory).

The brain would seem, at first blush, to be beyond the reach of adult stem cells. The problem, succinctly stated by Harvard Medical School researcher Evan Snyder, is collecting them: "If you're talking about the brain," says Snyder, "where would the adult stem cells come from?" In other words, who would donate their own brain cells? In 2001, Gage's group provided the first hint of an answer, and in doing so appeared to open yet another bioethical can of worms in the stem cell field. They isolated adult neural progenitor cells from cadavers — leading to the possibility that neural stem cells could be harvested from fresh corpses for medical use, much as hearts, livers, and kidneys are currently harvested from accident victims for organ transplants. Gage's group has shown that neural progenitor cells, when transplanted into an adult nervous system, migrate to the zone in the brain where new neurological cells are formed, and will also migrate to areas of injury, where they typically take on the function and morphology of cells in that spot. "Not only are new cells born, but they undergo synaptogenesis," Gage said at a recent Cold Spring Harbor meeting, meaning they form the connections, or synapses, that link nerve cells.

Brain implants of stem cells in human patients are not just around the corner, although a University of Pittsburgh doctor, Douglas Kondziolka, has treated a dozen or so stroke victims with embryonic cancer cells. StemCells is conducting preclinical studies with human neural stem cells of fetal origin, testing the cells in animal models of several neurodegenerative diseases as well as genetic disorders with a neural component, such as Huntington's and Parkinson's. Company scientists have published impressive studies in mice showing that neural stem cells, when injected into specific cell-forming locations in the brain, migrate through brain tissue to the site of injury. Evan Snyder's group at Harvard Medical School has shown the same capability with fetal-derived neural stem cells, which are being commercially developed by Layton Bioscience, so several biotech companies are pointing toward neural applications. "We envision that one can make cell banks of these human neural stem cells," Tsukamoto said, "to treat neurodegenerative diseases, genetic diseases with a neurological impact, eye diseases, and spinal cord injury."

StemCells scientists have also reported significant progress in the use of adult stem cells to regenerate the liver. It has long been known that animals can rebuild their own livers; when researchers remove two thirds of the liver of rats, which typically encounter many toxic food substances in the wild, the animals have the ability to regenerate a complete new organ in five days. Eric Lagasse published a promising study in 2000 showing that blood-forming adult stem cells in the mouse, once isolated from the marrow and expanded in a test tube, could be injected into mice suffering severe liver damage to seed the slow but steady regeneration of liver. This approach is being pursued, Tsukamoto said, to provide "another source of cells that can regenerate the liver without doing a liver transplant, because there are so few livers available."

To speed the growth of this regenerative tissue, StemCells has licensed a truly futuristic technology from Caltech biologist David Anderson, the company's third cofounder. This technology involves the use of a cancer-promoting oncogene called *myc* to goose the production of new tissue, conferring "the ability to immortalize, and then dis-immortalize, a cell," Tsukamoto explained. "So you can introduce a gene that'll help the cells grow. Once you've grown up lots of those cells, you can remove that gene so that you're not transplanting back cells that still contain that immortalizing gene."

~

Many of these futuristic techniques are far from clinical application, but the fact remains that a number of new adult stem cell therapies — that is, therapies other than bone marrow transplants — have taken the first tentative steps into clinical testing. The initial efforts have been small, modest, and, while intriguing, far from definitive. In 1996, to cite one of the earliest applications, Edward M. Horwitz and his colleagues at St. Jude Children's Research Hospital in Memphis, Tennessee, began testing bone marrow stem cells to treat young children with a devastating disorder called osteogenesis imperfecta. In this condition, the bones are so fragile that an infant can fracture a rib merely by rolling over in bed. Three of six young children in the initial trial temporarily improved after having bone marrow transplants — a procedure in which they presumably received adult bone-forming stem cells. In a paper published in 2002, Horwitz described preliminary results from a follow-up study in which these same six children were treated anew with an infusion of bone marrow stem cells collected from the original donors. The cells, expanded outside the body at Darwin J. Prockop's laboratory at Tulane University's Center for Gene Therapy, were then injected into the children. Horwitz reported that the cells have been safe to infuse, and that they appear to be incorporated into, and presumably help form, new bone. Four of the five children, he reported, can now stand and take steps, where previously they couldn't.

In another early trial at the University of Minnesota, doctors began testing Osiris's mesenchymal stem cells in 2000 in conjunction with umbilical-cord-blood stem cell transplants to treat children with leukemia. Again, the results are very preliminary, but the response in the first few patients suggested that the body rebuilds its blood supply following a transplant more quickly when adult mesenchymal stem cells are part of the treatment. "What we can say so far," said John E. Wagner, Jr., who headed the study, "is that we have seen no negative side effects, and we have the impression that it's faster, with several patients having the most rapid replenishment of neutrophils and platelets that we have ever seen — and we've done two hundred fifty cord-blood transplant patients now." Osiris scientists have reported at meetings that while it usually takes seventy-five to ninety days to reconstitute platelets in a recipient following a marrow transplant, it can take as little as thirty-five days with the addition of adult mesenchymal stem calls, and that the use of these stem cells may also reduce the incidence and severity of graft-versus-host disease. The recovery of platelets has an important economic dimension, too, because patients must usually remain hospitalized until platelet counts reach a certain level.

As promising as the adult stem cell research seems, it is still preliminary and uncertain — and that uncertainty has been further blurred by political interpretations of complicated scientific work. The proponents of adult stem cells do not always appreciate the fitful progress of science, and a cautionary tale is embedded in a rash of recent research reports suggesting that adult stem cells appear to be much more biologically versatile, and capable of adopting many more cellular fates, than anyone originally thought.

In May 2001, for example, Neil Theise of New York University and Diane Krause of Yale published a report in *Cell* claiming that an adult stem cell they had isolated from the bone marrow of mice could form virtually any organ tissue in the body. The cell, Krause said, can form not only blood, but lung, liver, stomach, esophagus, intestines, and skin. Theise said these adult stem cells were as flexible as the embryonic kind, and referred to them as the "ultimate adult stem cell." Krause added, "It suggests that there's a lot more plasticity in adult cells than we had previously thought." This was part of a cascade of research findings that made the stem cell field sound more like seventeenth-century alchemy than like twenty-first-century biology. Ira Black of the Robert Wood Johnson Medical Center in New Jersey reported turning bone marrow cells into neurons, and a Swedish-Italian team headed by Angelo Vescovi reported turning brain cells into blood. For sheer titillation, the story that seems to have tickled the public fancy to an unprecedented — and probably unwarranted — degree came in April 2001 from UCLA researcher Marc H. Hendrick, who published findings suggesting that fat tissue removed by liposuction could serve as a source of adult stem cells that could form muscle, cartilage, and bone. The research immediately became a topic of conversation on radio talk shows and on late-night television, and Hendrick and his main collaborator, Adam J. Katz of the University of Pittsburgh, immediately set up a company, Stem-Source, to commercialize the discovery. This was a development that any couch potato could understand. Not long after the research was published, I had a chatty cab driver in Cleveland who volunteered that he was ready to be "lipo'd" to get some stem cells. Opponents of embryonic stem cell research seized on all these results as further proof that embryonic stem cells were unnecessary. Yet in the spring of 2002, two reports in *Nature* cast serious doubt on the credibility of some of the earlier observations and sent everyone scurrying back to repeat their experiments.

Science mixed with politics almost always produces a sour cocktail, but this odd process of preliminary scientific results lighting a fuse on dogmatic political pronouncements reached its inevitable and illogical conclusion in January 2002, when *New Scientist* magazine — a newsmagazine about science, not a peer-reviewed journal — reported on a remarkably potent adult stem cell isolated from human bone marrow by Catherine Verfaillie of the University of Minnesota. Do No Harm, an antiabortion group of doctors founded by David Prentice, immediately issued a press release hailing the research. Antiabortion senators like Sam Brownback pointed to Verfaillie's findings as offering an alternative to embryonic stem cells. I even heard Leon Kass, head of the President's Council on Bioethics, confide to a fellow panel member that the Verfaillie research genuinely looked like the answer to the ethical dilemma posed by embryonic stem cells. All this, and *Verfaillie's work had not even been published in any peer-reviewed scientific journal.* Indeed, Verfaillie later sent a letter to U.S. senators to set the record straight. If, as many have argued correctly, the ethics of stem cell research are too important for ethical decisions to be left to the scientists, it is equally true that the science of stem cells is too important for scientific judgment to be left solely in the hands of politicians, bioethicists, and ideologues.

⚘

Truth be told, there's been a long-standing, vigorous intellectual rivalry between embryonic stem cell researchers and adult stem cell researchers ("There's definitely two camps, no doubt about it," one company research director told me). But among many scientists, it has become almost politically incorrect to speak with unguarded enthusiasm about the progress of the adult stem cell story — not because the research isn't exciting, but because such praise has inevitably provided ammunition to the right-to-life opponents of embryonic stem cell research. Again, an essential scientific dialogue has been polluted by politics.

In this regard, a particularly curious figure in the stem cell and cloning debates has been David Prentice, the biology professor from Indiana State University. He is a quiet, self-possessed, pleasant-looking young man with glasses, a sharp nose, and, occasionally, a self-satisfied smile. He looks for all the world like someone who enjoys the newfound attention, and he received plenty during the spring and summer of 2001. He testified at a House subcommittee hearing in July, for example, waving a long list of scientific citations that, he asserted, represented proven human therapies using adult stem cells (virtually all, in fact, were bone marrow transplants). The underlying message was that adult stem cell research had progressed so far that there was no need even to consider embryonic stem cells.

Prentice was a cipher to virtually all the scientists I've talked to — no one seemed to be familiar with his work. His university Web site describes him as "an internationally recognized expert on stem cell research," but there is not a single stem cell paper to his name in the vast PubMed database of scientific publications. He had applied for an NIH stem cell grant, he told me, but it was not funded. His breezy summations of research by other scientists triggered both anger and disbelief. "What do they know that we don't know?" Irving Weissman demanded at one point. It might be funny but for the fact that, far from being a disinterested scientific observer, Prentice served as the principal biomedical consultant to leading opponents of stem cell research and cloning in Congress, including Kansas senator Sam Brownback and Florida congressman David Weldon.

Brownback and other elected officials used the results of the Tulane and St. Jude Children's Hospital groups, for example, to argue that adult stem cells are so potent and versatile that there's no need to destroy embryos to get ES cells, and thus no need for the government to provide funding for embryonic stem cell research. Almost every major adult stem cell researcher with whom I have spoken — Ed Horwitz, Diane Krause, Neil Theise, and Darwin Prockop, to name a few — felt their research had been misrepresented by politicians.

Despite all the grousing from the scientists themselves, adult stem cell research became an important chip in the political debate over embryonic stem cell research. And as the argument over stem cells moved into the White House during the early months of 2001, adult stem cells were used as more than a tool of political leverage. They became a critical component in a legal argument waged as part of a lawsuit against the government filed in the spring of 2001 — a lawsuit that had roots in the antiabortion movement, a lawsuit with friends in high places, a lawsuit that, in seeking to have federal funds for embryonic stem cell research declared illegal, may have been a stalking-horse for a new administration hoping to dodge a controversial issue during its first few months in office.

14

THE SNOWFLAKE

INTERVENTION

ON JANUARY 26, 2001, one week after he took the oath of office, President George W. Bush engaged in a brief question-and-answer session with reporters at the White House. Bush had been asked a kind of apples-and-oranges question: What did he plan to do about federal funding for fetal tissue and embryonic stem cell research? The president said, as he had during the campaign, that he did not support the research, and White House aides later told reporters that Bush was signaling his intent to block federal funding. But block what exactly? There's a subtle difference between fetal tissue and embryonic stem cell research, at least to the people who do it, and the president voiced his opposition in a way that left room for biological confusion. "I believe we can find stem cells from fetuses that died a natural death," he said, "but I do not support research from aborted fetuses." Embryonic stem cells do not come from fetuses; in fact, they cannot be found in fetuses, aborted or otherwise. Presidential spokesperson Ari Fleischer later clarified, with proper precision, that Bush "would oppose federally funded research for experimentation on embryonic stem cells that require live human embryos to be discarded or destroyed." By the end of a remarkably polemical summer, millions of Americans, including Bush, would have a much firmer grasp of the issue.

Until September 11, 2001, the debate over stem cell research improbably became the defining issue of George Bush's first year in office. Following his decision to allow federal funds for limited research in August, the White House painted a portrait of a keenly interested, intellectually curious president who had been "agonizing" for months over the issue, who had picked the brains of scientists, bioethicists, theologians, and patient groups, who had weighed important competing claims — the destruction of potential human life versus the potential creation of new human medicines — and who finally arrived, at the end of a long summer of soul-searching, at a moral compromise worthy of Solomon, a decision both wise and uninflected with political considerations.

There is certainly some truth to that, but among those with their noses pressed up against the windows of this notoriously tight-lipped White House, there is another, less aggrandizing version of the events leading up to Bush's August announcement. The president's initial inclination, in this view, was to ban all federal funding for embryonic stem cell research, as his principal political adviser, Karl Rove, had repeatedly urged him to do, according to numerous press reports. Indeed, during its very first week in office, the Bush administration had immediately flashed its right-to-life credentials, curtailing foreign aid payments to family planning groups that promoted abortion and moving to review federal approval of RU-486, the "morning after" abortion pill. Patient-advocacy groups, scientists, and other proponents of stem cell research had held their collective breath in the face of persistent rumors suggesting that the president would in fact issue an executive order banning federal funds for embryonic stem cell research altogether. To conservatives, that would have offered a satisfyingly symbolic political symmetry to the actions of Bill Clinton, who during his first week in office in 1993 had issued an executive order lifting a long-standing ban on government funding for fetal tissue research. And although a preemptive Bush research ban would have prompted howls of protest from stem cell proponents, those proponents privately concede that it would have made political sense.

In fact, you could say the scientific community was primed for disappointment. Geron and its subsidized collaborators at the University of Wisconsin and Johns Hopkins had been free to experiment with pluripotent stem cells since the fall of 1998, to their considerable scientific and commercial advantage, but everyone else had essentially been sitting on their hands for two years, and the pace of the NIH review process left many scientists feeling intensely frustrated. Alan Trounson, Martin Pera, and Ariff Bongso, the Australia-based research team, had isolated several promising new embryonic stem cell lines, but literally couldn't *give* them away until NIH guidelines were in place. Roger Pedersen, the UCSF embryologist who had applied to the NIH to use the Australian cells, felt so thwarted that he quietly began to lay plans to move his entire laboratory operation to England. Other scientists bailed out on the NIH altogether.

In early 2000, for example, well before the presidential elections, Doug Melton, the Harvard developmental biologist, had turned to the Howard Hughes Medical Institute, an enormously well endowed and influential research foundation that funds top-drawer biomedical research at American universities, seeking support for his research on pancreatic cells derived from human ES cells. Melton felt doubly frustrated, as a scientist and as a patient advocate; his diabetic son Sam had endured another two years of insulin shots and other medical care, while minimal progress had been made on the possible role human embryonic stem cells might play in treating that disease. The Hughes institute began to formulate a plan about how it might fund human embryonic stem cell research outside the purview of the NIH. "Both we and Harvard thought that this was not only something that could be done," said Hughes director Thomas Cech, "but something that we really felt we had a responsibility to support. As a nonprofit, nongovernmental agency, our role in American biomedical science, we think, is really to be adventuresome, to do the things that are not yet easy to do, but that might have a huge health benefit down the line. And we came to the conclusion that this was the right thing to do from the ethical point of view — in fact, that it was the *only* ethically correct thing to do. That it would have been incorrect *not* to support Melton's work. And therefore we decided to fund his derivation of the cells."

Reflecting yet another variation on the theme of scientific frustration, Ali Brivanlou of Rockefeller University had traveled coast to coast, seeking some way of breaking the logjam in his own research. He talked to venture capitalists. He talked to Geron. He discussed a possible stem cell initiative with a private biomedical organization near San Diego called the Burnham Institute. He considered relocating his laboratory, and family, to England, where stem cell research was legal. All the while rumors continued to fly that Bush would ban all federal funding for the research.

One of the earliest skirmishes in the public campaign began when lawyers contacted leading stem cell researchers in the United States and Australia in the spring of 2001 and asked them if they wanted to sue the National Institutes of Health and the U.S. government. For researchers whose very reputations and livelihoods depend on government funds, this was like asking them all to bite, in unison, the hand that fed them. Still, with little hesitation, the scientists — James Thomson, John Gearhart, Douglas Melton, Roger Pedersen, Alan Trounson, and Martin Pera — said yes. One of the reasons they consented was their long-standing frustration;

in Pedersen's view, for example, this general avenue of research had been blocked for more than twenty years. But another reason was that the effort was being organized by a well-connected Washington lawyer named Jeff Martin. Martin served on an NIH planning committee and had done business with Tommy Thompson, a member of the Bush cabinet. Within a few weeks, he would be suing both of them.

≈

In February 2001, Jeffrey C. Martin was on an airplane headed to San Francisco, to deliver one of the most optimistic and, in some respects, most politically naive speeches to be heard during the annual meeting of the American Association for the Advancement of Science. Martin didn't belong to the tribe, but he sat on the Parkinson's Disease Implementation Committee at the NIH. It was there that he came to know Ron McKay, a biologist in the National Institute of Neurological Disorders and Stroke. Since the early 1990s, McKay's lab had published a series of groundbreaking papers demonstrating how mouse embryonic stem cells could be prodded into becoming powerful specialized nerve cells, including precisely the kind of dopamine-producing neurons that are destroyed in Parkinson's patients. But, like so many other government-funded stem cell researchers, McKay had been unable to translate this tantalizing body of basic research into a test of applied human medicine because of the continuing controversy about human embryonic stem cells. During their committee work, McKay came to appreciate Martin's understated intelligence. "He's very *clever*, Jeff," McKay told me one day. His voice was inflected with surprise as he said this, as though the remark included an unstated preface, something like "You'd never think so at first, but he's very clever, Jeff." I later realized that Martin used underestimation as a tool of his trade.

The previous fall, McKay had asked Martin if he would be willing to speak at the 2001 annual meeting of the AAAS. Martin agreed, and so it was that on February 16, Martin launched his PowerPoint presentation with a slide of Salvador Dalí's painting *The Persistence of Memory*. His talk was entitled "Solving Parkinson's Disease During the George W. Bush Administration: Stem Cells and Other Promising Therapies," and it sounded overly optimistic precisely because all signs back in Washington seemed to be pointing in the opposite direction. While most scientists remained deeply skeptical about the Bush administration's intentions, however, Martin had reasons, both political and personal, for holding a more sanguine view. First, he honestly believed that Bush would turn out to be the compassionate conservative he said he was. Second, he knew that Thompson, the new secretary of health and human services, shared a strong commitment to stem cell research. Indeed, Martin had joined a group of Wisconsin businessmen who had had lunch with the former Wisconsin governor earlier that same week, on February 14, and he emerged from that encounter convinced that at least one member of the Bush cabinet was an ardent supporter of stem cell research. "I made the case," he said of his speech, "that, although there was a lot of concern based on his statements in the campaign, ultimately he [the president] would be supportive of stem cell research."

He would soon have reason to change his mind. Even as Martin was delivering his remarks in San Francisco, a high-powered Washington law firm, assisted by a Virginia-based right-to-life group called Human Life Advocates, had mobilized to prepare a preemptive lawsuit against the NIH. Despite the election, and despite Bush's "culture of life" philosophy, the National Institutes of Health had continued to implement the Clinton administration's stem cell research policy as if nothing had changed — indeed, as if American biomedical research policy should proceed unaffected by religious passions or changing political winds. Much had happened, on paper, over the previous two years, and yet very little had changed in the laboratory.

In January 1999, Harriet Rabb, general counsel of HHS, had written a legal memorandum concluding that stem cells are not embryos and that therefore federal funds for embryonic stem cell research would comply with existing federal law against the destruction of human embryos, as long as researchers used stem cell lines that had been derived from embryos by other, privately funded researchers. This was a narrow legal definition that, rather than settling the ethical and social divisions, wriggled between them; as long as privately funded researchers destroyed the embryos, NIH-funded researchers were free to experiment with the fruits of that destruction. "No matter how much one hates disease and loves embryonic stem cell research," said Erik Parens, a bioethicist at the Hastings Center, "one can see that this ruling was a legalistic end run around the spirit of the congressional ban." Nonetheless, with that opinion in hand, NIH director Harold Varmus began to prepare scientific and ethical guidelines by which the

NIH would fund the work. This protracted and contentious process, buf-
feted and annotated by some fifty thousand citizen comments, resulted in
the official NIH guidelines, which were issued in August 2000 — just in
time to be politically irrelevant. The presidential campaign was heating up,
and federal policy would very much depend on who won. As soon as the
final guidelines were made public, Sam Brownback attacked the NIH policy as "illegal, immoral and unnecessary." That phrase would echo again
and again during the entire debate.

Nonetheless, throughout that fall of electoral uncertainty, the NIH solicited applications for embryonic stem cell research, with an initial deadline of March 15, 2001, and a meeting scheduled for late April to assess the
applications. These plans continued, even though Thompson had asked
lawyers at HHS to analyze the legality of the Rabb memo, and had also on
February 28 ordered the NIH to conduct a sweeping scientific review of
stem cell research. Exactly one week before the NIH deadline, however, a
group calling itself Nightlight Christian Adoptions filed a little-noticed
lawsuit in U.S. District Court in Washington, D.C., seeking to declare the
entire bureaucratic process — the Rabb decision, the NIH guidelines, the
imminent assessment of grant applications — illegal. The Nightlight lawsuit also sought an injunction preventing the NIH from funding embryonic stem cell research.

Based in Fullerton, California, the Nightlight group had initiated a
program in 1997 to find homes for the estimated 100,000 frozen embryos
stored, and essentially abandoned, in freezers at in vitro fertilization clinics
throughout the country. This effort, known as the Snowflakes Embryo
Adoption Program, identified infertile couples who were willing to "adopt"
a frozen embryo; in practice, however, since embryos are not "alive" in
terms of adoption law, the organization arranged a contract between the
biological donors and the recipients. Each contracted embryo would be
thawed and implanted in the womb of a would-be mother, where, with
luck, it would grow to term. By the spring of 2001, the organization had arranged approximately seven "adoptions." As JoAnn L. Davidson, director of
the Snowflakes Embryo Adoption Program, later explained, the program
"derives its name from the idea that snowflakes, like human embryos, are
frozen, unique, and cannot be re-created." The organization's Web site
shows a smiling, bearded figure embracing an infant, alongside a quotation
from Corinthians celebrating "the light of the knowledge of the glory of
God in the face of Jesus Christ."

Arranging the adoption of frozen embryos was an innovative way to help infertile couples, even though it gave bioethicists like Leon Kass pause. In this case, however, the help came in a political and religious package; indeed, conservative fingerprints were all over this lawsuit. Human Life Advocates was the Christian Legal Society's public interest law firm, and the lawsuit was funded by the Alliance Defense Fund, an organization that describes itself as devoted to "family values, religious freedom and the sanctity of human life." In stating the case for its grievances, the Nightlight group and its fellow plaintiffs trotted out some familiar conservative arguments against stem cell research. The Christian Medical Association, a group of fourteen thousand religious health professionals that joined the suit, claimed harm because the NIH policy would force it to divert funds from worthy educational and research purposes to a costly public campaign against stem cell research. Another plaintiff was David Prentice, the Indiana State University biologist, who claimed harm because his applications for NIH grants for adult stem cell research would unfairly face "increased competition for limited federal funding" from embryonic stem cell researchers. Perhaps the most ironic part of the entire legal action was that it purported to be a class-action suit on behalf of all adult stem cell researchers — even as those very researchers were privately complaining that their work had been recklessly misinterpreted by religious conservatives. All these claims were framed within a larger complaint as much political as legal: that the NIH policy on stem cell research was developed, in the words of the lawsuit, "in an arbitrary and capricious manner" and violated the terms of the Dickey-Wicker amendment, the congressional budget rider that forbade funding for research in which embryos were destroyed.

Although some have argued that the stem cell and cloning controversies are not an encore contretemps of the long-running abortion debate, the Nightlight Christian Adoptions lawsuit sets the record straight. The court filings specifically itemize the long history of moral tussling and legal confrontation over the status of the embryo in the Background portion of the complaint, a lineage running all the way back, in an unbroken thread, through the Dickey-Wicker amendment of 1996, the repudiated report of the NIH's 1994 embryo research panel, and even further, to the initial battles over in vitro fertilization in the 1970s and the phantom Ethics Advisory Board, and the 1973 *Roe v. Wade* decision of the Supreme Court. Abortion politics was the river running through this entire, contested domain. As if to underscore the lawsuit's political patrimony, the press conference held to announce it was "hosted," according to the newsletter of the Tennessee Right to Life Organization, by Brownback, the Kansas senator who had emerged as the leading Senate voice in opposition to both embryonic stem cell research and cloning. Samuel Casey, executive director of the Christian Legal Society and an attorney on the case, echoing word for word the comments Brownback had made about the NIH guidelines the previous August, said at that press conference, "Destroying living human beings for their stem cells is illegal, immoral and unnecessary." That was the conservative mantra — illegal because of Dickey-Wicker, immoral because destroying embryos was a form of killing, and unnecessary because adult stem cells were just as good.

ᑌᔑᑎ

On March 8, the day the Nightlight lawsuit was filed, Jeff Martin was on his way from New York to Philadelphia, accompanying his eldest daughter on a round of auditions at music conservatories she was considering for graduate school. It had already been a discouraging day — the *New York Times* had a front-page story that morning on some disappointing results of a fetal-cell transplant technique to treat Parkinson's disease. But when Martin received word of the Snowflake lawsuit by e-mail, he became especially concerned. It wasn't the plaintiffs that concerned Martin, or even the arguments made in the lawsuit. It was the attorneys.

They came from the Washington law firm of Gibson, Dunn & Crutcher — "a law firm," Martin noted with characteristic understatement, "which was very well connected to the Bush administration." Thomas G. Hungar, the lead attorney for the Nightlight group, was a highly regarded appellate attorney who had just played a major role helping his mentor, Theodore Olson, argue the Bush campaign's side of the historic 2000 Florida election dispute before the Supreme Court. *That* connected. But Hungar was also affiliated with Human Life Advocates, the Virginia-based right-to-life legal group that had been preparing the Snowflake lawsuit since the previous fall, in the event that Al Gore won the election. And at this point, perhaps, mixed messages coming out of the administration increased the odds of legal action. While stem cell proponents feared a preemptive Bush ban, stem cell opponents became increasingly restive about Tommy Thompson, who had continued to encourage applications for NIH research funds and had even suggested, in congressional testimony in early

March, that the Dickey-Wicker language was "troublesome." "Secretary Thompson, at least in the press, made statements that seemed in tension with the president's views," Hungar recalled. And that, he said, led to the filing of the Snowflake lawsuit.

"The identification of the law firm gave us cause for concern as well," Martin told me. "My concern was that the administration might in fact agree with the lawsuit. I wouldn't call it collusion, but if they in fact agreed with it . . ." His voice trailed off, but he didn't need to complete the thought. He wouldn't say collusion, but that's exactly what he and his legal colleagues thought. The fear was that the Nightlight lawsuit, as Martin put it in an e-mail to a scientist, might "be a vehicle for the administration to cave," an opportunity for government lawyers to simply agree with the Nightlight argument, in effect declaring the NIH stem cell policy illegal and sparing the president involvement in a no-win political controversy. In fact, Jeff Martin was surprised they didn't. "Based primarily on statements to Congress and press statements, there was sort of a fairly strong hint that the administration might decide that because of the Dickey-Wicker amendment, it really didn't have a policy decision to make, that it was going to take a different legal position from Harriet Rabb. And that might have been a politically astute thing to do, in that you could say, 'Well, we're just reading the law more correctly, or more literally.'"

But what really made the Nightlight lawsuit look ominous to Martin was how it seemed to dovetail with rumors about the legal thinking inside the administration. In the wee hours of that same Thursday morning, Martin had received a disquieting e-mail from a prominent member of the business community (Martin showed me the letter, but asked that the correspondent remain anonymous). The e-mail summarized a telephone conversation the business leader had had the previous day, March 7, with Tommy Thompson. "Friendly, frank, and sobering," Martin's correspondent said of the conversation. Thompson was "strongly in favor of ES research," but indicated "it is against the law. 1999 interpretation" — referring to the Rabb opinion — "was made under duress and in his lawyerly opinion is wrong." This was the same sentiment Martin himself had heard several days earlier when he had spoken to two midlevel White House officials. The e-mail concluded: "My take away: we are done this round."

Against the law . . . The phrase stuck in Martin's professional craw. "The fact is," Martin told me nearly a year later, "his conversation with

Thompson was so stark in saying that stem cell research was illegal that it really made me get off the dime and say, 'We have to do this.'" And so at 2:42 A.M. on March 8, in replying to the e-mail of his business confidante, Martin first hinted at the legal crusade that would consume his spring and summer, that would form a small but crucial tile in the mosaic of public support for stem cell research that built over the summer, and that, in concert with the efforts of many other grassroots soldiers, would improbably influence the most pitched public-policy debate about an area of scientific research since the recombinant DNA controversy of the 1970s.

Martin, a soft-spoken but, to his adversaries, no doubt irritatingly persistent man, decided to sue his friend Tommy Thompson, and his colleagues over at NIH. "The purpose of the lawsuit," he explained later, "was to try to take that legal issue away and make the president decide on policy grounds."

↬

When I first went to visit Jeff Martin at Shea & Gardner, where he occupies an eighth-floor office overlooking nearby Dupont Circle, I found myself traveling on one of the few untrampled paths in the stem cell story. During the social and political debate that unfolded during 2001 over the government's embryonic stem cell policy, precious little attention was paid to Jeff Martin's activities, both legal and lobbying, and it was easy to understand why. When I went to meet him, he was sitting at his desk, slightly hunched at the shoulders, without much expression. His low voice had a kind of quakiness to it, his face — plain, midwestern — seemed nearly without affect. At the time, I thought that his impassive demeanor was the physical embodiment of lawyerly caution.

Martin does not possess the bloodlines or résumé of your typical Washington powerbroker. He was born in 1953 in Columbus, Ohio, and raised in a middle-class milieu; his father — the first member of the family to obtain even a two-year college degree — was a technical engineer for RCA. "My father kept getting transferred; they kept closing down plants," Martin says of familial dislocations to Rochester, New York; Memphis, Tennessee; and finally Indiana. Martin studied philosophy and political science at Indiana University, and obtained his law degree in 1978 from the University of Chicago, where he was articles editor of the law review. He clerked for Spottswood W. Robinson III, a famous civil rights lawyer and later a federal circuit judge. By 1985, he had made partner at Shea & Gardner.

Aside from one youthful indiscretion in 1972, when he voted for George McGovern, Martin was a Republican through and through — he took a leave from the law firm in 1991 to serve two years as general counsel for the Department of Education during the first Bush administration. But his lawyering was only a half-time pursuit, because he wore an unusual second hat: Martin is a senior vice president in charge of government relations for Saks, the retail company that owns various department stores around the country, including the flagship Saks Fifth Avenue store in New York (Martin's brother Brad is the chairman and CEO of Saks, and his other brother, Brian, is the Saks general counsel). In other words, he was an experienced Washington lobbyist as well as a respected lawyer.

But what really qualified Jeff Martin to enlist in the stem cell wars roiling the capital was something involuntarily added to his résumé in the fall of 1997. "In November or December," he told me, "I first noticed a tremor when I put in my contact lenses. My left arm and my left hand would often shake." He asked his wife for a copy of the *Merck Manual*, a handbook of medical conditions and treatments, for Christmas that year, and reached a tentative diagnosis by himself. "I ended up reading about Parkinson's disease," he said, "and noticed the reduced arm swing, which also was something I had noticed, and the reduced facial expression, which is something my wife had noticed. And so in January, I went to my regular internist and I said, 'I think I have Parkinson's disease.' And after he examined me, he said, 'I think you have Parkinson's disease.' And then he sent me to a neurologist, who said, 'You have Parkinson's disease.'" Martin was stunned. He was forty-four years old at the time — unusually young for a disease that more typically strikes people in their late fifties or sixties. For a year, he absorbed the gravity of having a progressive, irreversible neurological illness. Then he flung himself into the "cause," as he puts it, joining the NIH treatment committee, assisting patient-advocacy groups, consulting with organizations like the Michael J. Fox Foundation, and lobbying. "He was told as a relatively young man that he has this disease," Ron McKay told me, "and he's dealing with it in a very interesting way."

Martin's work for Saks had introduced him to Tommy Thompson, and a series of back-channel contacts with the secretary throughout 2001 confirmed for Martin just how difficult the stem cell battle was going to be.

On Valentine's Day, for example, two days before his speech at the AAAS meeting, Martin received a call from the head of the Wisconsin Merchants Association, Chris Tackett, who was in Washington. "He asked me if I wanted to go see Tommy with him," said Martin, who eagerly joined the group for lunch. "Tommy — Secretary Thompson — was new to the office, and he knew me from the business world. So we talked fairly generally about stem cells, and it was clear to me that he was personally supportive, but saw it as a very difficult thing to win in the administration."

The lame-duck administration at NIH was proceeding with stem cell research as if the Clinton policy and guidelines were still in effect. But there were plenty of warning signs of a change in political weather. The Nightlight Christian Adoptions lawsuit was filed exactly one week before the NIH's March 15 deadline to receive applications for stem cell grants. Only three researchers had applied: the two Australians, Alan Trounson and Martin Pera, who wanted money to distribute several cell lines to fellow researchers, and Roger Pedersen of the University of California at San Francisco, who hoped to conduct research using the Australian cells. Continuing a pattern dating back at least twenty years, researchers interested in anything hinting at reproductive medicine or embryonic research simply didn't bother coming to the NIH for funding.

But in observing the bureaucratic rituals of public notification, NIH officials managed to divine some of the Bush administration's intentions. The NIH had scheduled a meeting on April 25 to consider the stem cell applications; according to federal regulations, officials had to place a meeting announcement in the *Federal Register* by April 10. By that time, Jeff Martin, already plotting legal action, had been in touch with acting NIH director Ruth Kirschstein, seeking to learn the names of people who had applied for funding in hopes of approaching them as potential plaintiffs. But as the draft of the *Federal Register* announcement wended its way upward through the HHS bureaucracy, it finally tripped a political wire. Word came back to NIH that the meeting had to be postponed. The "postponement" was never officially announced. Martin learned about it in a phone call from Kirschstein, and it became public on April 13, when the *Wall Street Journal* reported that the meeting had been abruptly canceled. The official reason, according to Thompson's office, was that HHS had not yet completed its review of stem cell policy.

The cancellation came as no surprise to the Australians. "We expected that to happen all the way along," Alan Trounson said, even though he and Martin Pera had gone through the motions of booking airplane flights and hotel rooms. "Martin and I just said, 'Well, *riiiight . . .*'"

ᘓ

While NIH administrators wandered in a policy fog, Martin began to assemble his lawsuit. In mid-March, he had approached the Coalition for the Advancement of Medical Research (CAMR), a newly formed group of patient advocates, academic institutions, and biotechnology companies, seeking funds for a lawsuit. "They didn't have the funds," Martin said, "and they didn't really want to get into the law. Some people actually pledged some money, but I ultimately decided that it was taking more time to raise the money for the lawsuit than I wanted to spend. So I went to my firm and said, 'Can I handle this on a pro bono basis?' And I got that permission."

Other lawyers, almost all with a personal connection to diseases that might be affected by stem cell research and all working pro bono, joined the cause. The mother of a diabetic lawyer in Chicago had written an unsolicited letter to the Juvenile Diabetes Research Foundation, volunteering her daughter for legal services on behalf of the stem cell crusade; Larry Soler, head of the foundation (and chairman of CAMR), passed the letter on to Martin, and soon Julie Furer of Schiff, Hardin & Waite was on the team. Four other lawyers from her firm with a special interest in diabetes pitched in, including Fred Sperling, best known as one of Michael Jordan's lawyers. Clarence T. Kipps, Jr., a longtime Washington lawyer at Miller & Chevalier, who had represented the Lockheed Corporation for many years in government contracts, agreed to help, too; his wife has Parkinson's.

Martin, Furer, and Kipps then rounded up a blue-chip collection of aggrieved parties to defend. "We looked for plaintiffs who were interested in participating in the legal issue, who had credibility as scientists, and who had standing," Martin explained. "And in particular the reason we went to the Australians was that they were in a rather unique position in that they had actually applied for approval of their stem cell lines under the prior guidelines. And that helped in terms of the potential standing objection to other researchers." So Trounson and Pera were on the list. So, too, was almost every groundbreaker in the stem cell field: Jamie Thomson, the University of Wisconsin scientist who first isolated human embryonic stem cells, and his colleague Dan Kaufman; John Gearhart, the Johns Hopkins biologist who was the first to isolate human embryonic germ cells; Roger Pedersen of UCSF, who had applied to use the Australian cells; and Douglas Melton, chairman of the Molecular and Cellular Biology Department at Harvard. Although Melton had independently obtained funding through the Howard Hughes Medical Institute to derive his own human embryonic stem cells, his participation was sought, Martin said, "because he added a lot of weight to the scientific team."

Trounson stressed a political point when he explained his motivation for joining the lawsuit. "There's a public interest that I think has to be considered," he said. "The current problems with IVF in the United States, and they're substantial, exist because there is no public funding at all, because all the NIH scientists were prevented from doing any work. Anyone who was funded by NIH was not permitted to do any work in this area. So currently, research in IVF is driven by private IVF clinics. It's market-driven research, and it results in large numbers of embryos being put back into patients, and a lot of multiple births, and a lot of fetal reduction. That's not the case in Australia and other places, where the public interest has been met by public funding. So we eventually joined that litigation against the administration basically because we felt that it was terribly important for NIH and the U.S. public researchers to be involved."

Rounding out the list of plaintiffs were several well-connected nonscientists: a Chicago executive named James Tyree, who suffered from diabetes, and a Pittsburgh businessman named James Cordy, a Parkinson's patient who made a lasting impression on Senator Arlen Specter of Pennsylvania, among others, by traveling around with an hourglass — a reminder that political and bureaucratic delays had real consequences for the ill. "Then, to get a little cachet into the case," Martin said with a sly smile, "we asked Christopher Reeve, and he agreed to participate." So, in the course of several weeks, they had lined up the Embryonic Stem Cell All-Stars, patients with political juice, and a man not unfamiliar with superhuman feats of strength, which is what they all thought it would take to change the mind of a White House on record against any research that would require the destruction of human embryos.

At first, Martin and his colleagues intended only to file what is known as an intervention in the Nightlight lawsuit. In an intervention, a third party essentially barges uninvited into an ongoing lawsuit, asserting that its interests are also at stake in the dispute and that the defendant (in this case, the government) might not be concerned about protecting those interests in any subsequent settlement. There was a precedent for concern: Tony Mazzaschi, vice president of the Association of American Medical Colleges, recalled that an animal rights group sued the Department of Agriculture, seeking to include rats and mice under the protection of the federal Animal Welfare Act, which would dramatically change the rules by which laboratory animals could be treated; the USDA agreed to an out-of-court settlement, threatening to paralyze scientific research. "We moved to intervene right after the cancellation of the April NIH meeting," said Martin, who filed papers on May 2. The intervention was "not a particular surprise" to Hungar, who said, "We knew there were people who had a financial interest in the Clinton policy continuing."

But a funny thing happened the very next day. The Nightlight attorneys and the government agreed to stay their case — essentially put it on hold. The government said it wouldn't fund any stem cell research until at least thirty days after the Department of Health and Human Services had completed its review of the stem cell issue. Antiabortion forces hailed the agreement as an important victory. "The stay prevented any funding for stem cell research to go forward," said Hungar, "and was important and helpful in providing some breathing room for the administration to reach a decision." Martin believes the timing was a coincidence, but his colleague Julie Furer interpreted the government's agreement to stay the case as "a fancy way of saying we agree to do nothing." The stay, approved by the federal judge Royce C. Lamberth on May 4, had two practical consequences. It marginalized stem cell proponents from any legal argumentation in the Nightlight lawsuit, since it was now on hold, and, more important, it turned off the clock. Without any ticking judicial deadlines, the administration could take as long as it wanted to complete its policy review, and as Martin said in his San Francisco speech, "Speed counts." "We believe it was a setup," Furer said. "We believe the department never intended to defend itself in that suit."

That little hiatus in activity lasted exactly one weekend. Martin and his legal colleagues, anticipating a response from the Nightlight lawyers to their intervention, had already prepared their legal rejoinder. When Judge Lamberth ordered the stay on a Thursday, they simply recast these arguments as an independent lawsuit over the weekend and filed their papers in federal court the following Monday, May 8. "That forced their hand," said

Furer. "They couldn't just sit back and do nothing." It also, Martin said, "gave us control over our own destiny in terms of making motions and proceeding with the case if we wanted to." The clock was ticking again. The government was now compelled to respond to the scientists' lawsuit within sixty days, or by July 13. On the docket now was a case with a rather coy title: *Thomson* [James] *et al. v. Thompson* [Tommy] *et al.* Maybe it was a coincidence, but Thompson had lunch with the president on May 8 and learned for the first time that Bush was open to something other than a complete ban. Thompson recalled in an interview, "I spun pretty hard that day!"

To put this in perspective, Harriet Rabb, formerly the top lawyer at HHS and now general counsel at Rockefeller University, pointed out, "The fact is, the government and HHS are sued so many times, every day, every week, that it is an extremely rare occurrence that litigation generates policy." Nonetheless, she added, "the lawsuit can become an organizing vehicle" for political groups. In this case, the "vehicles" took everyone on a wild summer ride.

‍ఌ

Behind-the-scenes lobbying and lawyering notwithstanding, the public debate over embryonic stem cells during the summer of 2001 was a thing to behold. Obscure embryological terms — blastocyst, primitive streak, inner cell mass — improbably became the lingua franca of congressional hearings, radio talk shows, and newspaper columns. The euphemistic inventiveness of the right-to-life movement transformed embryos into "preborn children," "embryonic babies," and even "microscopic Americans." The word that continued to come out of the White House to describe the president's deliberations was *agonizing*. Bush and a small group of advisers — usually Karl Rove, adviser Karen Hughes, legal counsel Jay Lefkowitz, and occasionally Andrew Card — met with a steady stream of scientists, bioethicists, patient-advocacy groups, and religious leaders, but the auguries were so grim that it seemed CAMR had committed itself to, in the words of one board member, "a real loser of an issue." Rove, according to a story that summer in the *Los Angeles Times,* presided over a weekly telephone conference call with Catholic intellectuals, who firmly opposed any form of human embryo research, including the investigation of stem cells. Lefkowitz, general counsel in the Office of Management and Budget, who played a key role in shaping stem cell policy, brought an unpromising résumé to the task; his nickname as a litigator was Viper, according to a profile in the *Washington Post*, and he cofounded (along with commentator Laura Ingraham) the Dark Ages convention, the conservative rejoinder to Clintonian Renaissance weekends. People who met with him say he took a keen interest in the issue but probably didn't stray too far from the president's "culture of life" position.

For both proponents and opponents of stem cell research, the uncertainty surrounding the Bush administration's intentions made for a vertiginous, unnerving summer full of lobbying, litigating, and politicking. While Rove continued to push for a total ban on the research as a way to satisfy the president's conservative base, he had to go toe-to-toe with Tommy Thompson. No one who spoke with Thompson during that period doubts that he was firmly in favor of stem cell research and that he pushed hard to keep the door open. "There's no question that I was out farther than the White House on this," he said. But some people later questioned how influential Thompson actually was in the internal debate. One observer, who asked not to be named, said, "This thing was run by the White House. It wasn't run by the Department of Health and Human Services, and explicitly so. The Department of Health and Human Services was not in the loop." But Jeff Martin, who was like a ganglionic relay station for Beltway rumors, gossip, and tidbits of intelligence, passed along a story about a meal Thompson had at the White House with Bush and Karl Rove. At one point, Thompson argued that the president's decision on whether or not to allow stem cell research would form a crucial part of his legacy. If he opposed it, Thompson warned the president, "virtually everybody in this country has a loved one who's got an illness that could be benefited by this, and they won't forget it, politically." When Rove attempted to interrupt Thompson and remind the president of his political base, the president held up his hand and said, "No, I want to hear this . . ."

During that summer, stem cell proponents "had the feeling that a decision could come down any day," recalled Lawrence Soler, head of CAMR. "That's what made it so difficult." Soler opened the weekly conference call of CAMR on June 7 with disquieting news: multiple sources inside the administration were telling CAMR that the White House was poised to ban federal funding for stem cell research outright. "We had heard from two or three separate sources within the White House that a ban was coming, and was coming imminently," Soler confirmed in an interview. "We really thought it was going to happen." It turned out to be a false alarm, but it served a useful purpose: CAMR immediately launched a massive grassroots advocacy effort, which resulted in so many telephone protests that it jammed the White House switchboard. People within the administration called CAMR, Soler said, and "we were asked to call off the dogs." Mazzaschi later boasted of an electronic grassroots network that could instantly send out ten thousand e-mails to patient groups.

In a quintessentially Washington kind of way, Jeff Martin — the well-connected, self-effacing, low-key broker of information, legal advice, lobbying expertise, and patient advocacy — had a finger in several pies. Working with several patient-advocacy groups, he identified stem cell proponents who were friends of key congressmen and then helped to set up meetings. He lobbied Republican congressmen he knew through his government relations work. On one occasion, he even urged fellow advocates to tell the staff of a prominent Republican senator that they better talk to their boss because he was telling friends that he was a lot friendlier to stem cell research than even his own staff suspected. If the president sent a hand-written note to a friend expressing his opinion on stem cell research, Martin not only knew the contents of the letter but was sharing it by e-mail with stem cell proponents within hours.

Perhaps the most unusual aspect of Martin's lobbying effort was his occasional contacts over the summer with a nominal defendant in the case, Tommy Thompson. On June 12, for example, Saks held a board meeting in Washington, D.C., which included a dinner that evening in a private upstairs room at Georgetown's 1789 Restaurant. "We invited Secretary Thompson to come and say a few words at the dinner," Martin said, "and he came." Now, it might seem a little odd that an attorney would speak to a defendant in an ongoing lawsuit, much less invite him to dinner. But Martin and Thompson had a history. Several years earlier, while Thompson was still governor of Wisconsin, Saks had plans to locate a regional headquarters in either Illinois or Wisconsin. Thompson had spent a good deal of time selling Martin on the virtues of Wisconsin, and Saks ultimately selected Milwaukee for its business center. So the two had done business before, but with Thompson as supplicant and Martin the petitioned.

Before he left 1789, Martin managed to buttonhole the secretary for a short but meaningful conversation. "He had said they were looking at a potential compromise," Martin recalled. "And I said, 'Well, limiting it to excess embryos could preserve the president's culture-of-life position.' He said, Well, in his view, it may be sufficient just to use the existing cell lines."

This first hint of a possible compromise wouldn't appear in the press for another three weeks, but Martin immediately e-mailed his scientist-clients, reporting the possible policy shift and soliciting reaction. He got an earful. At the time, there were only half a dozen or so human embryonic stem cell lines published in the scientific literature, and the biologists were nearly unanimous: that wouldn't be enough. Even at this early juncture, the number of viable cell lines was seen as crucial. Within a day, Martin was flooded with scientific objections, and on June 15 he sent a hastily composed two-page letter to Thompson, warning him — and the administration — of serious shortcomings in any policy that limited the research to existing cell lines. "There are serious problems that would be generated," Martin wrote to Thompson, "if the seven published human ES cell lines in the world today were the only ones that could be used in federally supported research." Martin's gang of scientists identified two main areas of concern — one arose "from the fact that different cell lines have different properties with different research and therapeutic potential," and the other from "bestowing an exclusive status of 'federally approved' to a few cell lines controlled by a few institutions who may have proprietary and commercial interests in those cell lines." Almost every objection enumerated in Martin's letter — the inadequate number of cell lines, their unfitness for clinical use, and severe constraints on their distribution because of intellectual property issues — anticipated by two months all the criticisms raised about the plan announced by the president on August 9. When asked if he recalled the letter, Thompson replied, "Sure do. Jeff is a friend and an adviser to me, so I take whatever he writes seriously."

I asked Martin if it was customary to have so much interaction with a defendant in a lawsuit. "Normally," he conceded, "you do not contact the adverse parties." But, he added, the contacts were few and brief. We later toted up all his interactions with Thompson during the time George Bush was wrestling with the stem cell decision. Aside from the Valentine's Day lunch, which took place before the lawsuit was filed, Martin estimated that his stem cell–related conversations with the secretary consumed less than five minutes over a six-month period. This was a Washington-style hourglass, and he made every second count. "We always had the sense for the first six months of the year that we were fighting an uphill battle," Martin said afterward, "but it was getting less uphill. We were making progress. But we were still behind."

❦

During the summer, the Juvenile Diabetes Research Foundation had arranged a series of meetings with administration officials, and the scientist they brought along was Douglas Melton, one of the plaintiffs in Jeff Martin's lawsuit. Both of Melton's children suffer from diabetes, of course, so he made an especially effective advocate. He met twice with Thompson, once with Vice President Dick Cheney, once with Karl Rove, and one time, on July 9, with the president and his advisers. Melton's thumbnail assessments of these meetings are illuminating.

Thompson, he said, made no secret of his support for the research. "In my two meetings with Tommy Thompson, he said point-blank he was very excited about stem cell research, was a very strong proponent of it, and wanted to do everything he could to let NIH investigators and others work on it aggressively," Melton said. "And in these conversations, he always presented it that I didn't have to convince him that this was an important area of science, that it was legally, morally, ethically justifiable, and that it was politically important to do. As the governor of Wisconsin, he was extremely proud of Jamie Thomson and had always been one of the most ardent supporters of this." So, Melton wondered, what is the problem? "The problem," Melton continued, "is the White House, not him. That's how it was always presented to me, though he would never have used the word *problem*. He would have said that the people in the White House have to be convinced, and he's doing everything he can, that he's on my side to get this through."

The meeting with Cheney was "really delightful." The vice president was "very sharp, a very curious mind, in the positive sense. And very interested in understanding the biology, and we had a very nice long discussion about the religious aspects, and how one could see creating stem cells as either a pro-life position or not." Melton's meeting with Rove was less encouraging; the president's political adviser seemed distracted and walked out of the room at one point. When he met with the president — with Rove, Hughes, Card, and Lefkowitz sitting in — Bush didn't tip his hand at all. "In fact, it was notable that I couldn't really understand what he thought the key issues were," Melton said. "And when I gently probed whether it was a moral-scientific decision or a political one, that got the biggest response from him. He said that it was absolutely not a political decision. That someone had come in and tried to show him polling numbers on this, and he threw them out of the office, that it had nothing to do with polling."

If Bush sneaked a peek at the poll results, he would have found little comfort in the numbers. Organizations like the Juvenile Diabetes Research Foundation had internal polling data as early as January showing that 65 percent of Americans supported federal funding for stem cell research. By midsummer, news organizations like the *Wall Street Journal* and ABC News did their own polling, "and their results supported ours," Soler said. In the ABC and Beliefnet.com poll, even a majority of Catholics — 54 percent — supported the research, while 35 percent were opposed. The public education campaign was going full bore, and the public seemed to be warming to the idea of the research. This is a point worth lingering on: when considering ethically nettlesome biomedical areas like stem cell research (or recombinant DNA in the 1970s, or "research cloning" in 2002), the initial public reaction seems to be a demographic version of the "wisdom of repugnance," with widespread revulsion and unease. But as concern over the issue assumes a certain critical mass of attention, and provokes a more sustained and in-depth public discussion, at least part of the public reaction typically evolves beyond repugnance and lands in a different place — usually a place that makes people feel more informed about, and thus more comfortable with, the new technology. It is probably no accident that efforts to ban research, such as the House of Representatives vote on cloning in July 2001, were hastily debated and rushed through; the more deliberate and reflective the conversation, the more likely people feel they can exercise social control over a controversial research enterprise.

During the summer of 2001, no newspaper was too small to editorialize on the subject. No columnist could resist sucking a thumb on the issue — William Safire in the *Times*, Richard Cohen in the *Washington Post*, Anna Quindlen in *Newsweek*, Pete Hamill, doctors, lawyers, merchants, lab chiefs. "The Stem Cell Wars" landed on the cover of *Newsweek*, and then graduated to *Hardball* and *Crossfire*. CAMR hired an op-ed ghostwriter, who created a "Swiss cheese" opinion piece that could be adapted by CAMR-recruited scientists and published in local papers; the organization placed more than two hundred newspaper editorials favorable to stem cell research. Howard Kurtz, the media critic for the *Washington Post*, officially blessed the arrival of stem cells as a Big Story by writing a column about all the other stem cell stories.

This growing din of public discourse accompanied another turning point in late June and early July: the emergence of a small bloc of influential antiabortion Republican senators, including Orrin Hatch, Strom Thurmond, Gordon Smith, and Bill Frist, as proponents of the research. More than any other development, these defections from party-line Republican positions made it clear that stem cell politics, although heavily influenced by the abortion debate, abided by a different calculus, at least in the Senate. The most influential of these proponents by far was Hatch. The fact that Hatch, who had led the Senate opposition to fetal tissue research in the 1980s, came down differently on stem cells had, according to one government insider, a "huge" impact on the public debate. CAMR hired a Republican lobbyist, Vicki Hart, to work the White House, and Jeff Martin joined in, using his access to many key senators — Frist, Jeff Sessions of Alabama, Max Baucus of Montana, and others. I asked Martin if he had lobbied Hatch or Smith. "I did," he said, "but it was by getting other people to talk to them, people they knew well. It wasn't that we had to persuade them so much," he added. "It was just that we'd learned early on that they were more supportive of it than we thought." Bush, too, was getting caught in the crossfire of all the lobbying. CAMR used its electronic grassroots system to mobilize seventy thousand "patient advocates" over the Fourth of July holiday to contact their elected representatives. Perhaps the most influential meeting of all occurred on the afternoon of July 9, shortly after Doug Melton left the White House, when two bioethicists, Leon Kass and Daniel Callahan, paid a visit.

Kass, a vigorous-looking sixty-two-year-old, had originally trained as a doctor and biochemist, but had switched to philosophy and was the Addie Clark Harding Professor in the College and on the Committee on Social Thought at the University of Chicago. He was genially old school, with an open and inclusive public manner, manifest intelligence, a ready wit, and, beneath it, a deep reservoir of humanistic intuition and scholarship. But he wasn't without opinions. When you examined his writings, you discovered elegantly worded and beautifully composed essays that invariably took a negative, almost extreme position on most practices of modern biology, indeed on modernity in general (in one infamous essay, dredged up by a *Wall Street Journal* reporter, Kass had decried the eating of ice cream in public). In the introduction to his 1986 book *Toward a More*

Natural Science, Kass spoke of science as a socially subversive and politically destabilizing force. He wrote that "science essentially endangers society by endangering the supremacy of its ruling beliefs." He went on, "Science — however much it contributes to health, wealth, and safety — is neither in spirit nor in manner friendly to the concerns of governance or the moral and civic education of human beings and citizens. Science fosters and encourages novelty; political society, governed by the rule of law, cannot do without stability. Science rejects all authority save the truth, and prefers skepticism to trust and submission when truth is unavailing; the political community requires trust in, submission to, and even reverence for its ruling beliefs and practices." A philosopher who viewed science as a dangerous social force suddenly had an audience with the president.

More to the point, Kass had published a long essay on cloning in the May 21, 2001, issue of the *New Republic* that was nothing short of a call to arms — a call not only to stop human cloning in any form, but more generally to regulate, arrest, and otherwise impose social controls on what he considered the "runaway train" of modern biology and biotechnology. He frequently professed the laudable aim of elevating the level of public discourse on these issues, but you had the nagging sense that part of the elevation he insisted on reflected a kind of genial elitism, a paternalistic, old-fashioned fondness for the civility and values of a romanticized nineteenth-century ideal of family, citizenship, courtship, and moral governance. He was the thinking man's Gary Bauer, and you can easily imagine Bush responding not only to Kass's moral seriousness, but also to his warm, engaging, and good-humored personality.

Bush had asked Kass to bring along a bioethicist who held different views on the stem cell question, according to several news reports. Kass showed up with Daniel Callahan. There were any number of bioethicists who had thought long and hard about embryonic stem cell research and had articulated important moral arguments in support of it — Arthur Caplan of the University of Pennsylvania, Thomas Murray and Erik Parens of the Hastings Center in New York, Ronald Green of Dartmouth College, Laurie Zoloth of San Francisco State University, to name a few. Callahan, it turned out, was not one of them. In 1969 he had cofounded the Hastings Center, the nation's first bioethics research institute, in New York's Hudson Valley. Although Callahan tended to be more liberal than Kass, he had become increasingly skeptical of the scientific enterprise and, unbeknownst to Kass, an opponent of stem cell research. Indeed, Callahan later told me that he, not Kass, was the one who strongly urged the president to reject federal funding for embryonic stem cell research at that meeting. "Leon was more ambivalent about the issue," he said, "while I came out and flatly opposed it." Later on, several news accounts quoted administration sources as saying that Bush had been especially "impressed" by such moral unanimity from two prominent bioethicists supposedly divided on the issue.

⌒

In early July, just as the first reports of a potential stem cell "compromise" were appearing in the press, Martin and his legal colleagues returned to their lawsuit — attempting to apply more pressure on the administration.

About a week before the government was due to respond to the scientists' lawsuit (that is, July 13), Department of Justice attorneys requested an extension of time. Martin was inclined to agree, but at the same time was getting anxious about the delay, which prompted a decision to prepare summary judgment papers — a motion asking the judge to decide the merits of the scientists' arguments on the basis of court filings right away, without waiting for a trial. Often a preemptive bid to gain swift resolution of a case, it included an unusual tactic in this instance. "What I ended up doing," Martin recalled, "was calling over to the Justice Department and talking first to the staff-level guy, Jim Gilligan. I said, 'Jim, we have summary judgment papers that we've prepared, and I'm prepared to share them with you, to see if you have any thoughts about them, see if they could be actually persuasive to the administration on a prefiling basis. But my one condition is that they must get to somebody at the decision-making level. And that's what I want to know. I'll send them to you if they're not just going to give you more time to prepare your opposition, but to actually be thought about at a senior level. Because we think they're persuasive.'" Martin said he received assurances from people in the deputy attorney general's office that "if I wanted to share the papers with them, they would get to decision-making levels. And so I shared the papers with them."

To cover his bases, Martin tried to contact Tommy Thompson on July 6 to let him know about the summary judgment motion. Thompson returned Martin's call from an airplane. Martin explained that he was going to share his legal filings with the administration. "Be sure and send them directly to me as well," Thompson told him. Martin even shared a letter with me in which he asks Thompson if he should file for summary judgment or not. In the end, Martin and his colleagues filed their motion on July 11. As even Clarence Kipps, one of the attorneys, admitted, the case against the government was "not a slam dunk." But if they got a federal judge to agree that it was illegal for the NIH *not* to fund stem cell research, it would indeed have been, as Martin put it, a "big story." Kipps insisted that people in the White House definitely knew about the lawsuit.

The conventional wisdom is that the scientists' lawsuit was not much of a factor in either the timing or content of the Bush decision. No one in the White House has ever made a comment about the suit, and as Martin admitted, "The bottom line is, we don't know what moved the process along. But my thought is that the lawsuit made strong arguments, and we got it to the right levels of the government, so that at least the administration had turned from a feeling that it was illegal to fund the stem cell research to a feeling that they could at least support whatever decision the president made to allow funding of the research. That legally, it was supportable. Highly supportable." If the president was indeed beginning to feel trapped, as some have speculated, it was probably the combination of Republican senators coming out in support of stem cell research, growing support among the public, and the possibility — small, but hardly negligible in potential impact — that a federal court might order the NIH to fund work on embryonic stem cells and open up the research much more broadly than the administration ever envisioned.

"We were starting to become impatient by late July," Martin said. "And we did not want an indefinite extension of time to respond to the summary judgment." So Martin and his colleagues asked the judge to compel the government to respond by August 13. The government lawyers, Martin told me, "fiercely" opposed an August 13 deadline. "I can't say that they told me they were concerned," Martin said. "But they clearly didn't want us to go forward. They opposed that vigorously. They wanted the president to have as long as he needed to decide his position. We were concerned about an indefinite delay." Judge Lamberth indicated that he would rule on Martin's request on August 9.

⳽

Not that there was ever much doubt, but the relationship between the Nightlight Christian Adoptions lawsuit and the political opponents of embryonic stem cell research became clear on July 17, when a subcommittee of the House Committee on Government Reform held a tumultuous, tear-jerker of a hearing on stem cell research. The headline witnesses were, in effect, the plaintiffs in the Nightlight lawsuit. They included JoAnn Davidson, director of the Snowflakes Embryo Adoption Program; the biologist David Prentice; a representative of the Christian Medical Association; and John and Lucinda Borden, a couple who brought two very cute nine-month-old Snowflake infants. Like so many of these congressional affairs, fact-finding and serious public reflection took a backseat to theater — and bad theater at that. In a long summer of hyperbole and invective, perhaps the most jarring moment occurred at this hearing, when John Borden held up his two adopted "snowflakes," Luke on one shoulder and Mark on the other, and shouted at the congressmen, "Which one of my children would you kill? Which one would you want to take?" A photograph of the parents and children appeared on the front page of the *New York Times* the following day.

The House hearing featured another "snowflake," three-year-old Hannah Strege, who squirmed in her mother's arms with a pacifier in her mouth, although that didn't keep congressmen from seeking her expert testimony. At one point, Representative David Weldon said, "I would assume Hannah is too young to have a comment at all?"

"Well," replied her mother, "for the last two nights, she's said, 'I want to say a prayer. For the snowflakes, amen.' So I guess that would be her comment." By the time Representative Chris Smith of New Jersey said, "I've been in Congress twenty-one years, and I'm not sure I've heard more compelling and heartwarming testimony," people in the press section and audience were actively suppressing laughs. Two weeks later, the House of Representatives passed a ban, cosponsored by Weldon, on all forms of cloning.

But still no word from the president. Bush had been heartened in mid-July when the NIH was finalizing its scientific report about stem cell research, because in the course of its preparation NIH officials had learned that there might be as many as thirty human embryonic stem cell lines already in development and potentially available for research. Although the president's initial moral impulse might have been to ban all federal funds for embryo-related research, that looked neither legally defensible nor politically feasible by the end of July, and the next best thing, the next *least* thing, was to use existing cell lines — cell lines already created through the destruction of embryos — for the research. At an August 2 meeting at the White House, the president and his advisers met with a delegation from NIH and learned that more than sixty embryonic stem cell lines in various stages of development were available — exactly the figure the president announced in his speech a week later.

Bush had been inching toward a compromise for weeks, according to subsequent press reports, but he may also have run out of options. The defection of antiabortion Republican congressmen had mussed the clean, partisan split in the politics. The patient groups had mounted an unusually successful public education campaign. The Thomson lawsuit, Martin believes, denied the administration an easy legal out. And the NIH stem cell report requested by Tommy Thompson, formally released on July 18, let the air out of the right-to-life argument that adult stem cells offered just as good an alternative; in scrupulously neutral prose, the NIH report said there were compelling scientific reasons to proceed with both embryonic and adult stem cell research. You might say the president was "agonizing," as White House sources continued to suggest, but you might also say that he'd run out of maneuvering room.

On the morning of August 9, Jeff Martin, like a number of other people in Washington, received a call asking where he could be reached by phone that evening. Secretary Thompson, he was told, would like to speak with him then. At 10:30 A.M., the White House announced that the president would address the nation that evening at 9 P.M. to reveal his decision.

15

THE BREATH
OF LIFE

ON THE MORNING of July 18, 2001, many of the leading dramatis perso-
nae in the stem cell debate found themselves taking their familiar places
again on the public stage, this time converging on a second-floor hearing
room in the Hart Senate Office Building in Washington, D.C. Michael West
had arrived early, lugging on his hip a large box brimming with stapled
copies of the testimony he planned to give. Richard Doerflinger, the ubiquitous and sharp-tongued representative of the U.S. Conference of Catholic Bishops, hovered near the witness table. David Prentice, the adult stem
cell proponent and self-described "ad hoc adviser" to politicians, took in
the scene behind his political patron, Senator Sam Brownback of Kansas.
Members of the public had begun lining up almost an hour before the start
of the hearing because it promised to be unusually lively. A few days earlier,
two private biomedical companies — the Jones Institute for Reproductive
Medicine in Norfolk, Virginia, and Michael West's Advanced Cell Technology in Massachusetts — had separately disclosed that they had already begun attempts to create human embryos with the specific intent of harvesting embryonic stem cells; the ACT work touched a particularly sensitive
social nerve because the company intended to use cloning technology to
create embryos. That controversial development was of obvious interest to
Senator Tom Harkin, the Iowa Democrat who headed a Senate subcommittee that oversaw the government's biomedical research enterprise, and to
his Republican colleague, Arlen Specter of Pennsylvania. Indeed, this was
the eighth in what would become more than a dozen hearings held by the
subcommittee since 1998 devoted to the issue of embryonic stem cells and
the controversies that swirled around that technology.

There was especially keen attention this morning because the first group of witnesses comprised four Republican senators, and each came bearing quiet-spoken rhetorical gifts. Orrin Hatch, reiterating a position he had expressed to President Bush in a letter of June 13, explained how he had to part company with some conservative colleagues and state that using leftover embryos to create stem cells was a worthy goal, that "you can't equate embryos in a freezer to embryos in a mother's womb." Bill Frist of Tennessee, having declared qualified support for stem cell research in an interview appearing in that morning's *Wall Street Journal*, arrived with an ambitious ten-point plan that would govern the ethics of biomedical research in general. Sam Brownback reiterated his long-standing opposition to embryo research; in a kind of plainspoken yet fervent midwestern mantra, he repeatedly said, "We all agree that this embryo is alive. The central question remains: Is it a life?" The implicit answer, of course, was yes, and Brownback argued that embryonic stem cell research and also human cloning, whether to make babies or to harvest stem cells, "must be stopped." But perhaps the most surprising and powerful witness that morning was Gordon Smith, a Republican from Oregon.

Of the four senators, Smith was probably the least well known and, frankly, the least influential. Like Hatch, he was a Mormon, and known to hold antiabortion views, but his political positions were hardly doctrinaire; he was related to the great western populist Democrats Morris and Stewart Udall. And he invoked their memory and spirit immediately, as he spoke in a low but surprisingly resonant voice. "I thought, in the spirit of trying to be helpful to your deliberations, I would share with you what I have experienced and what I believe," he told the subcommittee in a slow, inexorable cadence. "As a young boy, I watched my Grandmother Udall die of Parkinson's disease. Growing up, I watched my cousin, Congressman Morris Udall, literally die in public of Parkinson's disease. Last April, I buried my uncle, Addison Udall, of Parkinson's disease. Last weekend, my brother-in-law, Dan Daniels, informed me that he now suffers from Parkinson's disease." This was all delivered in a matter-of-fact voice, but no one sitting in the room that day failed to hear the universality of pain embedded in each sentence: the pain of a dreadful neurological visitation, of bedside family vigils, of irreversible declines, of the sequential loss of loved ones, one by one.

Now that he had everyone's attention, Smith continued — and I'm quoting virtually his entire testimony here — as follows:

In the experience of my life I have not been a stranger to hospitals and trying to provide care and comfort to those who suffer and seek to be well. So, for me, this debate presents me with the ultimate questions, the answers to which I believe I will be held accountable in this life and the hereafter. And that is, "When does life begin?" Mr. Brownback has stated it well. Some say it's at conception. Others say it's at birth.

For me and my quest to be responsible and to be [as] right as I know how to be, I turn to what I regard as sources of truth. I find this: "And the Lord God formed man of the dust of the ground, and breathed into his nostrils the breath of life; and man became a living soul." [Gen. 2:7]

This allegory of Creation describes a two-step process to life: one of the flesh, the other of spirit. Cells, stem cells, adult cells are, I believe, the dust of the earth. They are essential to life, but standing [alone] will never constitute life. A stem cell in a petri dish or frozen in a refrigerator will never, even in one hundred years, become more than a stem cell. They lack the "breath of life." As an ancient apostle once said, "For the body without the spirit is dead."

I believe that life begins in a mother's womb, not in a scientist's laboratory. Indeed, scientists tell me that nearly one half of fertilized eggs never attach to a mother's womb, but naturally slough off. Surely, life is not being taken here, by God or by anyone else.

For me, being pro-life means helping the living as well. So if I err at all on this issue, I choose to err on the side of hope, healing, and health. And I believe the federal government should play a role in research to assure transparency, to assure morality, to assure humanity, and to provide the ethical limits and moral boundaries which are important to this issue.

Those boundaries and limits must stop at a mother's womb, for again that is where life begins. The Puritans call it a Quickening, and the Scriptures, I think, speak to that effect as well.

In conclusion, I say — with deepest respect to those who have views different from mine, both theologically or scientifically — that I respect those views but share with you my own views and my own experience in seeking cures for the most dreaded diseases on this planet. We are at the confluence between science and theology. I believe that we must err on the side of the broadest interpretations to do the greatest amount of good.

I was sitting several rows behind Smith, in the press section, furiously taking notes as he began to speak; then, when he was about a third of the way through his testimony, I stopped writing and just listened, swept up in the momentum of the remarks. Like a lot of other people in the room that day, I suspect, I suddenly realized two things: that I was listening to a breathtaking piece of political oratory, and that it is an exceedingly rare thing to hear these days. Not everyone agreed with it — indeed, Catholic intellectuals immediately attacked its theological reasoning — but for me it was a high point of the stem cell debate, as much for what it didn't do as for what it did. Smith's testimony was deeply felt, without being crassly emotional; it was passionate, but did not deploy its emotional power to denigrate other positions in the debate; it was unusually thoughtful, even hesitant, in a sea of self-sure rhetoric; it rendered a complicated and cutting-edge scientific concept in plain language; and it breathed life into the conundrum of conception by reverting to an ancient text, the Bible, and using it for guidance in interpreting a modern biological phenomenon of immense complexity. It reflected honest internal struggle, one person weighing two competing principles: the destruction of early embryos (even if abandoned) versus the life-giving promise of the cells that might be retrieved. And it was accountable not only to Smith's internal moral standards but to the people who would be affected by his opinion. It was all delivered in a strong, plain voice, all the more impressive because I later learned that Smith had spoken extemporaneously. It was also the shortest peroration of the four speakers. Senator Harkin, whose instinctual courtesy as a committee chairman sometimes borders on hyperbole, nonetheless seemed swept up by the elevated rhetoric when he said, "In my twenty-seven years in the Senate, I don't believe I have heard testimony as moving and as well thought out as from you four gentlemen." He was addressing all the senators, but I think he really meant Gordon Smith.

As I thought about Smith's testimony later that day, and continued to think about it for months afterward, I asked myself why the senator's remarks had stood out in such sharp relief amid the ocean of commentary and argument and opinion surrounding the twin issues of cloning and stem cells. First of all, it was one of the few times when you couldn't predict what someone was going to say. Truth be told, Smith's position on stem cells was well known, but not to me, and my ignorance conferred a blissful willingness to be surprised. Perhaps that's another way of saying how wearyingly predictable the abortion debate has made our political discourse.

But the appeal of Smith's testimony went well beyond surprise. When I reread the speech, I was impressed by its civility toward those with whom it disagreed. It never raised its voice, rhetorically speaking. Unlike so many speeches by so many politicians over those months, in which everyone strove to establish their bona fides with an obligatory but gratuitous mention of a relative or friend with this disease or that, Smith's grief sounded deep, familial, and earned. It didn't pretend to have all the answers. Larry Soler, then chairman of CAMR, the patient-advocacy group, who was also in the audience that day, later said, "Here you have a guy up for reelection in a year, in a state where his hold was marginal, but he was there from the beginning. He showed a lot of courage."

All this made me wonder why there were so few similar moments. For all the commentary and all the hearings and all the position papers and newspaper stories, this was a public debate of immense importance that rarely rose to the level it deserved. The same point was made, in an offhanded way, during an exchange at a scientific meeting in December 2001 following a talk by Arthur Caplan, the University of Pennsylvania bioethicist, who had complained about the quality of the public discourse on cloning and stem cells and had singled out Mike West for particular criticism. A dissenter in the audience (someone, it turned out, who was an investor in ACT) raised his hand and vigorously defended West, pointing out that he had testified repeatedly in Congress, spoken to the press, and discussed the controversy ad nauseam.

"I agree," Caplan replied, "about the ad nauseam part," to widespread laughter. The joke touched a nerve. The debate had gone on and on, but just how illuminating had it been?

~

If Gordon Smith's four-minute speech was one of the high points of the public conversation, perhaps the nadir occurred at a symposium about human cloning at the National Academy of Sciences on August 7. Even at 7:30 on a muggy Washington morning, the TV vans and towering stalks of satellite relays had sprouted up on the street near the organization's elegant white-marble building on Constitution Avenue. Inside, more than two dozen camera crews trained their instruments on the stage of the academy's auditorium — the same auditorium where, nearly twenty-five years earlier, throngs of protesters had gathered to oppose the new and frightening technology of recombinant DNA, chanting songs and slogans, heckling scientists, and unfurling banners reading "No Recombination Without Representation." This time around, there were no protesters (indeed, like so many stem cell and cloning "debates," the event attracted sparse public attendance). This time, the most disruptive force in the room happened to be those cameras, and the people attached to them.

They had plenty of reason to be there. Several people invited to speak at the symposium had expressed their intent to clone a human being — that is, create an embryo through nuclear transfer, implant it in a womb, and bring a baby into the world that would be the identical genetic copy of whoever donated the adult cell for the cloning procedure. In fact, a story in that morning's *Wall Street Journal* reported that two of the scientists scheduled to attend, Severino Antinori and Panayiotis Zavos, planned to present "new information" about their "human-cloning coalition," which purported to have recruited two hundred couples eager to produce a cloned child.

It fell to the chairperson, Stanford professor Irving Weissman, to remind everyone that the purpose of the meeting was to hear from scientists experienced in nuclear transfer and to determine whether it was safe to clone human beings. The idea for a meeting about cloning had been inspired by conversations earlier in the year at the World Economic Forum meeting in Davos, Switzerland, and the National Academy of Sciences committee had lined up a stellar cast to present their views. There was Ian Wilmut, the Scottish biologist who headed the team that cloned Dolly. There was Jose Cibelli, the principal cloner at Advanced Cell Technology, Mike West's company, who had successfully cloned many cattle. There was Peter Mombaerts, a biologist at Rockefeller University, who had headed a team that had advanced the process of therapeutic cloning in mice. There was Rudolf Jaenisch of the Whitehead Institute, who had identified several significant genetic and developmental problems in animal offspring produced by cloning. There was Alan Trounson, the Australian scientist who had pioneered many developments in the field of in vitro fertilization (and who, at that very moment, was one of the plaintiffs suing the NIH and Tommy Thompson over stem cell research). There were, in addition, experts in cell biology, reproductive medicine, law, and ethics. It's no exaggeration to say that the National Academy had gathered the best people in their respective fields to provide the latest snapshot of scientific progress, prospects, and problems to its own panel, which would later issue a report. That's exactly what a National Academy panel, which advises Congress on technical issues, is supposed to do.

But the press — and especially the electronic press — came to see only three people: Panos Zavos, Severino Antinori, and Brigitte Boisselier. Zavos, a Kentucky-based "sperm expert" (as the papers described him), had the dark canted eyebrows and peaked ears of a *Star Trek* walk-on. His Italian collaborator, Antinori, ran an infertility clinic in Rome, and had earned the sobriquet Dr. Miracle in his native Italy for, among other things, helping a sixty-two-year-old grandmother become pregnant through in vitro fertilization. The two men had vowed to clone a human being as soon as possible. Boisselier, meanwhile, was the scientist in residence for the Canada-based cult known as the Raelians, who believe in UFOs and have committed considerable resources to creating human clones.

It became clear this was not an ordinary scientific meeting the moment Irving Weissman announced the first midmorning break. A stampede of press people, led by a cavalry of camera-wielding television people, rushed down the aisles, clambered onstage en masse, and converged on Zavos, Antinori, and Boisselier, a flying wedge of microphones and wagging notebooks and bobbing electronic booms so massive and oblivious to all in its path that it resembled a lava flow or a mudslide. Alan Trounson, an Australian of rather stout architecture and low center of gravity, was nearly bowled over. "Sort of like a rugby scrum, isn't it?" he mumbled, watching the spectacle with bemusement. Peter Mombaerts, wiping up water from a carafe that had been knocked over, wondered aloud about calling for added security.

The National Academy was later criticized for even inviting these three to the discussion. "There were two things I hadn't anticipated," said Wilmut. "One, that they would invite the nutters, and two, that they would invite the television cameras. The two things together were — well, *nightmare* would be too strong a word, but . . . the meeting was not tightly chaired." Arthur Caplan later referred to Zavos, Antinori, and Boisselier in a public talk as "the loony cloning element" and sent a letter with colleague David Magnus to *Science* complaining about the "puzzling collapse of standards" that allowed their participation at the meeting, conferring a scientific credibility they didn't deserve. But as I sat in the audience that day, watching their antics in disbelief, I came to a different conclusion: I thought it was a brilliant bolt of strategy to have Zavos, Antinori, and Boisselier there, because it quickly became clear that these "scientists" (with the possible exception of Antinori) should not be taken seriously. To anyone who attends scientific meetings regularly, it was instantly clear how transgressive, how delusional and marginal their scientific skills were, despite all the media attention that had preceded them. They were, however, a good story.

Antinori was a painfully spot-on parody of the Vesuvian Italian, all smolder and eruption: with his dark eyebrows, bristly gray hair, and an operatic and seemingly permanent scowl on his face, he wore indignation like an Armani suit, constantly shooting his hand into the air to object to some imagined slight uttered by another scientist at the meeting, and he spoke with such a thick accent that he was almost impossible to understand. Boisselier was simply not credible; she spoke with an eerie, otherwordly serenity, her scientific assertions drenched in a lilting self-delusion (as Caplan later pointed out, she did not have a single publication relevant to cloning to her name in the vast Medline and Biosis databases). When Rudolf Jaenisch, a leading expert on imprinted genes (genes whose activity is favored, or "imprinted," by inheritance from either the mother or father), described several such genes that could be tested in early cloned embryos, Boisselier asserted that Raelian scientists could routinely screen their embryos for ten such imprinted genes. The claim was absurd, and almost every scientist on the stage rolled his or her eyes.

Zavos took this buffoonery to another level. He opened his talk with a picture of himself, then one with his arm around Antinori. He had his own patois, a kind of rapid patter, and there was, in the pattern and urgency of his speech, a kind of egomaniac impatience that colored every word and phrase he uttered, including "Next slide, please." "After this morning," he began, "we are going on to bigger and better things. Human reproductive cloning. What are the prospects?" He showed a slide of the success rate of in vitro fertilization, then added, "As the scientists alluded to this morning, the beat goes on . . ." Next up, a slide of Dolly the sheep. "Let's say HELLO, Dolly!" he fairly shouted. Moving on to insult, he lashed out at Rudolf Jaenisch and Ian Wilmut for comments the two scientists had made in *Time* magazine, warning that an attempt at human cloning with current technology would be dangerous and unsafe. Although all the speakers had been explicitly warned not to attack each other, Zavos's next slide read: "Dr. Wilmut is self-contradictory!" (Wilmut, pen to mouth, gently rocked in his chair throughout this unusual, ad hominem assault.) "LET'S EXAMINE THE FACTS FURTHER!" Zavos thundered and then showed a series of slides whose style could only be termed tabloid (one of them bore the memorable title "Let's Get *Real*, Gentlemen . . ."). Several members of the National Academy panel could be seen suppressing laughs at this point. Leaving no constituency unoffended, Zavos concluded this remarkably bombastic performance by accusing the scientific community of a conspiracy to outlaw reproductive cloning by arranging to publish in *Science*, just two weeks before the House of Representatives voted to ban human cloning, a report by Jaenisch's group detailing the safety risks of cloning. "They're afraid, they're scared, and they're uninformed," Zavos said. "What does this paper say? Very confusing results." Wrapping up his talk, Zavos flashed a slide entitled the "Current Status of Human Cloning." It showed a doll from the *I Dream of Jeannie* television show with the legend "The Genie is out of the Bottle!"

During a question-and-answer period afterward, Mark Siegler, a bioethicist from the University of Chicago who was on the National Academy panel, asked Zavos if he and Antinori would pursue their attempts to clone a human in a scientifically transparent fashion, reporting bad outcomes as well as good. "If we cannot do it right, we will not do it," Zavos said hurriedly. "Please say you're satisfied."

"I'm *not* satisfied," said Siegler.

"Well," snapped Zavos, "that's all you're going to get."

Before the National Academy meeting, I had never seen Antinori, Zavos, or Boisselier in public. I had read about their intentions in the papers for months, but nothing had prepared me for the stunning amateurism of their presentations. (Subsequently, the Zavos-Antinori collaboration fell apart, and a month after the National Academy workshop, the International Association of Private Assisted Reproductive Technology Clinics and Laboratories voted to expel Antinori for "disreputable conduct.") Their appearance had been much debated by the academy panel ahead of time. Weissman later said that the National Academy in fact anticipated some of the criticisms that were ultimately voiced. "We talked it over way ahead of time, because we knew that the three would-be cloners that we invited might be very media-oriented," he said. "But we said, Look, if we only hear one side of this, then we're not hearing everything we need to know. And then the second reason, and really the most important, is that in our early discussions with them, they seemed to be totally unaware, all of them, of the animal cloning experience and the disasters that result." The National Academy wanted Antinori, Zavos, and Boisselier in the room so they could hear the overwhelmingly negative, indeed frightening, data generated by animal cloners.

Fair enough. But to the average citizen (or member of Congress, for that matter) reading the paper the next day, the "nutters" emerged with their respectability a bit battered but intact. You'd have been hard-pressed to understand just how over their heads as scientists they seemed to be, and when I mentioned to Weissman that I had expected them to get a rougher ride in the press, he leapt in, saying, "They got lionized!" That may exaggerate a little, but it underlines an important point. Those of us in the media do a much better job of reporting on controversy than interpreting it. We tend to be so focused on *intent* in covering a scientific controversy — in this case, on the expressed intent of this scientist or that to clone a human being — that we don't spend nearly enough time assessing the realistic possibility of achieving that intent and communicating it to the public. Just because a scientist (even a very good one) says he or she wants to do something, it doesn't necessarily mean it can be done. When Dolly was first cloned, a certain Richard Seed played a similar role: the bogeyman-cloner whose mere statement of intent titillated the public and foreshadowed some imminent, undesirable feat. Seed never succeeded, and there is a good reason for that: cloning is a very complicated procedure to do safely, at least in primates. "It's very easy to talk about, but it's very difficult," said Günter Blobel. "When you think about it," said Wilmut, cloner of Dolly, "I think it's still a surprise that this technique works at all." But that was not the impression conveyed by the media coverage of Boisselier, Zavos, and Antinori. In an effort to be evenhanded, journalists invariably described the three would-be cloners as "maverick," "renegade," and so on. Those are code words, of course, but in retrospect the code is too polite; it fails to convey to a justifiably concerned public a true sense of the ability of these people to pull off the outlandish acts they proposed. True, Antinori had a track record in reproductive medicine; as Mombaerts later put it, "He has the technical skills, and a penchant for the bizarre." But the other two served mostly as straw men in the public debate. Given a stage, they startle the public, inflame lawmakers, and sensationalize the discussion (it's no accident that

Zavos was invited in the spring of 2002 to testify before a House subcommittee loaded with opponents of cloning). These free radicals do nothing but catalyze the worst sentiments in an already emotional debate.

Given the complexity of the issue, traditional television journalism was rarely up to the task, either (film crews at the National Academy forum actually followed Antinori to the bathroom). Dan Rather, the CBS anchor, admitted as much in an astonishing aside following the network's coverage of George Bush's stem cell decision, announced two days after the National Academy forum. Rather told viewers that embryo research is "the kind of subject that, frankly, radio and television have some difficulty with, because it requires such depth into the complexities of it. So we can, with, I think, impunity, recommend that if you're interested in this, you'll want to read, in detail, one of the better newspapers tomorrow."

The take-home scientific message of the National Academy symposium was that somatic cell nuclear transfer is exceedingly difficult and dangerous. "We are seeing a great range of abnormalities in these animals," Wilmut told the panel. "In general, we should expect very low efficiency, and deaths occurring during pregnancy and after birth." In humans, he added, "what we should expect is late abortions and, perhaps worst of all, surviving but abnormal children." But for months, the press had recounted every twist in the story of Antinori, Zavos, and Boisselier and their attempts to create human clones. It's always possible that one of them, against all odds, will produce a healthy cloned baby. But given the overwhelming failure of animal cloning experiments, the entire scenario seems about as improbable as sports reporters hounding a minor league baseball player simply because he has declared his intent to break Henry Aaron's home run record. In this instance, however, mere attention can have important public-policy implications. With Zavos and Antinori in the foreground vowing to clone humans, the House of Representatives hurriedly debated and passed a bill banning all forms of human cloning. R. Alta Charo, the University of Wisconsin law professor, characterized the bill as "a nuclear bomb to kill a fly."

⌐⌐

While Michael West and Jose Cibelli were busy in Massachusetts, trying to create cloned human embryos in an effort to obtain stem cells, the House of Representatives held a floor debate on the afternoon of July 31, 2001, on a bill that, if the Senate agreed with its particulars, would land West and

Cibelli in jail for up to ten years and hit them with $1 million fines. House Resolution 2505, introduced by Republican representatives David Weldon of Florida and Bart Stupak of Michigan, banned all forms of human cloning and imposed criminal penalties on anyone who attempted the procedure, either for reproductive or research reasons in a private or public laboratory. If, as some have argued, human cloning is the defining biomedical issue of our time, what should we make of the fact that the House devoted only three hours of debate to the topic? We should, paradoxically, be grateful, because if the debate had gone on longer, we likely would have heard even more of the same kind of sentiment so cruelly preserved in the *Congressional Record* for future generations to examine. Even within the House, there was bipartisan dismay at the quality of the discussion. Peter Deutsch, Democrat of Florida, said the debate "may be the lowest level of knowledge I've seen for a significant piece of legislation." "Honestly," said Representative Stephen Horn, Republican of California, "I cannot say I remember much from my own school biology class, and I think a lot of us are in the same way. We were dealing with leaves and not molecular objects . . . Ultimately, the debate and science are too complicated to leave to a group of unsophisticated legislators with instruments too blunt to be effective. I am concerned that the House leadership has allowed this debate to proceed in this hasty, reckless fashion."

One of the members of the House Republican leadership, J. C. Watts of Oklahoma, reduced an important debate to insults, slogans, and sophomoric humor. "This House should not be giving the green light to mad scientists to tinker with the gift of life," he said. "Life is precious, life is sacred, life is not ours to arbitrarily decide who is to live and who is to die. The 'brave new world' should not be born in America. Cloning is an insult to humanity. It is science gone crazy, like a bad B movie from the 1960s." His "mad scientists" comment attracted the most opprobrium, but Watts could also be credited with either the feeblest attempt at humor, or the scariest thought of all, when he prefaced those remarks by saying, ". . . there is no greater group of people who would benefit from human cloning more than Members of the House of Representatives. What a Congressman or Congresswoman would not give to have a clone sit in a committee hearing while the Member meets with a visiting family from back home in the District, or the clone could do a fund-raiser while the Congressman leads a town hall meeting back home."

But several remarks during this discussion deserve closer scrutiny, because they illuminate the different standards of rigor applied to our scientific and political discourse. Here is one of the cosponsors, Bart Stupak: "Opponents of our bill have said embryonic research is the Holy Grail of science and holds the key to untold medical wonders. I say to these opponents, show me your miracles. Show me the wondrous advances done on animal embryonic cloning. But these opponents cannot show me these advances because they do not exist . . . The Holy Grail? The magic? How about the human soul? Scientists and medical researchers cannot find it, they cannot medically explain it, but writers write about it; songwriters sing about it; we believe in it. From the depths of our souls, we know we should ban human cloning." Quite apart from the Tin Pan Alley meets the Wisdom of Repugnance angle, Stupak's remark about nonexistent advances wouldn't stand up against the standards of a well-informed newspaper reader. In April, in a widely reported advance, researchers at Rockefeller University and Memorial Sloan-Kettering Cancer Center had created cloned mice through nuclear transfer, harvested the stem cells, and nudged them into becoming the kind of neurons that disappear in Parkinson's patients. And within eight months of Stupak's dismissive remarks, scientists at the Whitehead Institute had indeed used so-called therapeutic cloning to cure mice of an immunological disease. Those are not miracles, but medicine does not advance from miracle to miracle, as any sophisticated observer of science well knows.

And here is Representative David Weldon, the bill's other cosponsor: "We have had a lot of discussion about whether or not these embryos are alive, whether they have a soul," said Weldon. "The biological fact is, and I say this as a scientist and as a physician, that they are indistinguishable from a human embryo that has been created by sexual fertilization." It would require a nuanced scientific discussion, well beyond the technical ken of most congressmen, to explain why Weldon's statement is wrong, touching on such issues as imprinting, epigenetic methylation patterns, and mechanisms of reprogramming; indeed, at the National Academy symposium a week later, one biologist after another, in a parade of drearily pessimistic reports that stretched on for hours, detailed all the ways that embryos created through cloning were in fact biologically and genetically different. Rudolf Jaenisch of the Whitehead Institute, who has published extensively on animal cloning, said there are "major differences" between cloned embryos and those formed through sexual fertilization, adding, "The scientific evidence that these cloned embryos are not normal is overwhelming." What makes Weldon's statement particularly troubling is that he traded on his scientific expertise, in a room full of neophytes, to peddle this misinformation.

To be sure, truth is an equal-opportunity casualty. Proponents of therapeutic cloning and stem cell research invoked the word *cure* countless times to tout the promise of this unproven technology. Darwin Prockop of Tulane University told a Senate panel in 2000 that Parkinson's disease could be cured in four or five years; Jennifer Estess, founder of Project ALS, told the same panel that consulting scientists "tell us that stem cell therapy *will* work on humans"; even Tom Harkin, in a chart prepared to explain the process of research cloning, had an arrow pointing to a figure with the caption "Cured Patient." But *cure* is the rhetorical equivalent of funny money — easy to mint, painless to pass around, utterly unredeemable in the present (and, in this case, probably in the near future). My sense is that the scientists and their allies invoked potential cures so much because of the vehemence of the opposition and the absolute funding ban on research those opponents sought to impose. When opponents of these new technologies argue, as a midwestern pastor did at one Senate hearing, that "some diseases are better than their cure," the discourse has assumed an irredeemable irrationality.

A highly polarized debate in a sound-bite world inevitably creates a pressure that rewards exaggeration, however, and the rhetoric quickly hardens into a kind of hyperbolic staccato. You don't hear the process of thought and logic in this kind of debate, only a shorthand that asserts murder or cures. For the scientific community, there may be a steep price, yet to be paid, for insisting so vigorously on the rewards of stem cell research. The technology may pan out, or it may not, but the promise has loudly been made, and nobody likes broken promises.

Stem cell proponents fell into this trap on their own. They tumbled down their own private slippery slope of rhetoric, where the *potential* of stem cell research and research cloning began to sound like the certainty of therapeutic rewards. This is a subtle and mysterious process: in 1998 and 1999, when the world was first digesting the news of human embryonic stem cells, there was an almost boilerplate list of diseases whose sufferers might — *might* — benefit from stem cell therapy. By the summer and fall of 2001, that argument had transmuted in a significant way, so that if the president declined to approve federal funding for stem cell research, for example, or if Congress banned research cloning, sufferers of Parkinson's, Alzheimer's, diabetes, multiple sclerosis, and all the other familiar ailments were going to be denied a cure — not the *possibility* of a cure, but the cure itself. This fallacious reasoning reached its logically illogical conclusion in December 2001, when Michael West told a Senate subcommittee hearing on therapeutic cloning that "a six-month delay in scientific research would cost 541,800 lives that might be saved somewhere down the line." This utterly fluffy number apparently not only presumed that the 3,000 people who die each day from degenerative diseases might have been saved, but that all could have been saved from their grisly fate by stem cell therapy — a preposterous and indefensible assertion. Nonetheless, it bears repeating, as Senator Bill Frist never tired of saying: for all its magnificent promise, stem cell therapy may never work. "We need to make it very clear: this is untried, untested research," he said.

⸕

One final issue casts a long shadow on this debate: abortion politics. Regardless of one's views about abortion, the persistence and vehemence of the disagreement has had a devastating effect on American biomedical research over the past thirty years. When an antiabortion leader describes human cloning as "the *Roe v. Wade* decision of 2002," it's clear that our discussion of new biomedical technologies is entangled, perhaps fatally, in an old and intractable social disagreement. "My feeling is that the United States is incapable of a sensible discussion of these issues because of the abortion debate," said Alta Charo. As a result, individuals with fertility problems, children with birth defects, families with a history of inherited genetic disorders, and now people with a host of degenerative diseases have suffered materially from this debate; they have not benefited from the best possible medical research, because that research has been politically proscribed. With the isolation of embryonic stem cells in 1998, the future health care options of virtually every American, young and old, to some extent became hostage to abortion politics.

There has been a reluctance among scientists and academic institutions to take on this fundamentalism, even as it threatens free academic inquiry, and that is perhaps the subtlest reason the present debate has been unsatisfying. To understand how tame our conversation has been, consider these remarks from a prominent university president about those who seek to impose a certain version of morality on others. "From the maw of this 'morality' come those who presume to know what justice for all is; come those who presume to know which books are fit to read, which television programs are fit to watch, which textbooks will serve for all the young; come spilling those who presume to know what God alone knows, which is when human life begins. From the maw of this 'morality' rise the tax-exempt Savonarolas who believe they, and they alone, possess the truth. There is no debate, no discussion, no dissent. They know. There is only one set of overarching political and spiritual and social beliefs; whatever view does not conform to these views, is by definition relativistic, negative, secular, immoral, against the family, anti–free enterprise, Un-American. What nonsense."

What's surprising about these comments is not that they date back to 1981 (A. Bartlett Giamatti, then president of Yale, prepared the remarks for an address to the incoming class of 1985). What's surprising is that no university president in the year 2001 felt courageous enough to speak out with similar force and passion in defense of (among other things) the freedom of academic scientists to pursue basic research on their own campuses, an issue at the very heart of free intellectual inquiry. There have been petitions; there have been letters with dozens of Nobel signatories; but at the center of all this high-powered murmur, there has been a deafening silence from what might be called the bully pulpit of the American research enterprise. "I'm sort of small potatoes in the field," Doug Melton of Harvard told me one day. "I don't have a bully pulpit like the head of the Hughes or the head of Harvard or the head of a big medical school. I don't know why those people are unwilling to do it."

Perhaps one of the key junctures in the entire debate was one of quiet subtraction: the departure of Harold Varmus as NIH director in late 1999. Varmus later expressed concern that he had left the NIH too soon; he said he "worried" that his resignation was possibly premature. After he left to assume the presidency of Memorial Sloan-Kettering Cancer Center, there was a vacuum in scientific leadership on the stem cell issue that has been privately acknowledged by many scientists. Without Varmus's energetic efforts to push the research, both through frequent testimony before Congress and the nuts-and-bolts construction of a bureaucratic organization designed to enable, promote, and monitor the research, many researchers perceived a lack of focus and urgency in the scientific leadership on these issues. In taking more than a year to nominate a new NIH director, George W. Bush perpetuated that leadership vacuum.

There were probably several reasons for this leadership void, among them the unwillingness of anyone in the scientific community to stick their neck out over the issue. While the Howard Hughes Medical Institute decided to privately support embryonic stem cell research in the Harvard laboratory of Doug Melton, the process by which it reached that potentially influential decision had no influence at all on the public conversation because it took place behind closed doors. Thomas Cech, the director of the Hughes, noted in an interview that "it was the *only* ethically correct thing to do." In reaching that conclusion, the institute conducted a workshop on stem cell research in April 2000, attended by scientists, bioethicists, and public-policy experts. But the event was not public, and no publication emerged from the deliberations (when the institute formalized its commitment to research, the decision was reported in the *New York Times*, but again, there was no formal announcement). Several scientists involved in stem cell research with whom I spoke felt disappointed by what they perceived as the timidity of the Hughes in this area. Ron McKay, the NIH researcher, attended the Hughes workshop and felt the institute missed an enormous opportunity to play a leadership role. "They just walked away from it," he said. Similarly, Ali Brivanlou of Rockefeller University argued that the politics of stem cell research "was too important to be left in the hands of geeks." When he expressed exasperation that his own institution, Rockefeller University, appeared reluctant to embark on a private venture to support embryonic stem cell research by Rockefeller scientists, he was advised to adopt a lower profile. It's understandable that any institution would like to avoid a confrontation with the religious right, but universities shied away from the fight without quite realizing that the fight had already been picked.

The unwillingness of institutions to speak out unwittingly increased the influence of a separate segment only too willing to climb up on a soapbox: the private sector. Ever since the 1970s, public squeamishness about a federal role in overseeing reproductive medicine — a squeamishness actively promoted by the religious right — has driven this segment of science to industry. As a result, the science of human reproduction and gestation, up to and including embryonic stem cell research and human cloning, has tended to be pursued in a much more entrepreneurial, self-promotional, and frankly less rigorous manner (the rigor pertaining not only to the scientific standards with which the research is reported but to the ethical and moral framework by which the research is conducted). "The Americans removing themselves from IVF was kind of good for us," Alan Trounson, the Australian IVF expert, told the National Academy forum on cloning, "but what it did was create a situation where it was left to the private sector to develop. Sometimes there's a conflict between clinical service and basic science," he continued, a conflict that creates "a situation that is not in the best interests of the patient." As a result, new developments in this socially sensitive area have tended to arrive with a thud, and without much warning, in the morning paper.

Each announcement by Michael West or Panos Zavos or the Raelians — whether attempts to clone cow-human embryos, or create human embryos for research purposes, or create human embryos for reproductive cloning — each revelation, inevitably captured and amplified by media attention, has reverberated like a grenade in the halls of Congress, in the solitary studies where bioethicists ply their cerebral craft, and around the dinner tables where average families have struggled to make sense of these head-spinning developments. I don't believe the rat-a-tat-tat of company disclosures has caused a national panic, as Arthur Caplan has argued, but the news has certainly unsettled a lot of nerves. These bombshells have been dropped at the discretion of private entities, with motivations as varied as corporate breast-beating, fund-raising, scientific (and patent) priority, and sheer me-too egomania — the original Geron announcement describing embryonic stem cells, ACT's series of announcements on human cloning, the Jones Institute's revelation that it had created human embryos for research, Zavos and Antinori's vow to create human clones, and the Raelians' posting of a picture of a putative cloned human embryo on the group's Web site. In one of the few credible remarks she made at the National Academy of Sciences forum, Brigitte Boisselier said, "What is the purpose of human cloning? Why are there three people here who are doing it? Because there is a *demand*. There is a huge demand. There are lots of couples who want to have a child." Market economics have not only driven the science but, by extension, have controlled the tempo and pitch of the social conversation as well. The ultimate irony here may be that in three or five years we will look back and realize that much of the controversy was stirred by companies that no longer exist.

Michael West has flourished in this oddly muffled landscape, happy to play the provocateur. But there is another, largely unremarked component to West's prominence in this debate. It is the unwillingness of any scientific figure of national stature — a James Watson, a Francis Crick, or, to an older generation, a Jonas Salk — to step up and speak on behalf of the scientific community. Many have spoken out, but there has been a reluctance to speak out in public with the same firmness and resolve I've heard in private about the desirability of embryonic stem cell research and research cloning, precisely because of the way it inevitably drags one into the quicksand of the abortion debate. Perhaps that's what motivated a plaintive letter from former Representative Constance A. Morella, a Republican from Maryland, to *Science* in July 2001. "As a community," Morella wrote, "scientists have a great deal of influence, but only if they choose to wield it. Both politicians and the public need to hear from the scientific community; I urge you to speak out." Six months later, Arlen Specter felt compelled to make much the same point. "The scientific community is going to have to be activated," he said at a hearing. "Too often the scientific community is inert . . ."

If it remains inert, the scientific community may come too late to the realization that what is at stake is not merely the ethics of conducting research on human embryos, but the larger issue of a particular segment of society attempting to proscribe, through political influence rather than public consensus, areas of scientific inquiry on the basis of largely (though not exclusively) theological beliefs. The greatest secular glory of Western society since the Renaissance, and the especial glory of the United States in the years since the end of World War II, has been the incredible productivity of basic science. But every instance of corporate adventurism strengthened the hand of the people who sought to portray science as a "runaway train."

It was just such adventurism by Advanced Cell Technology that brought Michael West once again — for the fourth time in three years — in front of a Senate subcommittee interested in stem cell research. In mid-July, ACT triggered what seemed like its obligatory quarterly social controversy, this time revealing that it had embarked on a program to clone human embryos in order to obtain embryonic stem cells. The senators wanted to hear about it. This was the same hearing at which Gordon Smith spoke so eloquently, and the political deference accorded the four Republican senators diminished the time and attention allotted to other witnesses; West joined six other speakers crammed into an abbreviated session just before lunch.

What he had heard during an earlier session inspired a deep, unexpected fear. This fear arose, perhaps mistakenly, during a brief exchange between Arlen Specter and Bill Frist, one of the four Republican senators testifying that day. Frist, the only physician in the Senate, had repeatedly been described in the press as one of George W. Bush's most trusted advisers on biomedical matters (and perhaps distrusted in equal measure by patient groups, who, according to Daniel Perry of the Alliance for Aging Research, watched him "sit on the sidelines until the last minute during one of the most defining moments in the dawn of regenerative medicine"). He arrived at the hearing with a ten-point plan for biomedical ethics. Frist's plan restricted embryonic stem cell research to a limited number of cell lines and established ethical guidelines (including informed consent) for conduct of the research. It also called for the establishment of a presidential bioethics commission and, significantly, a ban on the creation of human embryos, even for research purposes. Frist, in other words, supported limited stem cell research but drew a firm line at human therapeutic cloning.

After Frist enumerated these issues, Specter ran down the list and said he agreed with every point except two: limiting the number of embryonic stem cell lines to as few as twenty or so and banning the derivation of any new cell lines. Specter had apparently agreed with Frist, in a kind of negative affirmation, on another very controversial point: that no human cloning, even for research purposes, should be allowed.

It was a subtle point, but not without ramifications for ACT. If the Senate agreed with the House and banned therapeutic cloning, Mike West's business plan, the entire superstructure of his dream of regenerative medicine, would go up in smoke. When I asked him about it some months later, he recalled the exchange between Frist and Specter instantly. "I heard the same thing you heard," he confirmed, "but [thought], What does it mean? I was disappointed. Because therapeutic cloning is a key piece of this thing . . . It's not a minor point. It's not minor, you know, as a bargaining chip. It's one of the two wings of the airplane of regenerative medicine, so to speak. It isn't going to fly without it." He claimed that Specter's apparent change on research cloning didn't affect the speed with which ACT continued its efforts, because the company was already working flat out. "We're working as hard and as fast as we can," he said, hammering the table for emphasis.

As it turned out, Specter had apparently misspoken at the hearing; his staff later clarified that he supported cloning for research purposes, and he in fact emerged as a leading voice supporting research cloning in the ensuing months. But the brief pang of fear that stirred the very marrow of West's scientific being was a reminder of how fragile the political foundation supporting his company's stem cell and cloning efforts really was, and there may have been something to fear. By mid-July, public sentiment was clearly running in favor of stem cell research, and the next convenient barricade for political retreat was the cloning issue; there was a sense that politicians could justify their support for stem cells if they banned cloning — all forms of cloning — even though the issue hadn't been thoroughly discussed and many Americans (including some congressmen who voted on it two weeks later) barely understood the difference between research cloning and reproductive cloning. West seemed to understand what was at stake, because he not only delivered a fervent defense of embryo research and therapeutic cloning; he invoked St. Paul in the argument.

"As the apostle Paul said," he told the senators, "'When I was a child, I spake as a child, I understood as a child, I thought as a child: but when I became a man, I put away childish things.' In the same way, it is absolutely a matter of life and death that policy makers in the United States carefully study the facts of human embryology and stem cells. A child's understanding of human reproduction simply will not suffice, and such ignorance could lead to disastrous consequences."

At the end of that July 18 hearing, West stood surrounded by reporters, talking about stem cells, cloning, and every variation on those themes. All the senators and all the other witnesses had left; the camera crews had broken down their equipment; the lights in the hearing room dimmed, and still he went on and on and on, quoting a parable from Matthew at one moment, discussing the intricacies of embryology the next. "So how do we get from this opportunity to helping people who are sick?" he asked at one point. "That's my job, and my role is to steer this little biotech company through rough seas . . ." You had the feeling that he would stand there and talk all day, if there was even one other person willing to stay and listen. Maybe that's the real definition of a visionary.

West returned to Massachusetts knowing that resolution of this kind of controversy often boils down to a race between positive scientific results and negative social constraints. If the scientists at ACT could show that therapeutic cloning worked, that it could provide benefit to humanity, before the door got slammed shut, they knew it would be much harder, and politically much riskier, for Congress to slam that door. And so it became a bit of an undeclared race, although most handicappers, when considering a stakes race between the greyhounds of Darwin and the world's greatest deliberative body, would know where to put their money.

16

FREE THE
BUSH 64!

ON AUGUST 8, 2001, George W. Bush joked with reporters covering his extended August vacation on his ranch near Crawford, Texas, hinting broadly that he planned to announce his much-anticipated decision on embryonic stem cell research later in the month, around August 21, during a trip to an unnamed city. The following morning, even as the *Washington Post* reported that the nation would probably have to wait another couple of weeks to learn what Bush's decision would be, word began to spread in Washington that the president would go on national television that very evening to announce his decision. His remarks to reporters were surely just good-natured teasing, but in retrospect it almost seems as if he didn't want anyone paying attention.

It had been a long, strange trip, and it ended up in a place the president may not have wanted to go. Not since the 1970s, with the discovery of recombinant DNA and the controversy over the technology of gene-splicing, had any scientific issue so dominated the national conversation, and never had responsibility for resolving the conflict weighed so heavily on an American president's shoulders. "We jacked up the visibility of this so much that it was imperative for the president to deal with it personally," said Tony Mazzaschi, a board member of the Coalition for the Advancement of Medical Research. If the waiting made it hard for all the outsiders, the months-long process of arriving at a decision may have made it harder for George Bush, too — may have made it impossible for him to do what he initially suggested he wanted to do, which was to preserve his "culture of life" philosophy and steer the nation's biomedical research enterprise down the avenue of adult stem cell research.

Bush's playful bit of misdirection notwithstanding, there were signs that something was afoot during the first week of August. Karen Hughes, the trusted presidential adviser, had arrived at the Bush ranch on August 8 — the same day the nation's newspapers offered accounts of the raucous National Academy of Sciences workshop on human cloning from the day before. Science and its discontents were already very much in the air. That afternoon, the president met for two and a half hours with Hughes and Jay Lefkowitz; it was then, according to the White House, that he reached a final decision. On the morning of August 9, as temperatures pushed into the 100s, phone calls began to go out to people with an interest in the stem cell decision, checking on their whereabouts later in the evening. Jeffrey Martin, the attorney suing the government on behalf of the leading stem cell scientists, got one such call from the Department of Health and Human Services; Tommy Thompson, the secretary of HHS, hoped to speak with him that evening. Members of Congress and other interested parties got heads-up calls, too. The reason became clear when the White House announced that the president would address the nation at 9 P.M. — his first televised address to the country since the contested election — to explain his decision on stem cell research. Word of the impending announcement roiled the admittedly small-pond market of biotech companies devoted to stem cell medicine; the stocks of Geron, StemCells, and Aastrom Biosciences shot up in the afternoon, based on rumors — or wishful thinking — that Bush would announce a compromise.

Over the previous three weeks, the president had gotten an earful on the issue from just about everyone, from the eighty-one-year-old pope to the nine-month-old "snowflake" babies who'd been featured at an emotional Capitol Hill press conference. He had been chastised by Republican members of the House of Representatives for looking the other way as an "industry of death" had begun to take shape, and right-to-life groups had held his feet to the fire on his campaign pledge not to allow embryonic stem cell work. He had received a letter from Nancy Reagan, urging him to support the research, as well as a letter written on July 20 by Senator John Kerry of Massachusetts and signed by fifty-nine senators, urging the president to permit federal funding (thirteen senators signed a separate letter written by Senator Arlen Specter, calling for an even more aggressive policy). The House of Representatives weighed in a week later; 202 members of Congress likewise urged the president to support the research. Pope John Paul II, hosting the president at his summer residence at Castel

Gandolfo on July 23, pointedly remarked that "a free and virtuous society, which America aspires to be, must reject practices that devalue and violate human life at any stage from conception until natural death." As Bush told reporters during the trip to Italy, "My process has been, frankly, unusually deliberative for my administration. I'm taking my time."

The penultimate moment of an intense six-week period of lobbying and deliberation occurred when Bush returned. The president had reached a turning point in his thinking the previous Thursday, August 2, in Washington, according to a White House version of events that subsequently appeared in several press accounts, when he had met late in the morning with three scientific experts from the National Institutes of Health: Lana Skirboll, director of the NIH's Office of Science Policy; Ronald McKay, the well-respected stem cell researcher in the National Institute of Neurological Disorders and Stroke; and Allen Spiegel, director of the National Institute of Diabetes and Digestive and Kidney Diseases. These three arrived at the White House bearing unexpected political gifts. In the course of preparing an NIH report on stem cell research, Skirboll's office had canvassed scientists all over the world and had identified a surprisingly large list of more than five dozen human embryonic stem cell lines already in existence or under development.

Here was news almost worth waiting for, even if the wait had lasted half a year. If the Bush policy was indeed about to land on a compromise limited to existing cell lines, as Tommy Thompson had disclosed to Jeff Martin in June, the compromise would be easier to sell to the scientists and patient groups if some sixty-odd stem cell lines existed, as opposed to the half dozen or so that had met the rigors of peer review in the scientific literature. That higher number — and the exact number became critical to all the decisions that followed — would allow Bush to ban any further destruction of embryos in order to obtain stem cells, at least placating (if not satisfying) his conservative constituents. Georgetown University bioethicist LeRoy Walters, who met with the president and his aides later that day, recalled that the group seemed visibly excited by the availability of some sixty cell lines. Walters expressed surprise that the number was so high and recalled remarking to the president and his staff, "I would want to know what intellectual property limits were attached to the cell lines." Nonetheless, the *New York Times* later reported that the August 2 meeting "crystallized" the president's thinking. "Administration officials said Mr. Bush was leaning toward the decision he made when he asked for the stem cell inventory," the story reported, "and that the number 60 sealed it for him."

By taking so long to arrive at this decision, however, Bush had unwittingly worked himself further into the jaws of a political vise. According to one person who followed the deliberation closely, "My impression of where he was was that he *really* didn't want to go there, but that this was really important science, and he couldn't close the door. There's no question that he was cornered. He appeared trapped." If merely making the decision left him feeling uncomfortable, he was destined to feel a lot worse in the coming weeks, as it became clear that the handoff of information from the NIH to the administration about the status of the sixty-odd cell lines had not been clean. "There had clearly been a misunderstanding by both Tommy Thompson and the White House about what NIH was saying," said one stem cell proponent. It was only a matter of time before it became clear to many people that there had indeed been a fumble on the exchange.

⌇

As Texas-size bugs bounced off the window behind him, George W. Bush appeared on television at 9 P.M. eastern time. He wore a dark suit and, as always, looked oddly trepidatious as he gazed into the camera, but you could hardly blame him for the discomfort. No American president had ever wrestled so visibly and so publicly with an issue of such great biomedical import, and none had ever been forced, for that matter, to give a lesson in embryology on prime-time television. Bush immediately launched into a review of his decision-making process, describing the pros and cons of stem cell research — a back-and-forth oscillation of claims, with a destination so uncertain that the bioethicist Thomas Murray later likened hearing the speech to watching a bowling ball careening first toward one gutter and then the other, never knowing where it was going to end up until the very last moment. Only at the conclusion of the ten-minute speech did the president reveal his decision: he would allow stem cell research to proceed, he explained, but only using cell lines "where the life-and-death decision has already been made" — in other words, only on the "more than sixty" cell lines that existed as of 9 P.M. that day. No federal funding would be permitted for research on embryonic stem cell lines created after August 9, 2001.

Watching the speech, I was struck by the fact that Bush didn't look like a man at peace with his decision. True, it was his first televised address to the nation since the inauguration, but he appeared stiff, a little halting, uncomfortable with his material; whether you agreed with it or not, the speech on paper is alive with ethical reflection and moral resolve, but somehow, in the president's delivery, it did not resonate with either moral authority or political courage. The *Washington Post* television critic Tom Shales may have been thinking the same thing when, writing of Bush's performance that night, he noted that "there is deep in his eyes something rather haunted, perhaps even fearful."

Bush looked especially discomfited when discussing egg, sperm, and other wet and wiggly minutiae of in vitro fertilization, which may have come as close as any presidential address to a national lesson in sex education. And although the speech sounded evenhanded, on subsequent reading you can detect a political and moral vocabulary subtly tilted toward the house metaphors of conservative religious thought. Echoing the Nightlight Christian Adoptions rhetoric, the president likened a human embryo to a "snowflake," adding that "each of these embryos is unique, with the unique genetic potential of an individual human being." In his blanket condemnation of human cloning, in his pointed remark that "even the most noble ends do not justify any means," in his explicit and cautionary reference to Aldous Huxley's novel *Brave New World* (the novel beloved by all slippery-slopers), in his vague but dark allusion to "the ends of science," and perhaps most revealingly in his approving mention of an unnamed ethicist who dismissed scientific arguments in favor of human embryo research as "a callous attempt at rationalization," the president's remarks echoed positions frequently articulated by Leon Kass, whom he named as head of his presidential council on bioethics during the speech. (I later asked Kass if he had had any input into the president's remarks, and he insisted he had not; Karen Hughes reportedly wrote the speech.)

There was something in the Bush policy to distress everyone. Conservatives and right-to-life groups professed great outrage at hearing that the president was going to allow even a little stem cell research. "The president has introduced the camel's nose into the tent," Kenneth Connor, director of the pro-life Family Research Council, told the *Washington Post*, "and inevitably we'll soon have the whole beast in there. Moral principles are not divisible. It's going to encourage members of Congress to advocate additional research and to kill additional embryos." Scientists and patient-advocacy groups were dismayed to hear that no more human stem cell lines could be created; this significantly foreclosed potential research at a juncture in the scientific story when much remained unclear. But the *Washington Post* got it right when it described Bush's decision the next morning as "essentially the most restrictive use of federal money the administration could have permitted short of a ban," and much the same point was made several days later by the bioethicist Arthur Caplan. Writing in the *Philadelphia Inquirer,* Caplan said, "By limiting research to existing cell lines, Bush in effect banned federal funding for human embryo stem cell research," adding, "The 'compromise,' if it stands, will produce the worst of all possible outcomes."

But the Bush speech accomplished at least one goal that went almost completely unremarked. Even a drastically restrictive policy that nonetheless allowed stem cell research took the air out of the lawsuits filed by Jeff Martin's group and by the Nightlight Christian Adoptions group. In the days following the announcement, Martin consulted with his scientist-clients and polled their desire to continue pressing the case. With one exception, the scientific consensus was not to continue. ("The only person who wanted to fight on, a little bit, was Roger Pedersen," Martin said. "And he'd already moved to the United Kingdom.") Everyone, however, agreed to keep the original intervention papers alive until Nightlight Christian Adoptions dropped their suit several months later.

Despite disgruntlement among social conservatives and patient groups, the Bush speech appeared to have been a home run with the press and the public at large. Political columnist George Will wrote a glowing review in the *Washington Post,* hailing Bush's willingness to take on the issue as a turning point in American social policy (and eerily likening the Democratic position on stem cells to "the 'science *über alles*' school of thought"). Public opinion polls indicated widespread support. Journalistic accounts of the behind-the-scenes decision-making process depicted a president who aggressively solicited all points of view, struggled with the moral dilemma, and steadfastly kept politics out of the algebra of his decision making. Even the professed "outrage" among religious conservatives seemed to evaporate overnight. Indeed, several months later, when Will wrote a column describing conservative disenchantment with the Bush administration, the stem cell decision barely merited a mention; after a brief cloudburst of indignation that lasted several news cycles, you didn't hear many complaints about the Bush policy from the conservative right. Another telling reaction came from Thomas Okarma, the president of Geron, who said the company was "very pleased" with the president's decision, as well it should have been. To those with the key patents in hand, such as the University of Wisconsin, and to those who possessed the exclusive commercial rights to products derived from research under those patents, such as Geron, the president had, in the words of bioethicist Tom Murray, reinforced "an oligopoly over the existing lines," eliminating potential competition and turning all of academic America into a farm team for either the University of Wisconsin or industry.

⟡

When the president made his announcement, Leonard Hayflick was home at Sea Ranch, listening on National Public Radio. "I thought it was a revelation of encyclopedic ignorance about the field of cell biology," he said later of Bush's decision. "Every cell biologist knows that each population of cells is different from the next, and to limit research to a certain number of cell lines is absurd. If the best cell line is one in a thousand, then your odds of finding the right one are just that. You need maximum freedom, and that was a crippling action." Douglas Melton was at his vacation home in Woods Hole, Massachusetts, listening on the radio too; when he heard the president mention "more than sixty" stem cell lines, he remembers thinking, "That's news to me." Michael West, watching TV at home with his wife, found vindication even in a flawed policy, "amazed that what once was this little project of no interest to anyone except four or five people on the planet was now the subject of a presidential address." Ali Brivanlou, also watching TV with his wife at home, said, "I was really afraid that he was going to say no, which I think he ended up saying, but in a way that is smarter than I would have thought. From the first few sentences, though, it was already clear to me that I am screwed." Elder statesman, academic scientist, private sector entrepreneur, hard-core embryologist: all of them reacted with shocked disbelief to the speech, and the epicenter of that disbelief resided in the stunning number the president had uttered: "more than sixty."

It took only hours, if not minutes, for a culture of professional doubt to set in among scientific experts about those sixty cell lines, and these concerns made their public debut in the next morning's newspapers. "I have no idea where this number came from," John Gearhart told the *New York Times*. "I am totally unaware of this and obviously this is of great concern."

Roger Pedersen, the expatriate embryologist, told the journal *Nature*, "I am mystified myself as to this number of sixty cell lines." "If there are sixty lines that are robust, grow well and have the properties of human embryonic cell lines, that's news to me, and it is good news," Melton told the *Times*. "First let's find out if these sixty lines really do exist. Secondly, are they going to be available without restriction?"

Melton's concerns essentially echo the warnings about the dangers of a "compromise" decision that he, Gearhart, Pedersen, Jamie Thomson, and other leading scientists had articulated in June, warnings that Jeff Martin had passed along to Tommy Thompson (and, presumably, the administration). So the instant criticisms of August 10 should not have caught the White House by surprise. At the time, it probably seemed like scientific nit-picking, and that's certainly how the administration sought to portray it because Bush's announcement appeared to have played well to other constituencies. But the administration's efforts to paint a morally heroic background to the president's action and to draw a firm line in the sand actually limited the president's maneuvering room when it became apparent that there might not be as many cell lines as advertised.

On Friday morning, the day after the president's announcement, Tommy Thompson appeared at a press conference in Washington and promptly waded into the quicksand that inevitably swallows any politician who plays fast and loose with scientific facts. Of the sixty or so Bush stem cell lines, the secretary declared, in no uncertain terms, "They're diverse, they're robust, they're viable for research." In plain language, they were good to go. In fact, they were not, and as of this writing, more than a year later, all but a handful are still unavailable. As Doug Melton later complained, the "cell lines" were, in many cases, nothing more than cells plucked from blastocysts, sitting in freezers and still awaiting their biological moment of truth — thawing, plating out, growing — that would determine whether they in fact had the right cellular stuff to self-replicate and differentiate. Even in Jamie Thomson's skilled hands, only one in three blastocysts yielded a cell line; those were the kindest odds of success that could be superimposed on many of the Bush cell lines.

Second, the White House felt compelled to announce, within days of the original television address, that the president would never sanction federally funded research on any newly created cell lines. Since scientific knowledge, by definition, never stands still for very long, the presumption that the existing cell lines would forever satisfy scientific demands placed an inflexible and unrealistic corset over a dynamic, growing field. More to the point, it made no medical sense. The prohibition on new cell lines vastly increases the odds that progress, if it occurs at all, will be slow. And if no more cell lines can be created beyond the lines that existed on August 9, then embryonic stem cells will never be used in human medicine, because all the pre–August 9 cells have been kept alive by a contaminating layer of mouse cells, which might pass disease-causing viruses to the human cells. Indeed, the Bush policy reinvented the war that Leonard Hayflick had fought thirty years earlier. Everyone in the field acknowledges that new cell lines will have to be created, sooner or later; this was the unenunciated lie of the Bush policy. "If it is the case that the cells could not be used for treatment," said Senator Edward Kennedy, "it would be a damning restriction."

Yet three days after the announcement, Thompson was on the Sunday morning talk show *Meet the Press*, insisting that the president would not reconsider the strict limits he had placed on the research, and would never fund research that destroyed existing embryos. "That is the real distinguishing line," Thompson said, "and that's a high moral line that this president is not going to cross." By Sunday, the threat of a veto had been invoked against any congressional attempt to pass less restrictive legislation, and the president himself reiterated that threat the following day in remarks to reporters. On Thursday evening, a humble Bush had told the nation he prayed he was right about his decision; by Monday morning, he was certain he wasn't wrong and threatened a veto to prove it.

Through its cabinet members and its proxies, the administration stressed, perhaps too insistently, how morally courageous and bold the president's decision had been. At that same Friday morning press conference, Thompson declared, "Make no mistake. This is a bold step." The efforts of key White House officials, including Karen Hughes and Jay Lefkowitz, to brief reporters on the president's decision-making process were so strenuous that an account in the *New York Times* on August 11 quoted an unnamed administration official who questioned this public relations strategy; speaking of the Hughes press conference, this official said, "I've never seen such a thing. They're trying to stem any questioning on how he arrived at this . . . They're positioning him. They're trying to demonstrate how thoughtful and careful he is."

Perhaps the most peculiar part of that positioning was an op-ed piece signed by the president that appeared in the August 12 edition of the *New York Times*. For some reason, the White House felt compelled to reiterate, only three days after the president's televised address, that "while it is unethical to end life in medical research, it is ethical to benefit from research where life and death decisions have already been made." As a precedent, the president cited "the only licensed live chickenpox vaccine" used in the United States, which was "developed, in part, from cells derived from research involving human embryos." In fact, most vaccines in current use were created with human embryonic or fetal material much further along in development than blastocysts.

If nothing else, this last observation closed a circle of scientific respectability forty years in the making, for among the most avid and interested readers of those lines was Leonard Hayflick, back in California, wondering what all the fuss was about. Again. "Chickenpox vaccine is indeed made in my cells, or copycats of my cells," he said, "but it was probably the last human vaccine that appeared using our cells." Hayflick pointed out that Bush could also have mentioned vaccines for polio, rabies, and rubella, and even the adenovirus vaccine, which was prepared by the NIH and whose use has been compulsory for those in the military. Every one of those vaccines was created with the use of human fetal tissue. But then the morality of embryo research had long since been warped by politics. In the president's curious moral universe, destroying leftover embryos to create more stem cell lines was unethical, but allowing fertilization clinics to create and then toss out excess embryos was simply a legitimate cost of doing business. By some estimates, as many as 600,000 embryos have been created in the last decade at IVF clinics in the United States alone, and countless thousands — perhaps even the vast majority — have been discarded or destroyed. If it is murder, as the "culture of life" argument has it, why does the president allow the killing to go on for even one more day?

༄

Skepticism about the Bush policy mounted over the next week. On August 14, Harold Varmus and Douglas Melton wrote a polite but deeply skeptical op-ed piece in the *Wall Street Journal*, warning of serious consequences if the sixty or so stem cell lines were not as robust or viable as advertised. On August 17, the American Association for the Advancement of Science sent a letter challenging the Bush administration to identify the mysterious sixty cell lines. On August 20, the headline on a front-page *Washington Post* story captured the growing skepticism: "Viability of Stem Cell Plan Doubted." The updraft of expert doubt created its own weather, a kind of cultural cold front during that hot, lazy summer in which popular suspicion eased effortlessly into parody and derision. Soon the Chatterbox column of the online magazine Slate.com began soliciting anonymous tips from readers about the identity and quality of the alleged cell lines, promising to rate them with Michelin-style stars. In this increasingly irreverent atmosphere, White House spokesman Ari Fleischer made the preposterous assertion that the "burden of proof" about the quality of the stem cell lines lay with the doubters. Burden of proof? No one even knew what they were, or where, and the NIH refused to disclose any information about the mysterious cell lines.

It is clear, in retrospect, that the White House sent Bush out on national television without having vetted (or even understood) the biological status of the cell lines he had embraced as the foundation of his compromise policy. NIH officials were said to have been flabbergasted on the night of August 9 when they heard Bush announce "the number" in his speech; they hadn't been consulted on specifics, and had had no opportunity to correct this misrepresentation. Some will argue that the number was irrelevant compared with the overall thrust of Bush's speech, but in fact it was that number that gave the compromise policy even a shred of credibility among the people who would have to live with it — namely, the scientists and the patients. If the number proved not credible, so too would the policy. Hence the intense scrutiny when, on August 27, after more than two weeks of unusual public clamor (unusual, certainly, for August), the NIH finally revealed the identity of sixty-four cell lines eligible for federal funding (the total later rose to more than seventy-eight). "The Bush 64" came from, in alphabetical order: Bresagen, an Australian company with offices in Georgia (4 cell lines); CyThera, a biotech company in San Diego (9); the Karolinska Institute in Sweden (5); Monash University in Melbourne, Australia (6); the National Center for Biological Sciences in Bangalore, India (3); Reliance Life Sciences in Mumbai, India (7); the Technion-Israel Institute of Technology in Haifa, Israel (4); the University of California at San Francisco (2); the University of Göteborg in Sweden (19); and the Wisconsin Alumni Research Fund (5).

Far from putting the controversy to rest, this list lit the fuse on an extraordinary episode of pack science journalism; one public affairs official at a prominent university later told me that reporters were "out for blood" in their hunger to unearth new details about the cell lines. Each day brought stories that defrocked the status of various cell lines, and with each story the cracks in the edifice of the Bush policy became deeper and wider. CyThera, for example, had "derived" its stem cell lines only weeks before the Bush announcement and would not finish assessing their biological quality for months (remember that when Jamie Thomson originally isolated human stem cells in 1998 at Wisconsin, his standard for determining that they were indeed robust was to maintain them in culture for at least six, and preferably twelve, months). A researcher at the University of Göteborg told reporters that, at most, only three of the nineteen lines could legitimately be regarded as stem cells. The status of the cells from the two labs in India was, to put it mildly, murky. When the cell lines were finally listed in an NIH registry and Doug Melton examined their characteristics, the reality became clear to him and to many other researchers. The administration had not actually listed stem cell *lines* — that is, well-established self-perpetuating colonies of pluripotent cells that could differentiate into some two hundred specialized tissues. What they had collected and listed, rather, was every instance in which the inner cell mass of a human blastocyst had been dissected, harvested, and tossed in the freezer before the White House deadline. These cells might — or might not — ultimately yield embryonic stem cell lines. In many cases, the answer wasn't known.

Even the more established sources came with question marks. A spokesman for the University of Wisconsin group, for example, claimed that "Wisconsin's cells are the gold standard." Melton had sent away for them and reached a different conclusion. "The ones that WARF is giving out are complete duds," he told me. "This H-1 line doesn't grow well at all. H-9 is the good one, but they don't give that one out." (Melton obtained a version of H-9 from an Israeli collaborator.) Now, it should also be said that one or two truly robust ES cell lines could sustain basic research by virtually every stem cell researcher in the world. Indeed, Alan Trounson has said as much about a couple of the cell lines distributed by his company, ES Cell International, and Tommy Thompson repeatedly made the point that a huge body of research on mouse stem cells had been accomplished with fewer than half a dozen cell lines. What he neglects to mention — and what Melton never tires of pointing out — is that those five highly serviceable mouse lines resulted from a long, collective winnowing process that began with *hundreds* of cell lines. Melton guesses that it would take perhaps a thousand human embryonic stem cell lines to identify the best half dozen or so, and he is backed up in that estimate by someone who should know. Martin Evans, who in 1981 became the first scientist to isolate mouse stem cells, told me that "probably hundreds have got as far as being reported in the literature and many, many more have been used (or not) internally in labs. We ourselves have isolated many lines — not all of which are normal. I'd generally support Melton's guesses."

Although it was not reported at the time, key scientific institutions in the United States began to bail out on the Bush policy within days of the announcement — if they hadn't already given up on the NIH months earlier. In the spring of 2001, for example, before he became a scientific refugee in England, Roger Pedersen helped establish a freestanding private research entity under the auspices of the University of California at San Francisco, perhaps the nation's premier biomedical research institution west of the Mississippi; that August, the off-campus facility's research was "actively begun," according to a university spokesperson. Meanwhile, Rockefeller University, Memorial Sloan-Kettering Cancer Center, and Weill Cornell Medical College quietly launched discussions about raising private money to jointly create a privatized stem cell research group in New York City. In Cambridge, with the blessings of Harvard and funding from the Howard Hughes Medical Institute, Doug Melton set up a collaboration with a Boston IVF clinic and planned to derive his own "nonpresidential" stem cell lines from frozen embryos; Melton and HHMI promised to offer cell lines that resulted to researchers for free. The Scripps Institute and the Burnham Institute, both near San Diego, explored plans to set up privatized stem cell research centers, according to one researcher who was recruited. Medical foundations increasingly went the private route, too: in September 2001 the Michael J. Fox Foundation for Parkinson's Disease issued grants exceeding $4 million to researchers, in part to support work on human embryonic stem cell lines that would advance potential treatments, and in March 2002 the Juvenile Diabetes Research Foundation finalized a $7.5 million collaborative agreement with scientists in Sweden to derive more stem cell lines. Within weeks of the Bush announcement, Melton and his former postdoctoral fellow, Ali Brivanlou, began to approach venture capitalists to fund a new company, tentatively called Io, to pursue embryonic stem cell research in the private sector. In other words, many of the nation's leading stem cell researchers and institutions were voting on the Bush policy with their feet within days or weeks of the announcement, and they were walking — without attracting attention to themselves, but very much in a hurry — to private funders. Praised as politically brilliant in the days after August 9, the Bush policy seemed headed for disaster.

What had gone wrong? Why did the Bush policy turn so suddenly sour in the estimation of the scientific community? In his attempt to reach a compromise that would permit a minimal amount of research to continue, the president based his decision on an unrealistic understanding of scientific research in general and, more fatally, on faulty and incomplete information about the existing cell lines. In the view of lawyer Jeff Martin, as generous and sympathetic an adversary as the administration could possibly want, "It's not clear to me that it was vetted as much as it should have been. I respect the NIH; I know they were doing an honest job. But they maybe should have tried to track things down a little more before they gave the president that number." It is also possible, however, that the president and his advisers made a calculated gamble: with his sixty-odd purported cell lines in play, he could prevent the destruction of any further embryos and thus maintain his "culture of life" credibility with religious conservatives. At the same time, even this highly restrictive policy would buy time for the science — *his* kind of science, that is, with adult stem cells and with the limited ES cell lines — to advance.

There remains, however, an even greater long-term problem with the Bush policy. By limiting federal funds only to the use of stem cell lines already in existence on the evening of August 9, Bush, perhaps unwittingly, strengthened the de facto privatization of American embryonic stem cell research. Many of those existing stem cell lines had been created either by biotech companies or with their private support, and virtually all of them came with commercial strings attached. Hence, the Bush policy funneled all prospective NIH-funded researchers into complicated commercial relationships. This is not unusual in biological commerce these days, but it was immediately recognized as an enormous problem by the scientific community. "I think it's a bigger hump than the biological quality of the lines, frankly," said Harold Varmus. Joyce Brinton, a technology transfer lawyer at Harvard, said that negotiations for the use of stem cells involve "a much more complex set of issues," and that in terms of difficulty, "this one has to be up at the top." George Daley, a stem cell researcher at the Whitehead Institute, said that the tight intellectual property constraints on stem cell research, at such an early juncture in the evolution of the science, has had the effect of "basically choking the field."

The swell of scientific complaints was déjà vu all over again for researchers familiar with Geron's distribution of the telomerase gene. "The MTA and licensing agreement they sent to us is absolutely a joke," said Brivanlou, who flew out to California to discuss a stem cell collaboration with Geron in 2001. "The technology transfer person here was furious. Even though I would go out there to do it, they wanted Rockefeller to put up all the money, they wanted 100 percent ownership of whatever I did, and they wouldn't let me take any of my materials back with me when I finished. No one in their right mind would ever sign an agreement like that." On August 13, four days after the Bush announcement, the Wisconsin Alumni Research Fund tried to modify its agreement with Geron by filing a lawsuit against the California company. In court papers, WARF's lawyers stated that Wisconsin "desires to license additional stem cell types to parties other than Geron, and has a reasonable apprehension that Geron will attempt to interfere with any such licensing attempts." The parties later reached a settlement, but the lawsuit only highlighted the litigiousness of groups maneuvering to control stem cell technology and its potential therapeutic fruit. "I think there was some real tension in the relationship," said Andrew Cohn, a spokesman for WARF, "so much so that there was a lawsuit to clarify the relationship. But I think there's been a noticeable change in their [Geron's] behavior in terms of sharing and being open from before the lawsuit and after the lawsuit." While the practical implications of Geron's exclusive license remain unclear, the company's attitude was unambiguous. As CEO Thomas Okarma told the *Boston Globe*, "Patents are what patents are. We funded the work, we have the rights."

The patent situation may turn out to be a bit more complicated than even the Geron-WARF lawsuit suggests. Several scientists told me that the University of Wisconsin and the Rambam Clinic in Haifa, Israel, which provided the embryos that accounted for four of the five human stem cell lines derived by Jamie Thomson, were involved in a quiet, behind-the-scenes dispute over the original Wisconsin research. Joseph Itskovitz-Eldor has told other researchers he believes the four cell lines are "his." This argument may have already affected overall research, since WARF had, at least through the end of 2002, distributed only its H-1 cell line to researchers; the other four, all derived from Israeli embryos, had not been made available by WARF, although Israeli researchers have distributed at least one other cell line to collaborators. Neal First, a Wisconsin researcher, said of his longtime friend Itskovitz-Eldor, "I think he is perceived by WARF as a thorn in their side. They thought they had captured the entire embryonic stem cell field, only to find that one corner had not been sewn up." To add even more spice to this bubbling patent pot, by the end of 2002, Douglas Melton's lab had created two "nonpresidential" embryonic stem cell lines, using as a recipe the technique originally published by Martin Evans's group in 1981 for the isolation of mouse ES cells. "So if it's just like the mouse," said Melton, "what's the discovery? I mention that because of the patent issue."

All these questions thickened the air of late August, a kind of oppressive political humidity that marked the three weeks leading up to the Labor Day weekend. The robustness and viability of the sixty-odd cell lines looked increasingly doubtful. The patent issues were more complicated than the White House or Tommy Thompson may have been aware. And the press, in reporting each problem and shortcoming, accelerated a process of unraveling that continued unabated, day by day, like a political version of telomere shortening, until it looked like the Bush policy would become, in the lingo of cell biology, "apoptotic" — so unstable that it would self-destruct. That was where things were headed when senators summoned Bush's secretary of health and human services to answer some questions following the Labor Day break.

༄

Even Tommy Thompson's harshest critics had to empathize with his dilemma when the cabinet member and his entourage swept into a large, crowded hearing room in the Dirksen Senate Office Building on the morning of September 5, 2001. Here was perhaps the principal administration insider who'd stood up to Karl Rove over the previous eight months and kept the possibility of research alive. Here was the dogged, blunt-spoken bureaucrat who'd happily been fed lines by Jeff Martin to argue the legal and medical case for stem cells in the White House. Here was the former governor of Wisconsin, who'd taken almost fatherly pride in the research at the University of Wisconsin, and who was determined to make the Wisconsin stem cells a model for how the Bush policy could work. With his broad, ruddy face and caterpillar brows, his thick black sheen of hair, and his brusque, earnest midwestern accent, Thompson brought a kind of blunt, guileless everyman enthusiasm to this dauntingly complex scientific topic, and no one who had spoken to him during his time in office doubted the sincerity and passion of his support for stem cell research. "I think we would never have gotten to where we got without Tommy Thompson," said one source who requested not to be named. "I think he went and pressed the White House very hard." He had, a friend confided, been "taken to the woodshed" a number of times by the administration for speaking his mind, yet the fact that an antiabortion Republican had voiced early enthusiasm for stem cells, said Tony Mazzaschi, "allowed a lot of people to rethink their position." And now, appearing before the Senate Committee on Health, Education, Labor, and Pensions, chaired by Edward Kennedy, he had to defend a flawed, almost disintegrating policy.

Thompson made a game effort. Picture this no-nonsense Wisconsin Republican sputtering out tongue-twisting scientific terms like "blastocyst" and "proliferation stage" and "hee-may-to-poi-etick cells" to a semicircle of skeptical senators, and fending off a succession of tough questions about the administration's policy, deflecting shots from left and right like a hockey goalie. At one point, Senator Barbara Mikulski of Maryland remarked, "You sound like you were defending a Ph.D.," to which Thompson replied, "I could almost write one right now."

Thompson's message that day was to characterize criticisms of the Bush policy as misplaced and needlessly distracting; in the manner of contemporary political communication, he delivered that same message about a dozen different ways. "Our challenge now is to move out of the halls of debate and into the laboratories of science," he told the senators at one point in his prepared statement. At another moment, he said, "My hope is that we can clear up the misunderstandings and recognize that the only way we're going to resolve the speculation is to do the research." Later: "We have consistently said these lines are at various stages of development . . . but unfortunately, and I believe unfairly, some are choosing to engage in word games . . ." And still later: "We need to move beyond the back-and-forth over the numbers and get to work on the science." With great fanfare, Thompson announced that the NIH had negotiated a so-called memorandum of understanding and agreement — a legal document that covers the transfer of biological reagents between labs — with the University of Wisconsin to streamline distribution of the cells and relax some of the conditions that scientists like Doug Melton had found so objectionable. He promised that the NIH would publish its registry of the cell lines within two weeks. He revealed that the NIH would begin accepting grant applications from researchers on October 1 and that "it will probably take eight to nine months to get the money out." He suggested that stem cell–related treatments could emerge within four years. "President Bush has opened the laboratory door," he concluded. "Now let's get our best and brightest scientists into the lab so they can go to work."

But the senators kept returning to a nagging question: Just how wide open was that door? Kennedy sharply questioned the wisdom of the August 9 cutoff date and hounded Thompson about how accessible the cell lines truly were. Senator John Warner expressed concerns that researchers in other countries, unaffected by the Bush constraints, would enjoy a competitive advantage. Senator Tom Harkin, echoing concerns in the scientific community, expressed doubts that the limited number of Bush-approved stem cell lines, raised on mouse cells, could be safely used in therapies ("There's a cloud over that," he said, "a very dark cloud"). Senators Christopher Dodd and Hillary Clinton expressed similar concerns and kept pressing Thompson as to whether President Bush would revisit the arbitrary date prohibiting the isolation of additional cell lines. "The administration will not reconsider its deadline for the destruction of embryos," Thompson insisted once again. Dodd, reflecting a sentiment probably much more widespread than he imagined, observed at one point, "You know, people talk about mad scientists. But nothing is more frightening to me than Congress trying to be a scientist."

The most telling exchange of a generally contentious morning, however, occurred when Dodd got Thompson to admit that the mysterious number sixty-four was not so robust and viable after all. Asked how many of the cell lines were essentially good to go, Thompson replied, "We feel there are roughly twenty-four or twenty-five lines fully established, and sixty-four overall — and we hope there will be even more." With that concession, the Bush stem cell policy — received with so much applause and good will a month earlier — stood revealed in all its naked insufficiency. Despite Thompson's repeated exhortations about moving into the lab, the next day's headlines conveyed a different message: "White House Cuts Estimate of Available Stem Cells" (*Wall Street Journal*), "U.S. Concedes Some Cell Lines Are Not Ready" (*New York Times*), and "Thompson: Stem Cells

Not Ready" (*USA Today*). As of August 9, in other words, nearly two thirds of the "cell lines" did not exist.

One of the other witnesses that day was Douglas Melton. Testifying after Thompson's group filed out, Melton noted in his low and gently insistent voice that the August 9 cutoff date "was not chosen for scientific reasons" and predicted that the intellectual property restrictions would "impede progress." He expanded on his thoughts about the Bush compromise when I spoke with him later. "It's not an idea that would have occurred to me," he said of the August 9 cutoff. "Whenever people ask me how many lines would be needed, since one didn't want to set up a never-ending production, my response was a thousand. And that was a scientific guess based on the number of lines that must have been developed in studies of mouse embryonic stem cells in order to achieve the twenty-some lines which are now routinely used. And so it seemed to me inconceivable that with the handful of lines that were available at that time [August 9] that one would get even a single robust line. Probabilistically, it was just unlikely to have occurred. And that doubt turned out to be true.

"But let's suppose that I'm completely wrong," he continued, "that I'm just being nitpicky here and there *are* sixty lines, okay? The bigger problem is, Are they available? Can a researcher obtain them and do experiments with them? And that seems to me extremely unlikely. In fact, the code word to look for, which is listed in the NIH registry, is that these are only available by 'collaboration.' Well, what does that mean? To some companies, that means that we'll give you the cells if we own all the rights to anything you discover, and tell you what you can do with them. Well, my university won't allow me to accept them under those circumstances, nor should they. We don't want, as part of the nation's public health policy, to have companies telling researchers what experiments they should and shouldn't do."

Not everyone shared those sentiments, of course, but misgivings ran deep among both scientists and patient groups. In the space of a month, the administration's stem cell policy had gone from political home run to, upon further review, a foul ball. The day after the Senate hearing, Thompson felt compelled to summon a reporter into his office for damage control. "There are plenty of cells available to do the basic research," he declared to the *New York Times*'s Sheryl Stolberg. "Let's get on with it and stop quibbling about the numbers."

But the credibility of the policy — indeed, the credibility of the government's social contract with Americans to pursue the most promising avenues of research to find new treatments for diseases affecting tens of millions of citizens — resided in the credibility of those numbers. As of this writing, a year after the policy was announced and four years after the initial discoveries were made, the general consensus among researchers is that the number of usable and available stem cell lines is no more than three or four, according to stories in both the popular and scientific press. As soon as the NIH registry was posted in November 2001, Brivanlou's lab created a computer program called Stem Cell Tracker and systematically contacted the owners of every cell line listed, seventy-one at the time. They succeeded in obtaining only the H-1 line from Wisconsin, along with the promise of another line from the Trounson and Pera group in Australia. "You should know," he told me months later, "this is a complete nightmare. The bottom line is, you can't get the cells." Even the government policy was confusing. When Brivanlou consulted four lawyers at Rockefeller University, asking each if he was eligible to apply for a round of government funding in the spring of 2002, two lawyers said yes and two said no. In September 2002, researcher Curt Civin of Johns Hopkins reiterated that message, telling a Senate committee, "Embryonic stem cell research is crawling like a caterpillar."

This grating collision between societal oversight and cutting-edge biomedical research was deadly serious to millions of people with mortal illnesses, but it had occasional moments of comic relief, too. Surely one of the most frivolous was the cover of the *Weekly World News*, a supermarket tabloid, which featured two face-to-face pictures of the president with the screaming headline, "President Bush Cloned . . . and there's an imposter in the White House, say insiders!" This memorable image hit newsstands over Labor Day weekend, at a time when momentum seemed to be building in the Senate to take on the president's increasingly tattered stem cell policy. Senator Arlen Specter had vowed (not for the first time), "You're going to see a legislative free-for-all." The cover date of the tabloid, however, inadvertently advertised the event that would let all the air out of the opposition and, oddly, protect the president's stem cell policy from attack: September 11, 2001.

17

BEATITUDE

ALTHOUGH IT IS NEARLY IMPOSSIBLE, and often foolhardy, to try to paint a detailed picture of the medical future, you can at least get a sketchy, pointillist draft of the possibilities at a good scientific meeting, and that was certainly the case when the Society for Regenerative Medicine convened for its second annual meeting in Washington, D.C., at the beginning of December 2001. You didn't have to subscribe to chairman William Haseltine's belief that the fountain of youth "is likely to be found within our own genes." It was enough just to sit back and listen to the parade of astonishments currently being pursued.

Eric Lagasse, a scientist at StemCells, reported on experiments with mice in which adult stem cells culled from the bone marrow could be manipulated to restore liver function in animals with severe hepatic disease (4 million Americans, Lagasse noted, suffer from hepatitis C, the leading cause of liver transplants). Annemarie Mosely of Osiris Therapeutics described initial human trials of the company's adult stem cells in bone repair, following experiments in baboons in which, over the course of eight months, these cells rebuilt bone across two-inch gaps; within a year, she added, the company hoped to begin human trials of adult stem cells to repair cardiac muscle damaged by heart attack. Hans Peter Zenner of the University of Tübingen in Germany showed off his "titanium ear" — an artificial, bioengineered device that mimics auditory physiology (91 million people in the United States and Europe, he explained, suffer from modest to severe hearing loss). Biologists from Massachusetts General Hospital and Harvard Medical School updated their progress in tissue engineering — building new intestines on biodegradable scaffolding, building new bladders and ureters (already being tested in children with congenital defects), building new kidneys, even penises (so far, only tested in rabbits). Steve Badylak of Purdue University showed a picture of the wounded foot of a fifteen-year-old boy who'd been shot; something called an extracellular matrix had been placed over the wound, and it recruited wound-healing cells to the site of injury and completely repaired the hole, an approach that is also being tested in humans suffering from rotator cuff, Achilles tendon, urologic, and dermatologic injuries. Lorenz Studer of Memorial Sloan-Kettering Cancer Center in New York described mouse experiments in which therapeutic cloning of cells snipped from an adult mouse's tail allowed researchers to isolate embryonic stem cells that were then nudged and nurtured into becoming the type of neurons missing in Parkinson's patients. Simon Melov of the Buck Institute in California, working with a Massachusetts company called Eukarion, described experiments in which antioxidant compounds seemed to extend the life span of mice. In what sounded like comic relief, Doros Platika, formerly the head of Curis, described the therapeutic potential of a gene called *hedgehog* to stimulate hair growth, although he noted, "There appears to be an even bigger market in *antagonizing* the growth of hair in women who don't like the pain and agony of wax treatments."

Some of these developments were years away from clinical application; others were already being tested in humans. Taken together, they represented a sliver of the spectrum of biological knowledge and human ingenuity that promises to transform our notion of healing. Nobody talked about immortality pills, but every biomedical innovation not only held the promise of extending life, but promised to make that life less painful and of higher quality.

Perhaps the biggest headliner at the conference was someone who had been invited to speak only a couple of days earlier, someone who, as Irving Weissman said in his introduction, "probably needs no introduction." Around 11 A.M. on that Sunday morning, Michael West stepped to the podium, opened a titanium notebook computer, and began to describe, to a scientific forum, the same experiments that had ignited an eruption of public opprobrium over the previous two weeks: the first tentative steps his company had taken toward creating a cloned human embryo, for the purpose of obtaining stem cells. Somehow, in this setting, with this audience, the public outcry seemed as quaint and old-fashioned as the vision of nature depicted in the landscape paintings lining the auditorium walls.

"I thought I would begin by having some fun," West said, launching into one of his deep-horizon reveries, musing on science's potential in the near future to reprogram human cells, to rebuild telomeres, to "continually rejuvenate cells." That aim, he added, sounding a theme that had resonated with audiences since Redwood Shores, was "making young cells for old people."

"As some of you know," West continued with a coy smile, "we recently published our preliminary data." Indeed, two weeks earlier, in a bottle-rocket burst of news that soared explosively and just as quickly plummeted back to reality, scientists at Advanced Cell Technology claimed to have created a human embryo by way of nuclear transfer. The experiments, and the way they were retailed through news outlets, generated far more controversy than scientific data. But on this quiet Sunday morning in the nation's capital, insulated from the high-decibel social debate and focused purely on the science, everyone sat, skeptical but intrigued, as West flashed a slide showing a glowing batch of orange-hued cells seemingly floating, like a clutch of balloons, on the end of a pipette. This was, the company had claimed, the first cloned human embryo. Up there on the screen, in this darkened and rapt room, it glistened with a kind of internal light, science's false-color, magnified version of a stained-glass window — iconic, aglow with some internal fire, inspiring awe at the handiwork and awe at the beauty of life itself, balanced fragile and protean on the tip of a needle.

As usual, West didn't shy from the big picture. "A lot of debate is in front of us right now," he said, "and it will likely be decided in the U.S. Senate. What is this?" he asked, nodding at the image up on the screen. "What should its moral status be? And how should it be treated in medicine and in society?" Gazing up at the picture, he repeated his frequent précis of embryological development, how structure did not even make an appearance until the primitive streak etched the developing organism at fourteen days, how individuation had not yet occurred, how personhood had not yet begun. "You're confusing cellular life with a human life," he said. "The idea that life begins at conception is logically flawed."

West was of course preaching to the converted, but there wasn't the faintest echo of the public furor that had greeted the news of these same "cloned" human embryos over the previous few weeks. No one in the audience questioned the premise of ACT's experiments (although they expressed eagerness to see more data); no one questioned the morality of experimenting with embryos (although privately there was considerable skepticism about what the ACT researchers had actually accomplished). "It was very preliminary data," observed Ian Wilmut, who sat in a back row through West's talk, "and it was presented in a very sensational way."

Later, when I had a chance to speak with West at the hotel, he placed the social controversy, including increased attention by the U.S. Senate, in a larger historical context. "There are a few people out there trying to be very careful and thoughtful," he said. "I sure hope so! This is . . . *such* an important debate. And it's actually a fascinating story. You know, this unique nexus of science and religion and politics and so on. All these stars don't normally align quite like this. But you know what's sad here is that the Christian church has got this nasty habit over the centuries, at least, of acting first and thinking second. You know, putting Galileo in prison because he had a different interpretation of the stars. If they'd thought about that carefully back then, they might have said, 'Wait a minute, the Bible doesn't say the earth is flat. The Bible doesn't say the sun revolves around the earth. And what harm does it really do?' The argument was, 'Well, that diminishes our place in nature, and will lead to, you know, *immorality*'" — here he gave a little laugh — "'if we have this view of the world.' But c'mon, any reasonable person would have known that, even back then. So why? Why this reaction?

"It's similar here. Any reasonable person would recognize that preimplantation embryos are not a pregnancy, and haven't individualized. I keep harping on that point, but the fact that they haven't individualized, that they're completely open to becoming even multiple people, tells you something very profound about what their moral status should be. And *they* know that. The church *knows* that, the Catholic Church. I've debated them on this point, and they don't argue against it. They just change the subject and then use inflammatory language. You know, 'embryo farms' and stuff. But it's interesting to see that the church hasn't learned this lesson, because they're making [the mistake] again in modern times."

West expressed hope that the Senate would act responsibly, but hints of another Christian tradition — martyrdom — crept into his speech. "What people who are in the know tell me," he said, "is . . . that there's not going to be a lynch mob. A lynch mob won't win. Brownback clearly wants to, you know — let's go hang this. No trial. And my understanding is, that's highly unlikely. But you never know," he said with a shrug. "Things happen."

If Congress went on to ban all forms of cloning, a lot of people — including a lot of scientists — believed that Michael West would have been a silent coauthor of that legislation because of the way ACT's cloning experiments agitated the public. Yet the most telling interaction between West and his peers at the meeting occurred outside the auditorium later that same day. West had stopped by the registration desk to pick up one of the perks of these sorts of conferences: a T-shirt. As he waited by the desk, a stem cell researcher edged up to him and blurted out, "I just want to thank you for pushing this debate." The scientist explained that as an NIH employee he was constrained in what he could say in public but wanted to express his gratitude. "I'm all for adult stem cells," he continued, "but we all know that the embryonic cells are the ones we need."

With a beatific expression on his round face, West thanked the man.

෴

It had been quiet, although not surprisingly so, during the months following George W. Bush's stem cell policy announcement. True, a national calamity of unprecedented dimension had unfolded on September 11, relegating the stem cell controversy and all other matters to the subbasement of political preoccupation. But Michael West did not assume a prominent role in this debate by deferring to distractions, even patriotic distractions. He allowed what he felt to be an appropriate amount of time to pass, and then he stirred the pot again. As someone affiliated with Advanced Cell Technology later told me, "I think he felt this debate was too important an issue to be ignored for so long."

Following the president's stem cell announcement on August 9, the traveling bioethics road show moved on to its next battleground: human cloning. In some respects, there wasn't much disagreement on this point. The closest thing to a social consensus on any issue of modern biomedicine seemed to have coalesced around cloning during the intense public conversations about this new technology after the birth of Dolly was announced in 1997: with few exceptions, no one wanted to allow what became known as reproductive cloning, or more colloquially, cloning for baby making. The fault line in the cloning debate lay just this side of baby making, however. It was the kindred technology known as therapeutic cloning, a term that, if not originally coined by the British scientist John Gurdon, was most authoritatively wielded by him. Gurdon participated in some of the early frog cloning experiments in 1960 and, with his colleague Alan Colman, had outlined the practical uses of this technology in a *Nature* article in 1999 called "The Future of Cloning." They described the technique and mentioned the obvious potential applications, including therapeutic cells and replacement tissues and organs. It was the closing statement that gave pause. "As soon as any new scientific technique works at all," Gurdon and Colman wrote, "it is almost always improved in both efficiency and ease of operation. This seems likely to be the case for cloning technology, too."

Opponents of "therapeutic cloning" used the term dismissively. What was so therapeutic about it? they demanded. Show one example where this technology had improved the medical lot of a single creature, man or beast. More to the point, the opponents of reproductive cloning vigorously opposed therapeutic cloning, because they were convinced that any refinement of this technology, regardless of intent, would bring society a fateful step closer to the birth of a cloned human being. Such an event, Leon Kass had frequently argued, would diminish our humanity and fundamentally change what it means to be human. Religious conservatives also argued that the mere act of cloning — that is, inserting an adult cell from a patient, say, into an egg cell stripped of its DNA, for the purpose of harvesting immunologically compatible stem cells — would create a human embryo that would ultimately be destroyed for purely utilitarian purposes.

But all this was merely the latest iteration of an irreconcilable social debate dating back thirty years, and it was right into the middle of this contested terrain that Mike West willfully waded in the fall of 2001.

⌒

In November 1999, I asked Michael West if the use of human oocytes — egg cells donated by human females — was under consideration at ACT, and he evaded the question. As it turns out, human cloning was not only on the table by then — the first steps had already been taken. In September 1999, Jose Cibelli had met with a fertility expert at Harvard named Ann Kiessling, "who," according to a subsequent press account, "agreed within five minutes to help set up a program to collect eggs from women."

The ACT scientists, meanwhile, busied themselves laying the groundwork for scientific and social acceptance of this controversial research initiative. West, along with his main scientific colleagues, Cibelli and Robert Lanza, wrote a kind of manifesto arguing the case for therapeutic cloning that appeared in the journal *Nature Medicine* in late 1999. ACT's most important and revealing publications on the issue, however, went unsigned; they took the form of advertisements that began to appear in Boston-area newspapers, including the *Globe*, in September 2000, according to a detailed and fascinating account by Antonio Regalado in the *Wall Street Journal*. "Research team seeks women aged 21–35 with at least one child to donate eggs for stem cell research," the ads read. "Compensation for time & effort." The company's search for human egg donors was already under way.

West's protestations of transparency notwithstanding, this form of self-serving corporate opacity thrives in an environment where reproductive medicine is divorced from federal funding and therefore from NIH oversight and monitoring. Nor was it limited to ACT. In the summer of 1999, several press reports, including a story in the *Washington Post*, suggested that Geron was creating human embryos for research purposes through cloning. Company officials vehemently denied these reports. Thomas Okarma, head of Geron, told the *New York Times* it wasn't true. When I asked Okarma about it myself a few months later, he said, "We are not now using, and not then, any kind of experimentation to do nuclear reprogramming with humans. Period. Full stop." When I asked a member of Geron's ethics advisory board about it, she professed ignorance. In May 2002, however, David Hamilton and Antonio Regalado of the *Wall Street Journal* obtained documentation showing that Geron had indeed supported a substantial effort to clone human embryos in Roger Pedersen's lab at the University of California at San Francisco — an effort that began in the first half of 1999. Okarma attributed his previous denials, according to the *Journal* story, to "semantics." If the public has a growing distrust of biotechnology, it may be precisely because of the industry's flexible definition of "transparency."

The first step on ACT's road to creating cloned human embryos was nonscientific: it was to establish an ethics advisory board (or EAB), as West had previously done at Geron. To West the self-styled "truth seeker" and amateur philosopher, this was as much a labor of love as an aspect of corporate good citizenship; it was also shrewd business. Although his successors at Geron dispute this, West claims to have been the motivating force for the creation of the bioethics board at Geron, which didn't formally convene until after he had left the company in 1998. But when he first described that history to me — how he had driven up to Berkeley in 1997 with

Andrea Bodner, a Geron scientist, to meet with prominent West Coast bioethicists like Laurie Zoloth, Karen Lebacqz, and Ted Peters; how they had discussed in general terms the ethics of stem cell research, as well as the framework for an advisory committee; and how, to hear West tell it, he had to battle Geron's management to preserve the independence of this group, in part by giving them the "power of the pen" to write anything they wanted about the company — he proudly mentioned an article that appeared in a biotechnology trade journal that heralded Geron in a headline as being "Ahead of the Bioethics Curve."

With West, as I had learned, there is always another version to the story, and I heard it from Karen Lebacqz. "At the time Mike West was still at Geron," she said, "they had gone to a person who invests in venture capital. This person knew some of us who were involved in the Graduate Theological Union, and told them, 'I will not fund you or contribute to this unless I am sure that you are attending to the ethical issues that the research raises.'" In this case at least, ethical review appeared to be merely a midwife to fund-raising.

West set out to create Advanced Cell Technology's EAB in 1999. He sounded out Arthur Caplan at the University of Pennsylvania about playing a role, but Caplan said he wanted a sizable compensation, in the form of a company donation to the Center for Bioethics at Penn. "Next I heard, he was sort of saying, 'Well, we've got an ethics advisory board in place,'" Caplan told me, "so I just backed away." Later in 1999, at a scientific meeting in London, West asked Caplan's colleague at Penn, Glenn McGee, to chair the committee, and McGee agreed. West subsequently secured the participation of Kenneth Goodman, a bioethicist at the University of Miami, Lee Silver of Princeton, Ann Kiessling of Harvard Medical School, and Ronald M. Green, director of the Ethics Institute at Dartmouth College. Among their immediate peers, they represent a kind of erudite, activist, almost radical fringe when it comes to issues of reproductive technology and regenerative medicine. Silver, for example, had become a fixture on *Nightline* and other network talk shows, publicly defending and even extolling the notion of enhancing children through genetic manipulations; Kiessling had devoted considerable effort and ingenuity to developing fertility techniques like "sperm cleansing," which allows HIV-infected couples to have children.

In many respects, Green was the key figure. He had served on the ill-fated 1994 NIH human embryo research panel and had authored one of the most important (and certainly the most controversial) working papers to emerge from that group — a consideration of the moral status of a human embryo. This argument, recounted in detail in Green's book *The Human Embryo Research Debates*, essentially concluded that a preimplantation embryo — an embryo before it implants in the womb, a week or two after fertilization — is not a person. "Because it is not sentient," Green wrote, "it is not physically harmed by its use in research, and its loss of opportunity to come into being is not something we appear to regard as a significant concern." Later, in 1997, he served on a special committee of the American Association for the Advancement of Science charged with examining the propriety of human embryonic stem cell research. Both experiences left him morosely convinced that therapeutic cloning, a technology he passionately believed was of great biomedical potential, would inevitably be sabotaged in the public sector by the political machinations of the right-to-life movement, and that scientific progress would therefore occur only in the private sector. Bearded, well-spoken, wearing his intellectual heart on the tweed sleeve of his professorial sports coat, Green had left a long paper trail not merely suggesting, but almost shouting, a well-articulated argument for the morality of embryo research, including human therapeutic cloning. When West attended a symposium on stem cells at Dartmouth, Green approached him and said, "If I can ever be of assistance to you, let me know, because I'm very committed to this research direction." So West knew exactly what he would be getting when he called up Green in August 2000 and asked him to head the board. When ACT's ethics advisory board took up the issue of research cloning that same month, the conversation concerned itself more about how, rather than whether, to do human cloning.

The eight-member group first met in late August, at a hotel at Logan Airport in Boston. But these bioethicists did not engage in a lengthy Socratic-at-the-Hyatt kind of dialogue, and the public relations value of an EAB can easily be inferred from the way *Scientific American* later described their activity. "Advanced Cell Technology," the magazine reported, "assembled a board of outside ethicists to weigh the moral implications of therapeutic cloning research." In fact, there was very little weighing. Green told me later that the board didn't spend a lot of time talking about whether therapeutic cloning was ethical or not, because there was already general agreement on that point. "What we said is, our purpose is not to determine whether this research direction itself is acceptable," he said, "but to do this right, in terms of all the human subjects and other issues . . . Now, if you wish, that is not the broadest possible social discussion. But I'm not persuaded that if we can't come to consensus on that at the national level, that individuals and individual companies should be prohibited from proceeding with this research. And then they need what I call internal ethical guidance on how to do this ethically." Indeed, this kind of preemptive private consensus accords with one of the criticisms that have been leveled at company-affiliated boards (or any ethics panels, for that matter); through careful selection of the membership, you preselect the outcome of any "debate" you ask the bioethicists to settle. A prominent bioethicist once said to me, "You tell me the composition of an ethical advisory board and I'll tell you whether they will give a green light, a red light, or a yellow light to a particular line of research." When I ran the membership of ACT's bioethics board past him and mentioned therapeutic cloning, he immediately replied, "They would give a green light." Even West conceded as much when, speaking of a government ethics commission, he said, "There is, I think, a misunderstanding that if you set up an ethics panel, there are some sorts of ground rules, like principles of accounting, and everyone knows what those rules are. In the field of ethics, there are no ground rules, so it's just one ethicist's opinion versus another ethicist's opinion. Really, all you're getting is a bunch of people's opinions. You're not getting whether something is right or wrong, because it all depends on who you pick."

ACT's ethics board focused its initial efforts on two worthy issues: the proper ethical protections for an egg donor program, and the proper financial compensation for board members. There has been a long-standing professional unease among many bioethicists about accepting money for moral counsel; a conservative bioethicist once remarked that company-affiliated boards of this sort represent "the best ethics money can buy." Green argued that ACT's board members should accept no fee at all; ultimately, the group decided to accept the same modest compensation as scientists who participate in NIH-sponsored peer review meetings, which is $200 per day plus expenses.

Much more involved were the discussions about the egg donor program. The board formalized a number of ethical rules involving informed consent, compensation (donors would be paid an hourly rate that amounted to about $4,000 per donation cycle), physical and emotional safeguards for the donors, and security measures to assure that no eggs — or, if they were lucky enough, embryos — could be spirited away. Green said there were "moments when things that the board recommended substantially slowed down ACT's research. *Significantly* slowed down the research." But in some respects, ethics was still chasing the science. "There had been an earlier effort at a harvest procedure in the autumn [of 2000], I think, following our early meetings, with a very preliminary consent form," Green said. "None of us were happy with that whole process, and so we went back and we looked very carefully at everything, and that slowed everything down." Green insisted, and West concurred, that the EAB's work on the egg donor program substantially delayed ACT's progress on this avenue of research.

Regardless of the rationale, the program instituted by ACT nonetheless provided ethical cover for the controversial experiments the company was about to commence. Arthur Caplan — who likes to categorize the role of ethics advisory boards as "hackers" (which identify every conceivable problem that might arise), "problem solvers" (which address specific issues as they come up), or "a shield" (which deflects outside criticism and provides cover) — said he thought ACT's board functioned mostly as a shield. "There's a model where you're hired to be a shield, to deflect the fact that someone's got criticisms," Caplan said. "And the company says, 'Well, we've thought this through, and we're kind of immune to criticism because we have the best minds thinking about this around the clock.' ACT's board sometimes looks to me to be playing that role, not hacking but deflecting."

ACT's bioethics board has not been without moments of controversy. A potential ethical conflict — and one that has received little attention in the press — involved the egg donor program. A major architect of the policy was Ann Kiessling, a doctor in reproductive medicine affiliated with Harvard's Beth Israel Deaconess Medical Center in Boston. But Kiessling also actively participated in the scientific research at ACT on those eggs — so much so that she was the second author on the cloning paper that ultimately emerged, an indication of major involvement. Moreover, the procedure to harvest human oocytes from female donors was performed by a team of doctors at a private clinic founded by Kiessling in Somerville, Massachusetts, so she also had a business relationship, no matter how tangential, with ACT. Green argued that having Kiessling on the ethics board allowed important issues surrounding egg donation to be incorporated into

ACT's policy, but when I asked other bioethicists about this, Tom Murray, Arthur Caplan, and others agreed that it represented, in Caplan's words, a "screaming conflict of interest" to have the people invested in research on those eggs simultaneously defining the ethical policy by which they are obtained and used. It is difficult to imagine a government agency like the NIH allowing a similarly problematic blurring of relationships. It's that kind of episode that causes people like geneticist David Cox, a member of the Clinton bioethics panel, to look at corporate ethics boards with great skepticism. "People either shove ethics under the rug," he said, "or shine a light on it where it doesn't matter."

⟡

If ethics should walk hand in hand with science, as West told Bill Clinton's presidential commission in 1998, it has to move at a pretty good clip, in Worcester and elsewhere. Scientific journals were increasingly filled with reports about experiments in animal cloning, but perhaps the most significant — although somewhat overshadowed by the stem cell debate in Washington — appeared in late April 2001, when a group headed by Peter Mombaerts of Rockefeller University published an experiment in *Science* that established the general feasibility of therapeutic cloning, at least in mice (a more complete demonstration would be published in April 2002 by George Daley and Rudolf Jaenisch of the Whitehead Institute). Mombaerts, a bustling Belgian-born scientist who likes to fly helicopters in his spare time, got sidetracked into the field of cloning as a way to explore his primary area of interest: the neurobiology of smell. Indeed, his dream project was to create a mouse from a single cloned neuron isolated from the olfactory bulb (and if his ambition to clone a creature from its nose sounds faintly like a plotline from an old Woody Allen movie, it should; Mombaerts himself refers to the experiment as "the Sleeper project").

The Sleeper project became a distinct possibility in July 1998, when scientists at the University of Hawaii successfully cloned mice for the first time. Mombaerts was soon on a plane for Hawaii. "I was the first one actually to contact them, and I was there three weeks later," he recalled. "I thought this was like the biggest thing to happen in biology for me." He eventually did more than collaborate with the Hawaiian group; he lured two of its key scientists, Teru Wakayama and Anthony Perry, to join his lab at Rockefeller, and by the winter of 2000–2001, they had their own mouse cloning project clicking along. Teru Wakayama, by painstakingly injecting the nuclei from single cells isolated from a mouse's tail into enucleated mouse egg cells, had created dozens of mouse embryos; as Mombaerts later said admiringly, "He manipulates a microinjector the way Tiger Woods manipulates a golf stick." From the resulting embryos, they isolated blastocysts, harvested mouse embryonic stem cells from them, and then walked the pluripotent cells across York Avenue to the laboratory of Lorenz Studer at Memorial Sloan-Kettering Cancer Center. Studer, who trained in Ron McKay's lab at the NIH, coaxed these embryonic stem cells into becoming neurons. Not just ordinary neurons, but the dopamine-producing neurons typically missing from the brains of Parkinson's patients. Mombaerts and Studer didn't take the final step — injecting these neurons into a Parkinsonian mouse to see if the cells actually had therapeutic powers — but the experiment was nonetheless a tour de force. It showed that, in principle, therapeutic cloning could be a source of presumably immunologically compatible stem cells that could be transplanted into a patient as a customized form of therapy. In this case, of course, the patient was only a mouse, but no one had to connect the dots for Mike West. Indeed, by the time the experiment appeared in *Science*, West had already lured both Wakayama and Perry to Worcester to work their cloning magic at ACT, and he had recruited Mombaerts to sit on ACT's scientific advisory board.

That was not the only eye-popping technology being explored by ACT, either. On the same day that President Bush was meeting with Doug Melton, Leon Kass, and Daniel Callahan about stem cells, I paid a visit to Lorenz Studer at his lab at Sloan-Kettering to discuss an even more futuristic technology that was well under way. Swiss by origin, Studer had done some early clinical experiments in Europe using human fetal cells to treat neurological disorders like Parkinson's disease, but he quickly realized the limitations in terms of material; cells from four fetuses were needed to treat a single patient, so he began looking for a different source of therapeutic cells. At this point, he turned on his computer and clicked on an astonishing series of slides. They showed embryonic stem cells that had been harvested from the early embryo of a rhesus monkey. ACT had sent the cells to Studer, and he had nudged them into forming several different tissues. "They can make heart cells," he said, clicking on a slide that showed them beating right there on his computer screen, "and ciliated cells, and brain cells that can switch into glial cells or neurons or dopaminergic neurons. So they are real, true embryonic stem cells." But the real kicker was how the monkey embryos had been created. They were not the product of the traditional union of egg and sperm, or even the nontraditional union of egg cell and adult nucleus, as in cloning. They were "parthenotes." That is, they had begun as egg cells only, eggs that had been tricked into thinking they had been fertilized (with a combination of chemicals and electrical shock) and then began developing. This research wouldn't be published until February 2002, nearly seven months later, but it was clear that Michael West had his cloning team at ACT hot on the trail of human embryonic stem cells in the summer of 2001, either by nuclear transfer or by parthenogenesis.

West's permanent sense of urgency only deepened when another political furor erupted during the summer. On July 11, the *Washington Post* reported that the Jones Institute for Reproductive Medicine in Norfolk, Virginia, had published a paper in the journal *Fertility and Sterility* in which scientists at the clinic described the creation of some forty human embryos strictly for research purposes; what's more, three of the embryos had yielded new embryonic stem cell lines. A day later, in another dubious episode of me-too "situational transparency," West admitted to several newspaper reporters that ACT had similarly embarked on a project to create cloned human embryos — again, not to make babies but to obtain embryonic stem cell lines. The back-to-back reports, only weeks before the House of Representatives would vote on a bill to ban cloning, poured gasoline on the already inflamed rhetoric of stem cell opponents.

West had no illusions about the shifting political winds. On July 31, the House of Representatives indeed passed its bill banning all forms of cloning — a bill that, if it became law with Senate passage, would criminalize the very research ACT was pursuing at breakneck speed in Worcester. On August 9, in announcing his stem cell policy, President Bush specifically expressed his moral contempt for all forms of cloning. Because the president's policy allowed a limited amount of embryonic stem cell research to proceed, the political terrain had shifted, at least temporarily. There seemed to be a more limited appetite on Capitol Hill to push for legislation that would expand the administration's stem cell policy, especially after September 11. There seemed to be very little appetite at all for legislation that would defend the legality of therapeutic cloning.

When West returned to Worcester after the July hearing, it was to work on the controversial technology that might allow ACT to create a human clone — an effort that, as a *U.S. News & World Report* account later put it, would "be hailed as the hugest medical breakthrough of the past half century . . ." Not quite. But it did make news.

உ

Jose Cibelli had begun human cloning in earnest during the month of July, when the first human eggs had been harvested in Somerville. According to procedures established by the company's ethics advisory board, ACT researchers would enter a small, locked room just off the main lab at ACT, and, under the constant gaze of a surveillance camera (to make sure no one attempted to filch the resultant embryos), they would perform the nuclear transfers. They were observed by more than the cameras; West had granted three magazines more or less "exclusive" access to follow ACT's human cloning experiments. This was either a daring form of corporate transparency, or a scenario worthy of Feydeau.

Contrary to the shorthand impression you sometimes get from newspaper headlines, the technique of nuclear transfer is not easy and certainly not efficient. Using seventy-one human eggs procured from seven different female donors, Cibelli attempted three rounds of microinjections, without a single success. The people who donated their somatic cells for these experiments, of course, did so with informed consent; under normal circumstances their names would be held in strictest confidence. In reality, the donors played a role in the public debate, and they could not have been more "photogenic" in terms of making a controversial technology more socially appealing. One of them, Judson Somerville, was an old friend of West's; they'd hung out together at the pool in the Houston apartment complex where they lived in the 1980s. He was a forty-year-old physician from Texas, Episcopalian, and a self-described conservative who had contributed to George W. Bush's gubernatorial and presidential campaigns. He had been confined to a wheelchair since 1990 because of a bicycling injury (in Worcester, of all places) that left him paralyzed. Another donor, identified by an article in the *Atlantic Monthly* as "Trevor," was a two-year-old boy who was dying from a terminal illness. Not to belittle their personal tragedies, but Central Casting could not have sent over more fitting symbols to represent the potential beneficiaries of research cloning.

Cibelli's first three rounds of experiments with the precious human oocytes failed to result in a single embryo. Only three of nineteen attempts

at nuclear transfer even began to divide, and all three of these fitful beginnings involved the use of so-called cumulus cells — these are cells that
hover, like tiny clouds, around the mother planet of a woman's egg cell.
This may be an extraneous point in the context of preliminary experimentation, but males don't possess cumulus cells. Indeed, the problem with cumulus cells — and with egg cells parthenogenetically tricked into becoming embryos — is that the resulting stem cells would hold out the promise
of immunological compatibility for only the distaff portion of the human
family. Men, in other words, would find this particular Holy Grail empty.

Cibelli finally got the technique to work, at least a little, on the evening
of October 10. Within two days, several of these eggs had begun to cleave,
or divide, and when the ACT scientists went to check on October 13, three
days later, a few of these man-made zygotes had indeed gone through several rounds of division. In its first-draft chronicling of this event, *U.S. News*
noted, with perhaps a slightly wide-eyed view of its historical importance,
that in them ACT's scientists "saw a revolution in medicine that will render
many of today's drugs and treatments obsolete." West, Cibelli, and Lanza,
in an only slightly less hyperventilating account in *Scientific American,*
wrote, "They were such tiny dots, yet they held such immense promise."

To get an idea of the speed with which West and his colleagues ran to
publish these preliminary results, consider that the ACT scientists worked
barely two more weeks after getting that first "embryo" before submitting a
paper to an electronic journal called *e-biomed: The Journal of Regenerative
Medicine.* "That snapshot," West said of the paper, "was toward the end of
October." And so it was that on the morning of November 25, 2001, during
Thanksgiving weekend, Mike West appeared on *Meet the Press.* As host Tim
Russert waved a freshly minted copy of *U.S. News & World Report,* its cover
heralding "The First Human Clone," West was breathlessly debriefed about
this stunning development. That kicked off a seventy-two-hour odyssey
of molecular ministry, scientific explanation, moral justification, and, of
course, corporate self-promotion on the airwaves. The news that ACT
claimed to have cloned a human embryo triggered an orgy of journalistic
coverage, social revulsion, scientific disgust, and editorial indigestion the
likes of which had not been seen since . . . well, no one could cite an appropriate precedent.

In protestations of innocence, West later took pains to explain to me
how the timing of ACT's announcement occurred completely beyond the

company's control. *U.S. News* had dictated the timing, he said. The company had no idea what the magazine would be writing, he said. The magazine arranged the appearance on *Meet the Press*, he said, not ACT; he just agreed to appear when they asked. ACT published its findings in an obscure electronic journal, he said, because he wanted to help the journal out. "I didn't think it was a *Science* paper," West said. "*Science* would say, you know, show us a blastocyst, show us stem cells." (They wouldn't have been the only ones, either.) "I like the *Journal of Regenerative Medicine*," he continued. "I'm trying to help, you know, submit papers to it. And I thought, Well, they have e-publishing, can get it out relatively quickly. I felt that this data should be out there, because it's important to people. Or at least people are always asking us for the data. But that's it.

"For some reason," he added, a look of bewilderment clouding his face, "no one wants to believe the simple explanation."

〜

On November 28, 2001, gossip mavens Rush and Molloy, of the *New York Daily News*, began their daily column with an item about a scheduling tussle between several morning talk shows over their efforts to snare the appearance of Judson Somerville as a guest. Within a minute of West's appearance on *Meet the Press*, Somerville later told me, the phone started ringing in his Laredo, Texas, home and didn't stop ringing for three days. Within a few hours of that first call, he and his wife were on an airplane to New York, headed for a network interview the following morning. Somerville had agreed to appear on NBC's *Today* show, which was footing the bill for his stay at the Essex House in New York. Nevertheless, Diane Sawyer of ABC's *Good Morning America* managed to reach him by phone in his hotel room, according to the *Daily News*, and tried to sweet-talk him into an interview. Representatives of CBS's *Early Show* had also come a-wooing. One of the networks even had his hotel room staked out, although Somerville told them he had committed to the *Today* show. "They said," he recalled, "that it was okay to screw the other guy, because 'that's the way we do it in New York.'" This little frenzy of unpleasantries was generated by Somerville's newfound — and, as it turns out, unfounded — status as "the world's first cloned person." He wasn't that at all, but the requirements of instant news — and instant "gets," in the parlance of the morning shows — help explain how a failed experiment conducted by a tiny private company could, if the claim were sensational enough, leverage national and indeed global attention.

Somerville had journeyed to New York with the blessings of Advanced Cell Technology and Michael West, who in fact appeared alongside his old friend on the *Today* show. Somerville seemed the perfect candidate to describe the promise of therapeutic cloning. In fact, he seemed too perfect. Did his public relations value inform his selection as a donor for these very preliminary ACT experiments? The company insisted not, but then West told me he'd also considered asking the actor Christopher Reeve if he would donate his cells for cloning; surely the company would not want to publicize *that*. In any event, Somerville seemed to entertain all interview requests (except Sawyer's), and was the living embodiment of therapeutic cloning's potential. But the frenzy to interview him also reflected the vacuum of critical judgment in which the cloning story took off. If the embryo created from Somerville's adult cells had matured to the blastocyst stage, and if embryonic stem cells could be harvested from it, and if those cells could be maintained in laboratory culture, and if ACT's scientists could then figure out the right way to propel them toward a neural cell fate, and if they could be reintroduced into Somerville's body in a safe and efficient manner, and if they turned on and began to reconnect nerves in his spinal cord, and if . . . Embedded in each one of those ifs was perhaps a year, or three, of intense scientific research. Those messy and deflating details were left out of most accounts.

In any event, it was all "silliness" in the opinion of Harvard stem cell expert Doug Melton, because ACT didn't "clone" Somerville at all. The company didn't produce even a single blastocyst through nuclear transfer. Worse, the researchers didn't even get a normal early embryo. The best they achieved was a six-celled "thing." "That alone suggests failure, a weird and aberrant development," Melton commented that Monday morning. That wasn't an observation heard on *Meet the Press*, when Mike West announced the news on Sunday, or on the *Today* show the following morning, or on the *NewsHour with Jim Lehrer* that evening, or on CNN's *Crossfire*, or any of the other television venues West adroitly used to describe experimental results that the scientific community almost instantly and universally recognized as a bust. Mike West had always been good at seducing the press, but in November 2001, three years after the fiasco of the cow-human hybrid story, he traveled from studio to studio like a coaxial Casanova.

"Science by press conference" made its debut as a phrase of contempt in the 1980s, coincident with the emergence of biotechnology. In those days, unlike now, the very notion of biotechnology was unproven; money was scarce, skepticism high, and credibility (both among the scientific and financial communities) hard to come by. The phrase was coined by Spyros Andreopoulos, a public affairs officer at Stanford University Medical School, who used it in a famous letter to the *New England Journal of Medicine* in 1980. He specifically referred, with ill-concealed disdain, to scientists at Biogen, but the indictment also included Genentech and other small start-ups, which during the recombinant DNA furor repeatedly presented research results at company-sponsored press conferences — sometimes as the results were about to be published in the scientific literature, sometimes when they'd merely been accepted for publication, always with an eye toward publicizing the company, and genetic engineering in general, during a time of social uncertainty about the new, unproven, and threatening technology. In a sense, Biogen and Genentech rewrote the book on scientific disclosure, and "science by press conference" entered the playbook of every biotech company struggling to score a touchdown in an overcrowded, dog-eat-dog financial environment.

Historians of science may someday conclude that the orgy of self-promotion and revelation reached some sort of zenith, or nadir, on November 26, 2001, when the reports about the "first human clones" dominated the news. The redeeming virtue of Genentech's scientists is that they consistently published superb, fully realized research of indisputable biological significance, and they did it in topflight journals, passing muster with the most demanding peer-reviewers — *Science, Nature,* and *Cell.* Until November 2001, the scientists at ACT seemed to have followed the Genentech game plan to the letter. They were aggressively self-promotional, but they backed it up with interesting research that routinely appeared in *Science, Nature Biotechnology,* and similar journals. Even Melton, a critic of the company's press relations, conceded that the science intrigued him. "Setting aside all this sort of silliness," he said, "what they're trying to do is quite interesting and important." Jose Cibelli, the company's director of scientific research and lead cloner, had an excellent reputation among scientists who knew his work.

But something happened in the fall of 2001 to alter Michael West's game plan. He orchestrated a full-fledged media circus, but failed to produce the elephant. He overscripted the coverage and persuaded normally conservative publications like *Scientific American* and *U.S. News & World Report* to produce sensational cover stories echoing the company's claims of "the first human clone" — claims that were instantly shot down. He published the results in, to be generous, a second-tier journal. And despite the infinitesimal scientific advance ACT achieved, the research provoked an enormous negative social reaction and increased the odds that Congress would pass legislation outlawing the very technology West viewed as crucial to his mission in regenerative medicine.

Everything about the exercise seemed ad hoc, frantic, overreaching. The paper, posted on the journal's Web site, had been *submitted* only a month before publication, and the results showed that not a single attempt at nuclear transfer resulted in a blastocyst, although ACT dubiously claimed a success rate of 27 percent in reaching the insignificant goal of a cleavage-stage embryo. So why the rush? Why had ACT felt compelled to publish such premature, inconclusive, and largely failed research, in an avenue of experimentation the entire world regarded with trepidation, when it might more prudently have waited to collect more impressive and persuasive results? Harold Varmus, in an op-ed piece for the *New York Times*, suggested the reason was money. "Although its executives claimed to be excited about the findings and said the information would promote educational debate," Varmus wrote of ACT, "the actual reasons may be more self-serving. Biotechnology companies are dependent on investors, and investors like publicity." At least two businesses working with ACT, I was told, had dissolved the relationship because of nonpayment of bills, so there may indeed have been a cash crunch. West, however, insisted that ACT had completed a new round of funding several months before the November announcement, and as several venture capitalists confirmed, controversy chases venture money away more often than not. There might have been a simpler, shabbier reason. About two weeks after the news was disclosed, a rumor circulated among scientists attending the regenerative medicine meeting in Washington that ACT had acted to preempt a rival. The company had learned that the Raelians, the Canada-based cult dedicated to human cloning, were claiming to have created a human embryo and planned to make an announcement (an announcement, it goes without saying, that never occurred).

Whatever the motivation, the reaction came fast and furious. West got it from all sides. In Varmus's words, the ACT work "showed little experimental progress and advanced no new ideas." Ian Wilmut told me he was "so angry about it" that he immediately sat down to write a letter of protest to *Scientific American*. Günter Blobel of Rockefeller University called it "one of those failed experiments that was converted into a successful experiment for the benefit of investors," adding that he felt West was "almost misleading the public." And those were among the more measured responses I heard from the scientific community. Melton, the Harvard biologist, heard the news on his car radio and offered one all-purpose reaction to describe both the mode of scientific disclosure and the media's complicity in it: "despicable." Brivanlou was furious: "I'm so pissed off, what he did. He really killed me and a lot of us who want to do this research. I think he destroyed the field, and he did it for public relations." Three respected scientists noisily resigned in protest from the editorial board of the journal that published the paper.

The response wasn't much gentler from the secular humanist side. Thomas Murray, head of the Hastings Center and a former member of Bill Clinton's ethics commission, said, "The announced findings are much more about the audacity of the people at ACT than any significant scientific finding. It was just a real flop." Conservatives demonized West; William Kristol and Eric Cohen, writing in the *Weekly Standard*, equated West, on a scale of human evil, with Osama bin Laden. Political critics, from Bush to Brownback, renewed cries for a total ban on cloning. Even West's old nemesis at Geron, Thomas Okarma, got in his licks. "My fear," he said, "is that this tiny step will inflame the conservatives in Congress into slapping a ban on this technology before we even know it's feasible."

That was a prescient concern. On the evening of November 27, two days after the announcement, Tom Murray got an urgent call at home, asking him to fly down to Washington by noon the following day to brief a group of Democratic senators on the cloning issue. Over lunch, Murray, Bert Vogelstein, and Andrew Kimbrell, an opponent of the technology, described to senators the promise and problems of therapeutic cloning. Following the meeting, Murray, Vogelstein, and the actor Christopher Reeve made the rounds of Senate offices, trying to put out the fires West had started. Brownback had already pledged to attach his proposed cloning ban to any bill up for passage, and the conservatives attempted to capitalize on the negative momentum. "The political ramifications of the West announcement were pretty strong," Murray said.

West pooh-poohed these criticisms when I asked him about them later. He claimed experimental failures were entirely normal in the exploration of a new scientific frontier. In fact, he said, the company had decided to follow the example of Robert Edwards, the pioneering in vitro fertilization expert, who had published each incremental step on the way to the first test-tube baby. Using any example from in vitro fertilization, a field of breathtaking self-promotion and sometimes shoddy science, as a model of scientific comportment didn't strike anyone as a particularly persuasive argument.

∽

Several months later, when the dust had settled a little, I went to Worcester to talk to West again for a magazine article. The company had expanded since my last visit; a poster on one wall showed a picture of a cloned cow and boasted, "Once a Dream, Now a Reality." West looked pale and weary, but it wasn't from the controversy; he'd just returned from a brief trip to England, where he'd been invited to participate in a four-person debate at the Royal Society on the issue of human cloning. His eyes were red with sleeplessness and his round face appeared unusually unanimated.

In the previous three months, the fracas had hardly died down. When I spoke to Art Caplan two months after the announcement, he was still fuming about it. "I think they've lost their marbles," he said of ACT. "Being a provocateur in this area is not good. Taking steps that wind up terrorizing people, frightening people, is not productive. Mike West has single-handedly been the person most responsible for the possibility of banning all forms of cloning, reproductive and therapeutic, by announcing the creation of human embryos, and the way it was announced, how it was announced, when it was announced — that is, post-Thanksgiving, not in mainstream journals, publicity campaign rolled out on *Meet the Press* with a, let us say, somewhat-out-of-his-element Tim Russert. So I don't understand what Mike West thinks he's doing, at all. Taking out a stick and poking it in the eye of right-to-lifers and pro-lifers is not the road to good policy." West even appeared to be reliving one of the more unpleasant chapters in the life of his longtime hero, Leonard Hayflick. At the urging of a Pennsylvania congressman infuriated by the cloning experiments, federal auditors from the Department of Health and Human Services had shown up at ACT's doorstep and launched an investigation into whether the company had inappropriately used any of its $1.9 million in NIH grants. The company ultimately agreed to pay about $150,000 in misspent funds. Meanwhile, a venture capitalist told me the company was staggering from month to month, desperately seeking funds.

Despite the money woes, despite having all that vitriol dumped on his fair-haired head, West appeared remarkably unrepentant, unapologetic, and eerily serene on this Friday afternoon. "It's just so frustrating to hear all that, because all we've been trying to do is to do things right," he said, sitting at the head of a long table in the company's conference room. "And at least I go to bed every night knowing that we did everything we could to get things right." They were just trying to be transparent with the public, he insisted again. They were completely at the mercy of the media. They believed their cloning results merited publication; they were only following the example of Bob Edwards.

As he spoke, there was an air of unreality to his remarks, as if there were an alternative reality in these rooms, among these scientists and technicians and entrepreneurs and their bioethicists, different from the one the rest of us experience. It would be overinterpreting the atmosphere to liken it to a sect, but there was the same sense of purpose, the same zeal to achieve a higher mission, and the same impervious faith of true believers who feel it is only a matter of time before the rest of the world comes around to their point of view. The serenity, the clarity of purpose, the unwillingness to slow down scientifically in the face of such overwhelming censure and opposition — the expression on West's round, optimistic face struck me as improbably but appropriately beatific. As Cynthia Kenyon had sent me back to Shakespeare's sonnets, West's righteousness sent me to the Bible, where I looked up the beatitudes, the Christian blessings that appear in the New Testament.

There was this passage from Luke: "Blessed are you when men hate you, and when they exclude you and revile you, and cast out your name as evil, on account of the Son of man! Rejoice in that day, and leap for joy, for behold, your reward is great in heaven; for so their fathers did to the prophets." Although West's spiritual allegiances had shifted from the Bible to science's book of life, he still appeared suffused by the radiance and righteousness of his mission, still felt convinced that any trips to the afterlife could, through the wonders of biology he relentlessly pursued, be indefinitely postponed, still believed in his role as a molecular prophet. As he liked to say, there was so much to do, and so little time.

By the end of the year, West's little entrepreneurial empire in Massachusetts, his platform for influencing the national debate on regenerative medicine and life-extension technologies, appeared on the verge of collapse. Jose Cibelli, Advanced Cell's chief cloner, announced his intention to leave the company for an academic post, and ACT was about to run out of money, according to people familiar with the situation. The company that had provoked so much societal debate (not to say outrage) seemed destined to be a footnote in everyone's memory five years hence. And yet Mike West would always have room to operate in a society reluctant to confront the vexing issues his work invariably aroused. On June 12, 2002, despite months of posturing and threats of action, the U.S. Senate announced that it would indefinitely put off a vote on whether to ban human therapeutic cloning. Once again, America's ambivalence about cloning and reproductive medicine provided Mike West with just enough breathing space to continue, and he resumed the work of his test-tube ministry. "We're working heavily on a different, related program that we're pretty excited about," he said that day. "But," he added with a little laugh, "I can't talk about it."

FINITUDE

IN THE LONG HISTORY of human apparel, the most pathetic garment ever created is undoubtedly the contemporary, standard-issue hospital gown. Halfheartedly floral, flimsy, and limp with serial launderings of God knows how many emergent bodily fluids, it is a kind of off-the-rack shroud that perfectly, if inadvertently, accentuates human frailty; to illness's conventional cosmetic torments of pallor and lassitude, it adds a final dollop of indignity. In the spring of 2002, these thoughts came to me, uninvited, while I was making a glum rite of passage for many people my age: staring down at an ailing, elderly parent wearing one of those cheerless gowns in the hospital. There is never a time when seeing a sick parent is not surprising, with its sudden snapshot of illness superimposed on a lifetime of more vigorous mental images, but it caught everyone in my family by surprise. My mother, seventy-four years old at the time, had always been reasonably healthy. She still possesses a fine, full head of genuinely black hair; she's rarely been sick; and, as the daughter of Italian immigrants, she'd been eating a Mediterranean diet decades before it bore a name or any cachet. But now she was in a hospital bed, wearing one of those flimsy gowns, and she looked terrible: face drawn and pale, smile weak, her lips swollen and chafed. She had been admitted to a hospital in Portland, Oregon, the week before Easter.

The immediate cause of her hospitalization was a severe gastrointestinal infection that left her dangerously dehydrated, but that was merely an acute overlay to an underlying illness that had been troubling her, and puzzling her doctors, for months: increasingly severe pain in her left foot. After a sophisticated, escalating barrage of diagnostic tests — x-rays, MRI, bone scan, CT scan, and CT-assisted needle biopsy of the bone — and an increasingly frustrating series of tentative and mistaken diagnoses, ranging from the benign (tendinitis) to the devastating (metastatic and incurable cancer), her doctors finally discovered that all the trouble stemmed from something unexpected: a tenacious bacterial infection that had worked its way into one of the bones in her heel. Her doctors eventually cleaned out the infected area surgically and pumped her full of antibiotics round the clock for six weeks. Confined to a wheelchair at first, she soon graduated to a walker and then a cane. But she has yet to regain the mobility and function she had before the infection, and it's not clear that she ever will. As I witnessed her pain and frustration and fear over many months, I found myself wondering: Is this an aberrant medical event? Is it an isolated incident, or is it the momentous beginning, imperceptible except looking back over your shoulder many years later, of a long, inexorable, and irreversible decline due to old age? I don't know. The point is, you never do.

But you can't help thinking about regenerative medicine (to say nothing of life extension), in a very practical and even acidly skeptical way, when confronted with an elderly loved one who may be on that downward slide of mortality. A hospital room is far removed from the abstract world of stock analysts blathering about telomerase inhibitors, or talking heads on CNN speculating that stem cells might be good to go in five years or ten maybe, or even philosophers opining that life extension would be wasted on the elderly because they're too set in their ways and have lost the ability to be outraged by social injustice. The reality is that the tests every patient, including my mother, typically undergoes — pressure, pulse, blood chemistry, and x-rays, to say nothing of the more sophisticated (and expensive) imaging technologies — are not simply indices of health but little observation windows looking down on all the things that can go wrong in the human body. And so many things can go wrong, especially as we get older, that sometimes that little window is a cruel variation on the portholes surrounding construction sites through which I used to peer as a child: now, looking in as an adult child, the view takes in the human edifice as it falls apart.

Hence, for many of us, the allure of biomedical innovation is inescapable, and personal. If the surgeons removed infected bone from my mother's heel, what would replace this crucial tissue in an essential weight-bearing part of the body? I thought immediately, of course, about some type of bone-forming stem cell, early clinical trials of which had begun in 2001. But consider the formidable obstacles standing in the way of the rapid adoption of even this straightforward new technology. First, the mere diagnosis in my mother's case proved extraordinarily difficult, which is a reminder that regenerative medicine is inevitably yoked to the many limitations of health care we already know all too well: access, affordability, timeliness of treatment, and, simply, good medicine, beginning with prompt and proper diagnosis. Second, even if the treatment of choice was adult stem cells, of the sort currently being tested, the general availability of such therapy would be years away, and I'm not sure I'd even want my mother to be among the first to try it. When you put living stem cells, no matter how "grown up" they appear to be, into a human body, you want to be very sure they stay where you put them, and grow no more than you desire, and become nothing other than what you want them to be. We'll probably have to wait a number of years to find out for sure that they're safe. If the treatment of choice was cells derived from embryonic stem cells, the wait might well be longer — their greater potential to form different tissue means they also might have greater potential to form unwanted tissue, in undesirable places. Unlike the high-tech aficionados of electronic gadgetry, "early adopters" of cutting-edge biomedical technologies do so out of medical desperation, not a temperamental willingness to experiment. Finally, considering the panorama of long-term chronic illnesses of the sort that typically afflict so many elderly people, and seeing the way these constellations of illness leach away the essence of a loved one's vitality and spirit, you begin to realize that the sheer complexity of age-related disease and the process of aging itself threaten to diminish, if not overwhelm, the ultimate promise of any single form of therapy.

In the same sense, the abstract promise of life-extending technologies must be weighed against the current realities of aging and the ageless realities of scientific research. The long-term promise of stem cell therapy is everything it has been cracked up to be: the potential clinical impact is staggering, on a par with the therapeutic importance of antibiotics or vaccines. But solving all the biological problems is a staggering task, too, and it is a task that has been largely assigned, by politics and happenstance, to a handful of underfinanced, understaffed, and scientifically overwhelmed boutique biotech companies. Günter Blobel, the Rockefeller University biologist who won a Nobel Prize in 1999 for his work in cell biology, considers the discovery of embryonic stem cells to be a "revolutionary" development in biomedical research. But the task of harnessing that power, he told me not long ago, probably exceeds the reach of any small biotechnology company. "It may be too big for Geron," he said. "It would be good for a very big company to dump a billion dollars on it over twenty-five years and do it in a very systematic way. It's something that should be done very systematically and, in my opinion, on a large scale." And this from a former member of Geron's scientific advisory board! As with germ-line cells, the patents and commercial rights to this technology are certain to live on, but the survivability of the companies currently holding them is not so clear.

A large-scale, systematic approach is not happening now, and probably will never happen in the United States, as long as the political terrain looks as uncertain and as controversial as it has for the past twenty-five years. As of this writing, in the summer of 2002, three of the companies most publicly associated with the field of regenerative medicine were suffering the financial bends. In June, Geron laid off 30 percent of its workforce and hunkered down for survival, with only enough cash on hand, it claimed, to last another two years. Osiris Therapeutics, the adult stem cell company, was also having financial difficulties, one venture capitalist told me. And Michael West's company, Advanced Cell Technology, was lurching month to month, barely able to meet payroll (again, according to a venture capitalist who declined to invest). Meanwhile, nearly a year after George Bush announced his restrictive federal stem cell research policy, the NIH had disbursed about a dozen supplemental grants, worth $50,000 each, to stem cell researchers — not even a microscopic drop in the agency's $27 billion budget — and agency officials privately complained that not enough researchers were applying for grants. And on November 18, 2002, the NIH quietly and drastically amended its Stem Cell Registry, rather than the seventy-odd cell lines so vehemently defended by Bush officials, it listed only nine accessible lines, the majority from Australia.

Research on telomerase, the "immortalizing enzyme," represents another variation on the theme of slow progress. The mechanisms of telomerase activity are now understood to be much more complex than the simple two-gene model that inspired several frenetic scientific races only five years ago. The prospects for activating telomerase as a method of cellular rejuvenation may be more promising than previously believed, according to some, but it is still years away from clinical testing. Geron, meanwhile, is close to testing a telomerase inhibitor in cancer patients, and the company has already been testing a cancer vaccine that has shown promising preliminary results in a clinical trial at Duke. But it will continue to take time, and it's an open question whether positive results can be achieved before the company runs out of money.

As for the longer-range technologies of genetic manipulation that may extend the human life span? The basic science has grown increasingly compelling, but the leap from laboratory animals to humans, especially in this instance, makes the Grand Canyon look like a sidewalk crack. Although the potential for extending the human life span is real, there are good biological reasons to argue that single-gene changes are not going to dramatically change longevity expectations any time soon, if ever. Leonard Guarente's most recent work on the *sir-2* gene illustrates how quickly the "promise" of life extension gets muddied by perplexing complications. When the MIT researchers inserted extra copies of *sir-2* into yeast, the yeast cells indeed achieved a longer life span — but they also exhibited a respiration rate three times higher than normal. Not to overinterpret, but who wants to live for 150 years if you're going to be panting the whole time?

The most persuasive reason for caution may be purely circumstantial: the recent "social" history of gene discovery. Since the invention of recombinant DNA and of gene sequencing technologies in the 1970s, a parade of medically important human genes have made their debut, often on the front pages of our leading newspapers. But rarely have those genes enjoyed as much medical promise as in their initial description in the lay press. Tumor necrosis factor, leptin, interleukin-12, endostatin, even hemoglobin (and its long-standing ailment, sickle-cell anemia) — all the genes were discovered with great fanfare, but few medical conditions have been cured despite years, and sometimes decades, of fervent research. Almost every time a promising protein is studied more closely, its effects become more complicated, its "benefits" more double-edged.

The more likely kind of progress, at least in the short term, will be attained by the sort of painstaking, nuts-and-bolts basic research conducted by scientists like Ali Brivanlou. In the summer of 2002, I visited Brivanlou in his lab at Rockefeller. It was a sunny day, and the scene outside his office window — like the prospects for his research — appeared considerably brighter than on that dark day in November 2000 when the NYPD pulled a body out of the river. But there were still grim moments. In February 2002, Brivanlou had become so frustrated by the university's lack of support for stem cell research that he told administration officials he planned to resign his tenured professorship. Several senior faculty members who learned of his impending resignation privately advised him to hold off on a final decision, and within a week Arnold Levine, the university's president, resigned for "health reasons"; his interim successor, Thomas Sakmar, immediately made embryonic stem cell research an institutional priority, setting up a privately funded institute within Rockefeller and aggressively pushing the research, to the point where university officials asked Brivanlou to take the unprecedented step of withdrawing a pending NIH grant application for $3.8 million. Here was a powerful vote for the potential of stem cell research, and simultaneously a vote of no confidence, from one of the world's preeminent research institutions, in the government's stewardship of this potential. Nor was Rockefeller alone. Stanford University ultimately announced a private stem cell institute, and at least one other major university hospital had a similar plan in the works. Indeed, one of the casualties of the stem cell controversy has been the longstanding trust between the NIH and academic researchers. The NIH feels it has taken extraordinary steps to facilitate the research, but academics have expressed distrust of the Bush administration's political agenda regarding scientific research, and NIH officials have privately expressed concern that not enough researchers are applying for government stem cell grants.

On the day I visited, Brivanlou sat at his computer, tapping a few keys. "So here it is," he said over his shoulder. "You want to see something that I've tried to accomplish for more than two years?" After years of waiting, his lab had managed to obtain exactly two human embryonic stem cell lines, one from the University of Wisconsin and the other from Australia, and had begun to compare them with mouse ES cells. The results were both exhilarating and sobering. It took Brivanlou five minutes to speed-scroll through some 22,000 genes whose activity had been assessed in these protean cells during their earliest instance of biological arousal. It will take years to sort out all the information, but several things were immediately clear. Not all the human cell lines available according to the Bush policy are the same, for one. More important, the difference between mouse embryological development at the molecular level and human development was "night and day," Brivanlou said. "Everything that we know about mouse, we already know is not true for humans." He was hotly pursuing an unexpected signaling system, different from the mouse pathway, in human stem cells, which conveyed not only a scientific but a political message. If we want to learn how to use human cells as human medicine, we are going to have to use human material — including human embryos — to make progress. If we are not willing to do that as a society, this "revolutionary" technology will incite no revolutions in care and healing, at least in the United States.

༄

In 1983, well before most people began to think about "practical immortality" as a pressing issue, Leon Kass published an essay called "Mortality and Morality: The Virtue of Finitude," in which he considered the ramifications of efforts to extend life. It isn't that Kass was "for" finitude; rather, he argued that mortality and our awareness of it — that unique human foreknowledge, among all creatures, that we are destined to die — is an enriching, ennobling, and even inspirational form of knowledge, deepening our appreciation of family and friendship, provoking our humanity to express itself, inspiring us to do good works of enduring value. "To praise mortality must seem to be madness," Kass wrote, but he nonetheless viewed life extension as an ominous possibility, one that would top-load society with the elderly, suppress the aspirations and mobility of the young, and create a huge population of bored and tedious people. In noting that such changes would likely have an impact on "work opportunities, retirement plans, new hiring and promotion, social security, housing patterns, cultural and social attitudes and beliefs, the status of traditions, the rate and acceptability of social change, the structure of family life, relations between the generations, or the locus of rule and authority in government, business, and the professions," Kass deftly suggested that the very fabric of civilized life would be stretched, if not tugged apart, by increasing numbers of the aged.

Fair enough. The problems begin when Kass places "the project to control biological aging," as he puts it, within a larger philosophical and social context. "Indeed, the prolongation of healthy and vigorous life — and, ultimately, a victory over mortality — is perhaps the central goal and meaning of the modern scientific project," he writes, "associated in its founding with such men as Bacon and Descartes." It is to Kass's credit as a skilled rhetorician that you need to peel away the skin of his prose to understand that what he says so reasonably is in fact not only unreasonable but profoundly reactionary: if you parse the sentence, you can see that Kass couples "the prolongation of healthy and vigorous life" — surely a goal all societies since the Renaissance have heartily embraced — with an arrogant, undesirable, and thoroughly theoretical destination of modern medicine. Prolonging a healthy and vigorous life, he implies, is dangerous precisely because it will also lead "ultimately" (or "inevitably," or "doubtless," or any of the stealth adverbs of clairvoyance that so often appear in these arguments) to a "victory over mortality." And there is no question in Kass's mind where this is headed. "For most of us," he writes, "especially under modern secular conditions in which more and more people believe that this is the only life they have, the desire to prolong the lifespan (even modestly) must be seen as expressing a desire *never* to grow old and die. However naive their counsel, those who propose immortality deserve credit: They honestly and shamelessly expose this desire."

In a sense, Kass is a merchant of immortality, too. He's not selling a drug, of course, but he's pushing the idea in order to sell an ideology. The "victory over mortality" is a canard. No serious scientist believes victory over mortality is possible. We may someday be able to slow down the aging process, but reversing it is unlikely, and engineering the germ line to attain greater life span represents a risky form of human experimentation few if any societies on earth would be willing to undertake. We in the media are merchants of immortality too. *Wired* magazine, for example, was happy to push the longevity envelope in a recent article, predicting that for people born in 1990, "immortalizing therapies will be available by pill and injection by the time you hit 40 — but you'll have aged, and the drugs will be pricey and imperfect. The wealthiest of you will live to 150." For those children born in 2020 — my grandchildren, say — *Wired* predicts that genetic engineering of the germ line will extend life, and that some people could conceivably live for five hundred or a thousand years. In a recent publication, the International Longevity Center suggested that "large increases in life expectancy to 150 years or more in humans, as often suggested by antiaging entrepreneurs, may be unrealistic without major new insights into the molecular mechanisms of aging." Needless to say, those "major new insights" have yet to appear. Leonard Hayflick and a number of other prominent gerontologists have become so annoyed by the claims of life-extension enthusiasts that they recently prepared a manifesto. The document — written by Hayflick, S. Jay Olshansky, and Bruce Carnes and signed by a number of scientific luminaries, including Robert N. Butler, Steven Austad,

Tom Kirkwood, George Martin, Carol Greider, and Andrew Weil — flatly asserts that dramatic increases in life span are unlikely. "The prospect of immortality," the scientists state, "is no more likely today than it has ever been, and it has no place in a scientific discourse."

Using immortality as a straw man in a social debate therefore attaches a doubtful destination to current scientific research that admirably seeks to cure diseases and prolong healthy and vigorous lives. But framing the problem that way begs an even bigger question: Who possesses the wisdom, the foresight, the sheer futurological chops to decide which prolongations of healthy and vigorous life are socially (or morally) acceptable and which should be forsworn? Distrust of the scientific enterprise is the subtext of current debates over embryonic stem cells and research cloning, and is likely to come up again and again.

Trying to predict medical advances, much less segregating desirable medical progress from undesirable attempts to extend life, is an exercise fraught with scientific uncertainty and moral capriciousness, and it reminds me of another, earlier episode in my mother's life. She was among the first Americans to receive a first-generation, experimental life-extension drug that worked remarkably well. In January 1945, as a seventeen-year-old senior in high school, she lay in a hospital bed in Cleveland, Ohio, one night, surrounded by the entire clan — not just her father and mother, but all her uncles and aunts and cousins. Their mere presence was a bad omen, she realizes now, because the hospital had been very strict about allowing no more than two visitors; all of a sudden, her room was full of the extended Albanese clan. When her father pulled out a pack of cigarettes and offered her one, it appeared to be the gesture of an executioner. She'd never smoked a cigarette in her life.

She was on her deathbed, as it turns out, due to an acute infection of the kidneys. Perhaps because one brother had died at nineteen of complications from scarlet fever and her other brother was off fighting in World War II, Dr. Joseph Brady — the family doctor who had delivered her into the world — contacted every hospital in town to gather precious amounts of this experimental drug, and then treated her with it. She didn't know it that evening, but Dr. Brady had told her parents that if she survived the night, the new medicine would probably save her life. The drug was called penicillin. Only by the most permissive of definitions would penicillin be considered a life-extension drug, and yet it extended my mother's life by more than half a century, and counting. It was another of medicine's sub-versive efforts to "prolong a healthy and vigorous life," another socially destabilizing attempt at "victory over mortality."

Any reasonable person would argue that when we talk about life ex-tension and other problematic medical technologies of the future, we're not talking about things like penicillin. But how can we be so sure? Many observers, technological enthusiasts as well as people like Kass, assume that some sort of life extension will definitely happen; many assume that we'll all have access to the medicine; many among us predict vast social disrup-tions on the basis of defeating mortality. But all those assumptions and predictions are about as credible as if someone, in 1927, had predicted that within a year a mold spore would blow unbidden into Alexander Fleming's laboratory window, and that it would land on an uncovered culture dish in his characteristically cluttered and messy London lab, and that the soft green halo of growth in that dish would forever transform twentieth-cen-tury medicine. The gifts of medicine come to us in mysterious and blessed ways, certainly beyond the ken of our predictive skills, and perhaps the most important lesson we can take from the penicillin story is to keep the windows, and doors, open to a future we can only faintly imagine.

The futurologists of immortality like to imagine that life extension will be accompanied by the kind of medical wisdom and skill that will keep everyone vigorous and healthy during their extended lifetimes — certainly a laudable goal, lest we all end up like Jonathan Swift's decrepit and crabby Struldbruggs, who had "not only all the Follies and Infirmities of other old Men, but many more which arose from the dreadful Prospect of never dy-ing." But that exceedingly improbable scenario would essentially require the simultaneous medical cure of virtually every age-related ailment that we know about. What good is curing heart attacks with stem cells, for ex-ample, if it merely preserves more elderly brains for Alzheimer's disease to steal? What good is the ability to bioengineer organs like livers and bladders if we implant them in bodies chronically incapacitated by arthritis and os-teoporosis?

Kass brings an elevated thoughtfulness to these issues, and I'm always glad to hear what he has to say, because he is invariably the most eloquent voice on the cautionary side of things. But it is equally important to point out that the edifice of his speculation, like that of biological optimists and techno-groupies, is built upon a landfill of assumptions and that these assumptions are often an amalgam of guesswork, omission, and obliging simplifications. The reality is that human biology is so complex, so prone to gerontologic mishap, that even with tomorrow's knowledge, it's much likelier that we'll see incremental improvement in health care than a pill that cures death and achieves "victory over mortality" — although in that less grandiose vision many lives will nonetheless be extended and enhanced, and needless human suffering will undoubtedly be reduced. There's nothing wrong with that. What's potentially very wrong is to proscribe certain areas of biomedical research, for ideological or political or even "ethical" reasons, on the basis of shaky and dubious assertions of inevitability. Even stem cells and other forms of regenerative medicine, despite their potential to rejuvenate certain precincts of our anatomy, are not likely to have such a global impact (anatomically speaking) as to be able to turn back the clock altogether. They might relieve an enormous amount of suffering and allow people to live longer in much better health. But immortalize the soma? Not likely.

These thoughts actually echo a robust debate currently consuming experts in demography and gerontology. Leonard Hayflick is fond of pointing out that even if you cured Alzheimer's, heart disease, stroke, cancer, diabetes, hepatitis, Parkinson's disease — if you completely cured every one of those diseases — you would still add only about fifteen years to the current average life expectancy, and he, along with many other gerontologists, believes that average longevity will probably not exceed ninety years throughout the twenty-first century. But there's plenty of room for disagreement just this side of immortality, even among the professionals. In a recent article in *Science*, demographers Jim Oeppen and James W. Vaupel noted that the straight upward climb of record life expectancy since the mid-nineteenth century "suggests that reductions in mortality should not be seen as a disconnected sequence of unrepeatable revolutions but rather as a regular system of continuing progress." They point out that over the past 160 years, what they call the "best performance" life expectancy has increased by three months *per year*, and that expert predictions of a ceiling on life expectancy — they pointedly cite Olshansky and Carnes here — have "repeatedly been proven wrong." So there is clearly vigorous disagreement among prominent demographers about the shape of this future. The real point is that this future would, in both views, look a lot like the world we already have, and the future we have already been busily creating. Without even factoring in the dramatic changes that regenerative or life-extending medicines might confer, some experts predict that there will be a million centenarians in the United States by 2050, a twentyfold increase over the roughly 50,000 identified in the 2000 census.

In that sense, all new and successful medicines extend life expectancy, and the short-term promise of regenerative medicine fits more comfortably into this larger picture. If this sounds like mealymouthed futurism, don't take it from me; take it from the people whose professional livelihoods depend on these calculations. In January 2002, the Society of Actuaries held a meeting in Florida whose theme was "Living to 100 and Beyond: Survival at Advanced Ages." In a talk entitled "Plastic Omega," Gene Held, vice president of SCOR Life U.S. Re Insurance Company in Texas, reviewed the entire history of aging research, from the Hayflick limit through the telomere story to life-extension genes and new medicines on the horizon. "Can we stop aging?" he asked at the end of his survey. "The honest answer is, 'Right now, we don't know.' Things are still in a rudimentary stage with respect to understanding the aging process, much less arresting it." But he added, "While there is a disagreement among scientists as to whether it will be possible to slow the aging process, I believe deeply it is something we will do." That is why, he suggested, the actuarial view of "omega" — the maximum life span in all these tables and calculations — may need to be revised and perhaps even be viewed as "plastic," or changeable, in light of current medical research.

Unlike many bioethicists, Held did not dwell on the devil we don't know. As he wrapped up his talk, he showed several simple graphs. One depicted what happens when a population of animals expands too much and exceeds the "carrying capacity" of its environment; the population crashes, and although it eventually restabilizes, it is at a lower level. Then he showed a slide documenting the rise of world population over the last ten thousand years. Since about 1800, the human population has risen in a nearly vertical spike; it now surpasses 6.2 billion people. "Whatever the ideal population for this planet is," Held told his audience, "it seems obvious we exceeded it a long time ago. And this is without extending the human life span."

࿉

Whenever I think about this future, this "plastic omega," I try to imagine it through the lives of my children, who will be in their twenties in 2020, who will enter middle age around 2050, and who may retire (if current standards obtain) around 2065. But when I consider the potential of life extension, I can imagine — with a little effort — a scenario in which various medical interventions, both preventive and remedial, could drastically extend their life span. This is actually easier than it might seem. If you were born around 1925, like my parents, and you're still alive today, your average life expectancy is about eighty-six, with a possibility of living to one hundred. If you were born around 1950, like me, you will probably live on average to about eighty-five, although perhaps a million people will surpass one hundred — some might even live to one hundred thirty with the help of regenerative medicine. If you were born around 1995, like my elder child, your average life expectancy will be a little longer — close to ninety for my daughter, according to some projections, and close to eighty-five for my son. In other words, my daughter might see the year 2085. Just writing the number causes a catch in my throat.

It also causes a reality check. Who in the year 1912 could imagine the world we live in now, and who among us is foolish enough to venture anything more than a sheer guess about what the world will be like ninety years hence? War, terrorism, environmental disasters, diminishing resources — will organized medicine even have the cultural luxury to pursue life extension, or will resources be diverted toward the development of vaccines against emerging diseases and antidotes to germ warfare agents? Will the relative safety of the developed West be sufficient sanctuary from HIV, cholera, bad water, *E. coli,* malaria, dengue, West Nile? Several African countries have already seen average life expectancies dip below forty years because of HIV infection, and demographers predict that between now and 2020, the global AIDS epidemic will cause life expectancy to *drop* in fifty-one countries — "a demographic effect," David Brown wrote recently in the *Washington Post,* "essentially without precedent in modern times." In view of nature's restless, ever-changing indifference to human survival, fighting mortality to a draw over the next century might be victory enough.

But let's say, more for the sake of entertainment than out of any rigorous prediction, that the developed nations of the world will indeed have the social, political, financial, and medical wherewithal to slow aging and extend life. Will the remedy be affordable to all? Doubtful. Will the possibility that others have equal access to the same medicines translate into more people vying for a fixed number of social and personal prizes, creating a kind of unprecedented culture of competition in daily life — for jobs, for mates, for a place to live, for a vocational dream to pursue, for a seat on the subway? Will my children have time to do all the things we thought we'd get around to and never did or couldn't afford: Reading all the great books, twice or thrice. Taking up watercolors at age one hundred, with still-steady hands. Starting piano lessons in their 120s, with still-nimble fingers. Finally settling down and getting a "real" job in financial services, bringing all the wisdom and skepticism of their 150 years to the task? These speculations bring a little smile to my face, and perhaps to yours, because futurology of this sort is an effortless form of whimsy. While there are many entertaining (or grim) answers you might provide to these questions, the only honest answer, as actuary Gene Held suggests, is, It's impossible to know. And that ultimately is the point. We are living the experiment.

Having said that, let's engage in a little harmless dishonesty. Let's stipulate that within ten or fifteen years, scientists will be able to pharmacologically tinker with the telomerase gene, temporarily rejuvenating all our tissues, or boost the activity of genes that seem to extend life span, as Elixir Pharmaceuticals would like to do. And let's say these advances will be paired with the ongoing progress in disease diagnosis, cellular therapies of regeneration, bioengineering of entire organs, and the like. What might *that* world look like? How would we spend time, spend money, and live the lives that would be so much longer?

Aside from the obvious economic burdens of social programs and health care for the aged, Kass has speculated about a kind of psychological and demographic displacement of young people, who will find fewer opportunities for work, for growth, for intellectual freedom and maturation. Francis Fukuyama, a political scientist at Johns Hopkins University, in his book *Our Posthuman Future,* advances the interesting notion that an older population in the developed world would probably be dominated by women (who tend to outlive men), leading to a gray matriarchy that could in turn have unusual sociopolitical implications. Women, he writes, "are less likely than men to see force as a legitimate tool for resolving conflicts." The age gap between the developed and developing worlds could lead, he adds, to a world divided "between a North whose political tone is set by elderly women, and a South driven by . . . super-empowered angry young men." Perhaps even more provocatively, Fukuyama suggests that because older generations may be more reluctant to make way for the young, "life extension will wreak havoc with most existing age-graded hierarchies." He adds, "It stands to reason, then, that political, social, and intellectual change will occur much more slowly in societies with substantially longer average life spans." Although he doesn't pursue this thought, history is rich with tales of societies that were reluctant to change — a reluctance that often inspired either swift, bloody, and violent repression or, conversely, swift, bloody, and violent overthrow of social and political structures. A culture top-heavy with the aged, however wise and refined and pacific it might be, would seem doomed.

Perhaps the most important point is this: even without dramatic life-extension medicines, even without "practical immortality," this is where we're headed already. The genomics revolution, personalized medicine, stem cell therapy, and other advances in regenerative medicine — the technologies may take decades to ripen, but they are already in train. If Oeppen and Vaupel's projections hold true, by midcentury we'll have added a decade to average life expectancy simply by doing what we're already doing — using our finest minds to prolong healthy and vigorous lives. It may exaggerate the link, but the economist Peter Drucker makes essentially the same point when he observes that the rapid growth in older populations already under way and especially the shrinking of the younger population "will cause an even greater upheaval, if only because nothing like this has happened since the dying centuries of the Roman empire."

✍

A more immediate and tangible question is whether society, in America and elsewhere, will even allow itself to pursue the medicine that might bring us to one of these less fanciful futures. To get there, we're going to need to resolve some intractable problems that have bedeviled domestic politics in the United States and elsewhere for many years — differences about when life "begins," whether it is ever morally defensible to destroy nascent human life, where the new technologies fit within this debate, and even the very words and tone with which we talk at each other over this ideological divide. Most of all, we need to understand that politics, not science, will ultimately determine whether we have the opportunity to get there or not.

The bioethics panel appointed by George Bush is the most recent, but hardly the only, example of how politics can trump scientific or even expert lay counsel in such matters. When the President's Council on Bioethics began meeting in January 2001, its chairman, Leon Kass, delivered an unusually thoughtful oration on the challenges that lay ahead and promised that the panel would achieve a "richer and deeper public bioethics." The group immediately tackled the controversial issue of human cloning and quickly reached unanimity that cloning for reproductive purposes should be banned. Over the course of several months, as the discussion increasingly focused on cloning for biomedical research, a subtext of the conversation was that scientists had to be reined in — they had to be regulated because they couldn't resist tampering with nature, with molding instead of beholding. Kass himself had written, in language with an odd Strangelovian resonance, "Now may be as good a chance as we will ever have to get our hands on the wheel of the runaway train now headed for a posthuman world and steer it toward a more dignified human future." Anyone who read those sentiments, including longtime friends, wondered whether he could keep his hands off the steering wheel of the bioethics council if it headed in a direction other than the one he'd spent three decades pointing toward.

Despite a good deal of cynicism on Capitol Hill and in the scientific community, the Kass council diligently and openly tackled these questions and performed its duties with distinction. When all was said and done, however, a narrow majority on the panel appeared to reach the conclusion that research cloning should be permissible, as long as the work was strictly regulated. The record of the panel's last public discussion of research cloning, as well as personal statements in the final report, is unambiguous evidence that as of June 20, a majority of Bush's bioethics council supported cloning without need of a ban or moratorium. But both President Bush and Leon Kass had taken very public positions against *all* forms of human cloning. In fact, when a member of the Bush ethics council requested a straw vote on the issue of therapeutic cloning, Kass reportedly declined, saying he "didn't want to embarrass the president." Several weeks later, when a draft of the council's report was circulated to members, two members had privately changed position and a new "majority" had emerged — a manufactured majority suddenly calling for a four-year moratorium on the research, the length and terms of which had not even been discussed in public sessions. A number of panelists expressed outrage at the change, claiming the moratorium recommendation misrepresented the true sentiments of the council (a more detailed account of this episode appeared in the journal *Science*). Even if Kass did not personally engineer the change of votes behind the scenes, he certainly allowed it to happen on his watch.

Much later, it occurred to me that there was an important social message embedded in these events. For more than two years, Ali Brivanlou, the Rockefeller scientist who told me that "nobody can stop scientists from doing the experiments they want to do," had been agitating to begin research on the embryos in his freezer, and yet they remained untouched the entire time; he understood that his responsibility to society was greater than his professional desires. For more than two years, Kass, the secular humanist who had exhorted society to grab the wheel of the "runaway train" of modern biology, had been consumed with the desire to block any form of cloning, but when the commission that unofficially bore his name reached a different conclusion, it appeared to outside observers that he couldn't keep his hands off the wheel. Who behaved with greater respect for the community at large, the scientist or the ethicist?

✍

The fact of the matter is that these little skirmishes temporarily delay, but do not ultimately derail, worthy science. If research cloning is shown to be a valuable avenue of research, it will happen, in Great Britain or China or Israel, if not here. The technology comes with problems, not least the business of egg donations, but if the medical potential becomes more tangible, public pressure will build to allow further exploration of cloning, here or elsewhere. The social and medical case to proceed with the research will become too compelling for either religion or ideology to block it. Of course, countless human lives may be negatively affected by the delay in research, but we'll have had a good debate.

At least the actuaries back up their arguments with charts and numbers. One of the most insidious words in the social debate about new biomedical technologies is *inevitable*. Conservative intellectuals such as Charles Krauthammer have argued that if therapeutic cloning is allowed, it is "inevitable" that human cloned babies will result, through surreptitious and illegal embryo transfers, and that crossing this threshold even a single time will diminish our collective human dignity. These arguments sound to my ears too reliant on a philosophical form of clairvoyance; in both political and ethical debates about cloning, opponents of the technology have wielded the future tense as a cudgel, stating — with a confidence and assurance to which real life, in the lab or on the street, rarely accedes — that if therapeutic cloning is allowed, human cloning for baby making will surely follow, will *inevitably* follow and develop into a thriving industrial activity. Despite this assertion of certainty, despite the fears it knowingly incites, these arguments fail to rise above the level of alarmist futurology. The odds of philosophers or politicians correctly predicting the contours of our biomedical future are no better, and probably much worse, than the odds of predicting the winners of next Sunday's football game, which are based on immeasurably more concrete information and involve wagers worth considerably less than actual lives. Constricting our biomedical opportunities in the face of speculative and often ideologically inspired fears is timid, reactive, and bad public policy.

When we can all agree that something goes against the essence of social norms, such as cloning for baby making, it's not inappropriate to proscribe that activity through legislation and enforcement — at least while we feel our way through the new terrain. In the absence of such consensus, however, in the face of an uncertain future about which we might respectfully have different interpretations and expectations, the social and political impulse to ban represents not moral decisiveness, as some would have us believe, but a form of moral insecurity. It nourishes itself on a kind of pessimism about the human condition, a lack of faith that we can understand and use our newfound powers wisely, a lack of faith that we can discriminate between desirable uses and undesirable misuses a lack of social and political faith that we can, if necessary, adapt to unpredictable consequences that might inadvertently ensue. This long-standing American impulse to control and prohibit reproductive technology is not a universally shared human imperative. The British, for example, have invented and exercised a completely different — and, in the view of many, more effective — way of dealing with the future of cloning, and other nations, including Sweden, Korea, Malaysia, and China, are actively pursuing it in one way or another. In considering cloning and other reproductive technologies through the lens of a "culture of life," the United States government is increasingly pursuing a moral isolationism that is out of step with the rest of the world. One culture's repugnance is another culture's opportunity to cure repugnant diseases.

And this leads to a thought experiment any reader can try. Let's say that in 2003, as some have predicted, a human clone will crawl into our midst. Imagine waking up one morning and hearing on the radio that Severino Antinori or another of cloning's "loony element" had successfully created a child through human cloning. How would you feel? Leon Kass has written that such a technological intervention "would surely effect fundamental (and likely irreversible) changes in human nature, basic human relationships, and what it means to be a human being," and no dearth of talking heads would cue the national psyche to feel this emotional reaction. But how would you, in your daily, domestic, familial circumstances, feel about this development? I understand that some people would share Kass's revulsion, but for the life of me, I can't imagine how the birth of a human clone would mark an irrevocable blow to my, or anybody else's, sense of human dignity, or irreversibly change human nature. It would probably surprise me, given the long scientific odds against success, and certainly sadden me, because of the human desperation or delusion that would drive such an experiment at this point in time. But would I feel any less human? No. Would I love my wife and children any less? Hardly; if anything, their love would seem more precious to me. Would relations between generations begin to crumble? Hardly. I'd still call my folks every week, talk to my mother about theater (which would still go on), and commiserate with my father about the fortunes of our favorite baseball team (a sport which would still go on), while silently hoping that the research potential of cloning could be explored, so that we might continue to have these conversations even longer. If I thought that cloning would become routine and that there would "inevitably" be embryo farms, I might be concerned. But I don't believe cloning will ever become widespread. I cannot imagine a single way in which my life or humanity would change, much less be diminished, by the birth of a human clone. Indeed, the Oz-like din of anticipatory fearmongering that surrounds this event will, I predict, swiftly evaporate if and when we confront the rather less frightening reality, although biologist George Daley offers an interesting prediction: "I think the scientific community is going to look at these cute little neonates and say, 'Prove it!' You think the O. J. Simpson trial was a joke? This is going to be a circus."

Based on the experience of animal cloners, the odds are high that such a birth, were it to occur, would produce a developmentally aberrant infant; perhaps this "Baby Igor" would tragically focus all of society's attention on the inherent risks of human reproductive cloning. I think it is possible, however, to pursue the many potential benefits of human therapeutic cloning in a way that will minimize its dangers. If, as its critics sneeringly maintain, the "therapeutic" part of therapeutic cloning is purely speculative, so too are the threats of reproductive cloning to human dignity and human essence.

While it's important to ponder these philosophical ramifications, there is ultimately a very practical — indeed, shamelessly pragmatic — aspect to this debate and its resolution, and it was brought home to me during a revealing exchange at one of the sessions of the Bush bioethics council. Paul McHugh, the former head of psychiatry at Johns Hopkins, described what it was like to have a waiting room full of patients with Huntington's disease, Alzheimer's disease, and Parkinson's disease — severe neurological disorders that, in a nontheoretical way, strip away human dignity and human essence. They are also diseases that might benefit from research cloning, and McHugh said he couldn't just walk away from the potential for new therapies without grave reasons. Leon Kass turned to Gilbert Meilaender, a respected bioethicist who was known to oppose any form of cloning, and asked, "What are you going to say to Paul McHugh's patients if the moral argument you are upholding prevails?" It was a brilliant question, perhaps more brilliant than even Kass suspected, because it rendered explicit the social danger of asking an ethicist — or politician or philosopher, for that matter — to be a doctor and make a medical decision based on something other than medicine.

Meilaender hesitated for a moment, and then said he would try to convey the following: "We owe you sort of a firm commitment that we will do everything within our moral power, everything that we think we can morally do, to try to find ways to, if not cure, at least relieve your condition." But, he continued, "We may arrive at a moment when we think that here is something that could conceivably be done or tried, but that we ought not do, and what I would say to them then is that the reason I would not do it is because I would not want to do something that helps create a world in which neither you nor I would want to live."

Everyone will have their own reaction to Meilaender's remarks, but as I sat in the audience, I imagined myself as a patient hearing those words, and I couldn't help hearing in them the sound of a doctor withholding a form of treatment on the basis of his morality, not mine. "Whose morality are we talking about?" I wanted to shout. "Are you truly saying you would refrain from possibly lifesaving research for me on the basis of a morality we don't share?" Americans tend to think of themselves as a moral people, but they are also pragmatic; just as there are no atheists in foxholes, there are probably very few idealists in hospital beds. No one, I believe, desires a treatment or cure based on the wanton and self-interested instrumentalization of another human life, and medicine should be happy to be reminded of that on a regular basis. But we don't all agree on the moral status of that developing organism only days after fertilization, and it's pretty clear we never will. Speaking only for myself, and especially as a parent, I don't like the idea of being deprived of the *possibility* of a medicine, for myself and especially for my children, on the basis of a moral belief I don't share. To return to penicillin, it would be as if a doctor refused to prescribe it to a patient in the sincere moral belief that to cure a person would contribute to overcrowding of the planet. Ultimately, we go to our doctors for medicine, not a moral worldview, and allowing bioethicists or anyone else to intervene in that relationship changes the medicine we're likely to get. I can't predict how this disagreement will be resolved, but reconciling the moral and pragmatic impulses of our national character will be the issue — multiplied a million times, one hundred million times — that Americans must confront in terms of the health decisions that will be made for them, for generations into the future, during this debate.

అ

Allow me to close the circle with another weekend in the Catskills. I was sitting next to my four-year-old son during a ride on a horse-drawn wagon as it made its way through the small town of Roxbury. At one point, the wagon driver made a turn into the village cemetery. It was one of those country graveyards whose appearance instantly accorded with the poetic vision of Thomas Gray, its residents "each in his narrow cell for ever laid," some tombstones as elaborate as Napoleonic monuments, others weather-worn slabs of stone, some surrounded by fresh flowers, others barren. My son wondered if the graves without flowers meant that those people were not loved or missed. He has only recently become aware of the meaning of death, ours as well as his own, and the realization has spawned many questions (and some truly impressive existential tantrums).

Our quick circuit of the necropolis conveyed a multitude of silent messages: the deaths in the 1860s and 1918 and during the 1940s reminding me that longevity will forever be subservient to disease and war, the dates of birth and death attesting to the rise in average longevity over the past two centuries, the quick subtraction leading me to point out to my son how long this person or that had lived. As we completed the loop and were about to make our way back to the road, Sandro broke out in an improbable smile and said, "I know how long I am going to live."

"Oh," I said, "how long is that?"

"I am going to live to . . . infinity." At the age of four, he had already devoted some conceptual energy toward forestalling the inevitable, and it expressed itself in a kind of precocious vocabulary of avoidance, in words he had picked up and applied to a very human form of evasion.

I'm still scanning the heavens, looking for a shooting star to which I might append the wishes for a long and healthy life. I'll concede that immortality, even of a practical sort, probably won't be in the cards for him. But I'd be happy if modern science has a chance — an *unfettered* chance — to find better ways to relieve potential suffering and allow him and his older sister as many years as possible, to allow them to discover who they are, to express fully the gifts they might bestow on the rest of us. More time to live, yes. But also more time to give, to appreciate, to love, to forgive, to repay the world.

While the destination remains unclear, the current debate on what seem to be tangential questions relating to regenerative medicine is in fact critically important for the future of our health, and deserves the close attention of a public, here and abroad, whose stake in the outcome will play out in hospital rooms and doctors' offices for decades to come. "We're never going to agree really, I think, about the particular moral status of gametes and eggs and embryos," Laurie Zoloth, a bioethicist at San Francisco State University, once told me. "Partially because of the way we talk about this in terms of reproductive language. And because it's reproductive language, it's heavily freighted by women, sex, erotic passion, control — all of the language that religion has thought about for a long time. So we have very different views that come from very specific places in the tradition. That being said, we then have to think of some new language for the things we do. But probably much more important, we have to figure out who we become when we do this work . . . If we do it, and if we don't do it. If we turn away from it, that will have implications for the work that we do, and for how the work will shape us or not shape us. The chance *not* taken will be as powerful a factor in human history as the chance taken."

The medicine available to our children fifty years hence, the medicine available to us perhaps twenty or thirty years from now, and the medicine available to our parents and ailing loved ones of all ages even five or ten years from now, may depend very much on how society, here and internationally, grapples with the delicate ethical and political questions that vex the field of regenerative medicine. We are no longer innocents in the garden. We have knowledge, including the knowledge to use nascent life, under strict conditions, to derive great potential benefits. That knowledge will ultimately deprive us of immortality, if we believe the Bible, but it nonetheless may be used to reduce a great deal of human suffering. It would be a tragedy if a nineteenth-century metric of theology and morality were applied to the twenty-first-century secular promise of treatments to heal and extend life for millions — not just for the biomedical implications, but because of what it would say about the confidence of our society to walk into a future that is never certain, but that always has the potential to be as beneficent and moral and good as we want it to be.

Notes

Unless otherwise noted, all quotations from people mentioned in the list of interviews for each chapter are from my interviews with them.

Prologue: The Never-Ending Life

Interviews: Leonard Hayflick, professor emeritus, University of California at San Francisco, Sea Ranch, Calif., Dec. 7, 2000; Cynthia Kenyon, Dept. of Biochemistry and Biophysics, University of California at San Francisco, Dec. 12, 2000.

PAGE

1 my Darwinian warranty was about to run out: for the views of evolutionary biology on aging, see Austad, *Why We Age*, pp. 94–122; Ridley, *Genome*, pp. 201–02; and Kirkwood, *Time of Our Lives*, pp. 63–80.

4 For most of . . . humans could expect: for average life expectancies of previous historic epochs, see Hayflick, *How and Why We Age*, pp. 86–87. The current average life expectancy for Americans is 76.9 years, according to a report issued by the Centers for Disease Control and Prevention, and that figure may not change dramatically in the future. According to the Centers for Disease Control (CDC), the average life expectancy of a female born in the year 2000 is 79.5 years and of a male, 74.1 years (see A. M. Minino and B. L. Smith, "Deaths: Preliminary Data for 2000," *National Vital Statistics Reports* 49, no. 12 [Oct. 9, 2001]: 24). Other estimates are higher.

5 Some of those massive and long-lived creatures: for the age of redwoods, see Hayflick, *How and Why We Age*, pp. 34–36.

7 I can't claim to have come: S. S. Hall, "The Recycled Generation," *New York Times Magazine*, Jan. 30, 2000, pp. 30–35, 46, 74, 78–79.

8 In his remarks to the group, William Haseltine: see W. A. Haseltine, "The Emergence of Regenerative Medicine: A New Field and a New Society," remarks at the First Annual Conference on Regenerative Medicine, Washington, D.C., Dec. 4, 2000. "This medicine will be transforming," Haseltine said.

"It can help everyone who ages: that is, all of us." Haseltine claims to have coined the phrase "regenerative medicine," which he defines as "the use of human genes, proteins, and cells to regenerate tissues and organs damaged by disease, injured by trauma, or worn by age." In a talk at the Aspen European Dialogue, Cernobbio, Italy, May 1998, he said, "The outlines of a strategy to alter the fundamental fact of aging itself is in view. If individual cells, tissues, and organs can be regenerated from within via the specific and controlled action of genes, then we can envision a time when individual and younger cells can be introduced and subtly integrated into existing structures. Cellular replacement may keep us young and healthy forever. The fountain of youth is likely to be found within our own genes."

8   "I am now working on immortality" and other quotes: B. Alexander, "Don't Die, Stay Pretty: Introducing the Ultrahuman Makeover," *Wired*, Jan. 2000, pp. 178–88.

10   a manifesto decrying the misguided messages: Hayflick shared with me an early draft of the manifesto: S. J. Olshansky, L. Hayflick, and B. A. Carnes, "Position Statement on Human Aging." A version of this document later appeared as "No Truth to the Fountain of Youth," *Scientific American*, June 2002, pp. 92–95. The draft manifesto states, "Even with precipitous declines in mortality at middle and older ages from those present today, life expectancy at birth is unlikely to exceed 90 years (males and females combined) in the 21st century."

1. THE HAYFLICK LIMIT

INTERVIEWS: Clayton Buck, director, Wistar Institute, Philadelphia, Aug. 16, 2001; Vincent J. Cristofalo, president, Lankenau Institute for Medical Research, Wynnewood, Pa., June 5, 2002 (telephone); William Fenwick, partner, Fenwick & West, Palo Alto, Calif., June 5, 2002 (telephone); Leonard Hayflick, Dec. 1, 1999 (New York), Dec. 7, 2000 (Sea Ranch, Calif.), Aug. 16, 2001 (Philadelphia), Apr. 29, 2002 (telephone), Nov. 19, 2002 (telephone), and Dec. 2, 2002 (telephone); Maurice Hilleman, director, Merck Institute for Vaccinology, West Point, Pa., July 16, 2001 (telephone), and Nov. 20, 2002 (telephone); Hilary Koprowski, professor emeritus, Thomas Jefferson Medical College, Philadelphia, May 31, 2002 (telephone); David Kritchevsky, Caspar Wistar scholar, Wistar Institute, Philadelphia, Oct. 22, 2001 (telephone); Ronald Lamont-Havers, senior associate for special programs, Cutaneous Biology Research Center, Massachusetts General Hospital, Charlestown, Dec. 16, 2002 (telephone); Joshua Lederberg, Rockefeller University, New York, May 28, 2002 (telephone); Paul Moorhead, former Wistar researcher, Baltimore, Aug. 9, 2001 (telephone); Stanley A. Plotkin, professor emeritus of pediatrics, Philadelphia, Oct. 30, 2001 (telephone); Theodore Puck, Eleanor Roosevelt Institute, Denver, Nov. 8, 2002 (telephone); Woodring Wright, Dept. of Cell Biology, University of Texas Southwestern Medical Center, Dallas, July 5, 2001 (telephone).

16 He has had an amazingly productive career: for background on Hayflick's scientific career, see J. Shay and W. E. Wright, "Hayflick, His Limit, and Cellular Aging," *Nature Reviews: Molecular Biology* 1 (2000): 72–76; and V. J. Cristofalo, "Leonard Hayflick," *Contemporary Gerontology*, in press (courtesy of the author).

19 The Wistar Institute occupies several buildings: for the history of the institute, see *Wistarabilia: A Centennial History of the Wistar Institute* (Philadelphia: Wistar Institute, 1994). I am also grateful to Clayton Buck, Nina Long, and Frank Hoke of the Wistar for providing additional background information.

19 the parade of beaming mothers: annual reports of the Wistar Institute frequently refer to Farris's work in assisted reproduction (N. Long, personal communication). In addition, Farris wrote two books on infertility, describing dozens of successful "conceptions" based on a precise prediction of the woman's ovulation cycle.

20 Koprowski, a flamboyant Polish-born biologist: see Hooper, *The River*, pp. 445–47, 480–88.

21 recent accusations, subsequently disproved: on the polio vaccine controversy, see Smith, *Patenting the Sun*, pp. 127–31. The "Hooper hypothesis" was not supported by DNA analysis of some surviving vaccine stocks; see H. Poinar, M. Kuch, and S. Paabo, "Molecular Analyses of Oral Polio Vaccine Samples," *Science* 292 (Apr. 27, 2001): 743–44.

23 [the cells] possessed a "quilt-like appearance": L. Hayflick and P. S. Moorhead, "The Serial Cultivation of Human Diploid Cell Strains," *Experimental Cell Research* 25 (1961): 586–621.

25 Alexis Carrel, a prominent doctor: see J. A. Witkowski, "Alexis Carrel and the Mysticism of Tissue Culture," *Medical History* 23 (1979): 279–96; "Dr. Carrel's Immortal Cells," *Medical History* 24 (1980): 129–42; and "Cell Aging in Vitro: A Historical Perspective," *Experimental Gerontology* 22 (1987): 231–48.

26 "It is now clear that the errors": Austad, *Why We Age*, p. 64. Kirkwood, *Time of Our Lives*, pp. 83–87, describes how the welfare of these chick cells became a recurrent feature in New York newspaper articles. Titia de Lange, a biologist at Rockefeller University, notes that Carrel was a colleague of Peyton Rous, the discoverer of the Rous sarcoma virus in 1911. This virus readily "transforms" — or immortalizes — chick cells through cancer. "I always thought that could be an alternative explanation," she said. "We'll never know" (personal communication, Sept. 27, 2002).

27 represented a "finite limit" and "may bear directly upon problems of aging": Hayflick and Moorhead, "Serial Cultivation."

28 "Oh yes, that happens often": Puck told me he did not recall the meeting or the exchange ("That's a long time ago, I don't remember specifically") but said that his lab had been able to restore replication to senescent cells "by addition of a nutrient molecule." For Puck's arguments, see T. T. Puck, C. H. Waldren, and J. H. Tijo, "Some Data Bearing on the Long-Term Growth of

Mammalian Cells in Vitro," in *Topics in the Biology of Aging*, ed. P. L. Krohn (New York: Interscience, 1966), pp. 101–23.

29 "The largest fact to have come out": P. Rous to H. Koprowski, Apr. 24, 1961. The 1961 paper by Hayflick and Moorhead became a "citation classic" in 1978 (*Current Contents*, June 26, 1978, p. 2), and Hayflick's follow-up 1965 paper achieved similar status in 1990 (*Current Contents*, Jan. 15, 1990, p. 14).

30 SV40, which could potentially infect human vaccine recipients: on viral contamination of monkey cells, see D. Bookchin and J. Schumacher, "The Virus and the Vaccine," *Atlantic Monthly*, Feb. 2000, pp. 68–80.

30 "For those who fear — a Thurber moral": H. Koprowski, "Live Poliomyelitis Virus Vaccines: Present Status and Problems for the Future," *Journal of the American Medical Association* 178 (Dec. 23, 1961): 1151–55.

32 he ended up inventing a powdered medium: see L. Hayflick, P. Jacobs, and F. Perkins, "A Procedure for the Standardization of Tissue Culture Media," *Nature* 204 (Oct. 10, 1964): 146–47.

32 the organism that causes primary atypical pneumonia: see R. M. Chanock, L. Hayflick, and M. F. Barile, "Growth on Artificial Medium of an Agent Associated with Atypical Pneumonia and Its Identification as a PPLO," *Proceedings of the National Academy of Sciences* (hereinafter *PNAS*) 48 (1962): 41–49.

32 WI-38 would be a much safer starting material: see L. Hayflick, S. A. Plotkin, T. W. Norton, and H. Koprowski, "Preparation of Poliovirus Vaccines in a Human Fetal Diploid Cell Strain," *American Journal of Hygiene* 75 (Mar. 1962): 240–58. See also L. Hayflick, "The Limited in Vitro Lifetime of Human Diploid Cell Strains," *Experimental Cell Research* 37 (1965): 614–36.

33 "WI-38, or its imitators": Hayflick is referring primarily to a cell line similar to WI-38, called MRC-5, which was developed in the United Kingdom some years after WI-38. These two cell lines have been used as substrates for vaccines targeted against rubella, chicken pox, polio, rabies, adenovirus, and hepatitis-A; many of the vaccines were made in MRC-5, however, because of uncertainty about the ownership of Hayflick's cells, as Maurice Hilleman put it. The impact of the rubella vaccine has been considerable. In 1969, around the time the Plotkin-developed vaccine became available, there were almost 58,000 cases of German measles in the United States alone, with many miscarriages and birth defects associated with the disease in pregnant women; in 1999, there were 272 rubella cases in the United States. See "Eradicating Rubella in U.S.," *Washington Post*, Jan. 23, 2002, p. A-18.

34 he performed a clever experiment: see W. E. Wright and L. Hayflick, "Nuclear Control of Cellular Aging Demonstrated by Hybridization of Anucleate and Whole Cultured Normal Human Fibroblasts," *Experimental Cell Research* 96 (1975): 113–21.

34 Wistar . . . offering the WI-38 cell strain: Koprowski said in an interview that he did not recall the details of the negotiations over WI-38. Another Wistar employee familiar with the WI-38 controversy said, "I don't know the details

. . . but it is believable to me that Hilary made arrangements for money to be paid to the Wistar, and that Hilary would distribute that money as he saw fit."

36 Hayflick would retain ten ampules: the ampules were at "population doubling level eight," meaning that the cells had already replicated eight times but could be thawed and expanded into hundreds of additional ampules.

37 During a meeting in Bethesda: in an interview, Lamont-Havers said he could not recall all the specifics of his interactions with Hayflick regarding the WI-38 dispute; of the directorship for the National Institute on Aging, he said, "He certainly was a very serious contender for the job, and I know he was not offered the position."

37 "I *asked* for that investigation": for Hayflick's version of the WI-38 legal saga, see L. Hayflick, "A Novel Technique for Transforming the Theft of Mortal Human Cells into Praiseworthy Federal Policy," *Experimental Gerontology* 33 (1998): 191–207.

38 Soon he was in federal court: see "Scientist Accused of Selling Lab Cells," *San Francisco Chronicle*, Mar. 29, 1976, p. 3.

38 selling cells "that were the property": H. M. Schmeck, Jr., "Investigator Says Scientist Sold Cell Specimens Owned by U.S.," *New York Times*, Mar. 28, 1976, p. 1. Six years later, an article in the *Times* acknowledged that Hayflick's long ordeal "appears to be over," but also stated that the out-of-court settlement ended the legal battle "without shedding much light on who was right and who was wrong in this bitter, tangled affair" (P. M. Boffey, "The Fall and Rise of Leonard Hayflick, Biologist Whose Fight with U.S. Seems Over," *New York Times*, Jan. 19, 1982, p. C-1).

38 "A personal tragedy which also raises": N. Wade, "Hayflick's Tragedy: The Rise and Fall of a Human Cell Line," *Science* 192 (Apr. 9, 1976): 125–27. On the issue of contamination, see N. Wade, "Vaccine Cells Found Mostly Contaminated," *Science* 194 (Oct. 1, 1976): 41.

38 Schriver was . . . "widely respected for honesty": Boffey, "Fall and Rise."

39 "I went from full professor at Stanford": Hayflick has published a number of repetitious, impassioned, self-pitying retrospective accounts of the WI-38 episode in various journals. Among them are L. Hayflick, "A Brief History of the Mortality and Immortality of Cultured Cells," *Keio Journal of Medicine* 47, no. 3 (1998): 174–82; "The Coming of Age of WI-38," *Advances in Cell Culture* 3 (1984): 303–16; "The Use of Human Cells for Production of Human Biologicals," *Tissue Culture Research Commun.* 16 (1997): 147–56; "Mislabeling, Contamination, and Other Sins of Cultured Flesh," Joint WHO/IABS symposium on the standardization of cell substrates for the production of virus vaccines, *Develop. Biol. Standard* 37 (1977): 5–15; and "History of Cell Substrates Used for Human Biologicals," *Develop. Biol. Standard* 70 (1989): 11–26.

40 Dozens of prominent scientists: the settlement of Hayflick's lawsuit was not reported in the news pages of *Science*, but rather in a letter to the editor

signed by eighty-five scientists, including prominent biologists like Robert A. Good, Robin Holliday, Robert J. Huebner, Henry S. Kaplan, George M. Martin, Zhores Medvedev, and Charles Yanofsky. See B. L. Strehler et al., "Hayflick — N.I.H. Settlement," *Science* 215 (Jan. 15, 1982): 240–42. In reply to a letter from Hayflick published in *Science*, Wade was singularly unrepentant, writing, "There is nothing I wish to change, add to, or subtract from the article as then written, including its title." N. Wade, *Science* 202 (Oct. 13, 1978): 136.

2. "A Circle Has No Ends"

INTERVIEWS: J. Michael Bishop, chancellor, University of California at San Francisco, Dec. 19, 2001 (telephone); Elizabeth H. Blackburn, Dept. of Biochemistry and Biophysics, UCSF, Nov. 12, 1999; Titia de Lange, Laboratory for Cell Biology and Genetics, Rockefeller University, New York, Oct. 13, 1999, Nov. 12, 1999 (telephone), and Aug. 22, 2000; Carol Greider, Dept. of Molecular Biology and Genetics, Johns Hopkins University School of Medicine, Baltimore, Oct. 20, 1999, Nov. 21, 2001 (e-mail), and Nov. 29, 2001 (telephone); Calvin Harley, vice president for science, Geron Corp., Menlo Park, Calif., Nov. 15, 1999; Leonard Hayflick, Dec. 7, 2000; Victoria Lundblad, Dept. of Molecular and Human Genetics, Baylor College of Medicine, Houston, June 25, 2002 (telephone); Alexey M. Olovnikov, Institute of Biochemical Physics of the Russian Academy of Sciences, Moscow, Oct. 22, 2001 (e-mail); Woodring Wright, July 5, 2001 (telephone).

42 Olovnikov was "simply thunder-struck": Alexey M. Olovnikov, "Telomeres, Telomerase, and Aging: Origin of the Theory," *Experimental Gerontology* 31 (1986): 443–48. Hayflick was the journal's editor at the time, and he confirms that he smoothed out Olovnikov's English.

42 the director of the Gamaleya Institute . . . was rumored: it is impossible to ascertain whether Olovnikov's allegation of a KGB-Gamaleya connection is true, but I was told by David Chudnovsky, a prominent Soviet mathematician and computer scientist who relocated to the United States, that such a claim was not at all improbable.

43 his father "somehow contrived to read": letter from Olovnikov to the author, Oct. 22, 2001. All information about his early childhood is from this letter.

44 the answer lay at the very ends of the chromosomes: for the early history of telomeres, see H. J. Muller, "The Remaking of Chromosomes," *The Collecting Net* 13 (1939): 1181–98; and B. McClintock, "The Stability of Broken Ends of Chromosomes in *Zea mays*," *Genetics* 41 (1941): 234–82. Elizabeth Blackburn published an interesting historical appreciation of McClintock's work in the essay "Broken Chromosomes and Telomeres," in a volume in honor of McClintock's work: *The Dynamic Genome: Barbara McClintock's Ideas in the*

Century of Genetics, ed. N. Fedoroff and D. Botstein (Cold Spring Harbor, N.Y.: Cold Spring Harbor Laboratory Press, 1992), pp. 381–88. Blackburn says McClintock's work "anticipated and laid the groundwork for our more recent, molecular, view of telomeres."

46 Olovnikov published his "theory of marginotomy": A. M. Olovnikov, "Principles of Marginotomy in Template Synthesis of Polynucleotides," *Doklady Akad. Nauk SSSR* 201 (1971): 1496–99.

46 Medvedev . . . suddenly disappeared: see R. Kaiser, "Intrigue Enlivens Conference in Kiev," *Washington Post,* July 9, 1972, p. A-30. Hayflick said that the mystery of Medvedev's disappearance was resolved when a scientist attending the meeting in Kiev received a telegram in which the Russian scientist apologized, in perfect English, for failing to attend because of another commitment. It was signed, Hayflick recalled, "Professor Kidnaper."

47 when he finally gave his talk: language difficulties clearly impaired Olovnikov's ability to describe his "theory of marginotomy." Vincent Cristofalo, who sat in the front row for the talk, recalled: "Olovnikov's English was abysmal, and when he was finished, everyone was relieved that he was done, because no one understood what the hell he was talking about."

47 Olovnikov later received approval to submit: A. Olovnikov, "A Theory of Marginotomy: The Incomplete Copying of Template Margin in Enzymatic Synthesis of Polynucleotides and Biological Significance of the Phenomenon," *Journal of Theoretical Biology* 41 (1973): 181–90.

47 Watson . . . had published similar speculations: J. D. Watson, "Origin of Concatemeric T7 DNA," *Nature New Biology* 239 (1972): 197–201. This paper, unlike Olovnikov's, confined itself to an aspect of viral replication (Watson, personal communication, Oct. 8, 2002).

49 "I loved the names and structures": for Blackburn's background, see *ASCB Profile: Elizabeth H. Blackburn* (American Society for Cell Biology, Dec. 1997).

50 the chromosome tips repeated the same short sequence: see E. H. Blackburn and J. G. Gall, "A Tandemly Repeated Sequence at the Termini of the Extrachromosomal Ribosomal RNA Genes in *Tetrahymena,*" *Journal of Molecular Biology* 120 (1978): 33–53. For a brief summary of the early history of the telomere field and more recent developments, see J. Marx, "Chromosome End Game Draws a Crowd," *Science* 295 (Mar. 29, 2002): 2348–51.

51 a graduate student named Carol Greider: see *ASCB Profile: Carol Greider* (American Society for Cell Biology Newsletter, May 1999); and R. Skloot, "The Marvels of Telomerase," *Hopkins Medical News* (Winter 2001).

52 Blackburn and Greider published their findings: C. W. Greider and E. H. Blackburn, "Identification of a Specific Telomere Terminal Transferase Activity in Tetrahymena Extracts," *Cell* 43 (Dec. 1985): 405–13.

52 number of times those early papers had been cited: on this point, and for an excellent up-to-date survey of telomere biology, see C. W. Greider, "Cellular

Responses to Telomere Shortening: Cellular Senescence as a Tumor Suppressor Mechanism," *The Harvey Lectures,* series 96 (New York: Academic Press, 2002), pp. 33–50.

52 telomere research began to assume greater visibility: see H. J. Cooke and B. A. Smith, "Variability at the Telomeres of the Human X/Y Pseudoautosomal Region," *Cold Spring Harbor Symposium on Quantitative Biology* 51 (1986): 213–19. On the first yeast gene related to short telomeres, see V. Lundblad and J. W. Szostak, "A Mutant with a Defect in Telomere Elongation Leads to Senescence in Yeast," *Cell* 57 (1989): 633–43.

54 a small army of researchers: the DNA sequence of human telomeres was reported by about half a dozen groups almost simultaneously, many of them presenting papers one after another at what has become an annual spring telomere meeting at Cold Spring Harbor Laboratory. The first published report, however, came from Robert Moyzis's group at Los Alamos National Laboratory.

54 By 1990, de Lange . . . had also published: see T. de Lange et al., "Structure and Variability of Human Chromosome Ends," *Molecular and Cell Biology* 10 (1990): 518–27.

57 [they] published a much-cited paper: C. B. Harley, A. B. Futcher, and C. W. Greider, "Telomeres Shorten During Aging of Human Fibroblasts," *Nature* 345 (May 31, 1990): 458–60.

58 "Find a chemical that can block": S. Russell, "A Discovery of Her Own," *San Francisco Chronicle,* May 8, 2000. This is hardly an isolated example. In her Harvey lecture (cited above), Greider pointed out examples of what she called "the hyperbole in the popular press about the potential of telomeres to predict or even alter lifespan" from many leading publications, including the *New York Times,* the *New Yorker, Newsday,* and *Time* (p. 33).

3. THE BORN-AGAIN DARWINIAN

INTERVIEWS: Vincent Cristofalo, June 5, 2002 (telephone); Caleb Finch, professor of gerontology and biological sciences, University of Southern California, Los Angeles, Aug. 21, 2001 (telephone); Calvin Harley, Nov. 15, 1999; Alan Mendelson, Axiom Venture Partners, Hartford, Conn., Aug. 2, 2002 (telephone); Daniel Perry, executive director, Alliance for Aging Research, Washington, D.C., Aug. 21, 2001 (telephone); Miller Quarles, president, Cure Old Age Disease Society, Houston, June 12, 2002 (telephone); R. J. Shmookler Reis, University of Arkansas Medical School, Little Rock, Oct. 28, 2002 (telephone); Eugenie Scott, director, National Center for Science Education, Oakland, July 24, 2002 (telephone); Jerry Shay, Dept. of Cell Biology, University of Texas Southwestern Medical Center, Dallas, Nov. 2, 1999 (telephone); James Smith, Dept. of Cellular and Structural Biology, University of Texas Health Science Center, San Antonio, Aug. 20, 2001 (telephone); Judson Somerville, Pain Management Clinic of Laredo, Tex., June 8, 2002 (telephone); Alan Walton, Oxford Bioscience Partners, Westport, Conn., July 6, 2001 (telephone), and June 18, 2002 (telephone); Michael West, president and CEO, Advanced Cell Technology, Worcester, Mass., Oct. 8, 1999 (telephone), Nov. 5, 1999, Dec. 7, 1999 (telephone), Dec. 15, 1999, Aug. 3, 2001 (telephone), Dec. 2, 2001, and Mar. 8, 2002; Woodring Wright, July 5, 2001 (telephone).

60   "I worry that I'm pathological": telephone conversation, Dec. 7, 1999.

62   the oldest living human: claims have been made that individuals have lived longer than Jeanne Calment, but many leading gerontologists question the accuracy and documentation of these claims; see Olshansky and Carnes, *Quest for Immortality,* pp. 206–7.

62   the myth of the Egyptian god Osiris: in the standard version, Isis and her niece Nephthys found all the dispersed pieces of Osiris, except for the phallus, and then buried him, giving him new life in the afterworld, where he then ruled. "The process of becoming Osiris, however, did not imply resurrection, for even Osiris did not rise from the dead. Instead, it was the assumption of immortality, both in the next world and through one's descendants on Earth." *Encyclopaedia Britannica Micropedia,* fifteenth ed. (1992): vol. 8, pp. 1026–27.

64   West's journey to the center: details about West's upbringing and education come from interviews and from a copy of his curriculum vitae provided by Advanced Cell Technology. A number of profiles and business biographies of West have been published. Among the best are D. Stipp, "The Hunt for the Youth Pill," *Fortune,* Oct. 11, 1999, p. 199; J. Alper, "A Man in a Hurry," *Science* 283 (Mar. 5, 1999): 1434; S. G. Stolberg, "Cell Biologist Traded Religious Fervor for Scientific Zeal," *New York Times,* Aug. 13, 2001, p. A-11; A. Zitner, "Cloning Advocate Fights the Clock on Capitol Hill," *Los Angeles Times,* Dec. 4, 2001, p. A-1; and M. McCullough, "Repro Man," *Philadelphia Inquirer Sunday Magazine,* Jan. 13, 2002, p. 12.

66   "the flagship educational institution": Andrews University Web site, www.andrews.edu.

66   "I made a fair amount of money": Alper, "Man in a Hurry," p. 1434.

66   he participated in antiabortion demonstrations: as West has testified, "It may be useful to point out that I think of myself as pro-life in that I have an enormous respect for the value of the individual human life. Indeed, in my years following college I joined others in protest of abortion clinics. My goal was not to send a message to women that they did not have the right to choose. My intent was simply to urge them to reconsider the destruction of a developing human being." Written testimony submitted to the Subcommittee on Labor, Health, and Human Services, Education and Related Agencies, of the Senate Appropriations Committee, July 18, 2001.

67   The Institute for Creation Research: see the organization's Web site, www.icr.org.

69   some notable findings about aging: R. J. Shmookler Reis and S. Goldstein, "Loss of Reiterated DNA Sequences During Serial Passage of Human Diploid Fibroblasts," *Cell* 21 (Oct. 1980): 739–49. West's CV lists no publications from his three-year period in the Goldstein lab.

69   He worked in the laboratory of James Smith: West published one paper while working in the lab; it had to do with "replicative senescence," or the Hayflick phenomenon, but involved the way skin cells overproduce a substance called collagenase (M. D. West, O. M. Pereira-Smith, and J. R. Smith, "Replicative Senescence of Human Skin Fibroblasts Correlates with a Loss of Regulation and Overexpression of Collagenase Activity," *Experimental Cell Research* 184 [1989]: 138–47).

72   "There were some biotech companies": Greider confirmed that she was approached about a possible academic-industrial collaboration based at the University of Massachusetts Medical School in Worcester.

73   Quarles . . . takes fifty vitamins: T. Mowatt, "Oil-Seeker to Youth-Finder," *Life Extension*, Oct. 1999.

75   he took the word from . . . the New Testament: Stolberg, "Cell Biologist Traded Religious Fervor."

75   National Conference on Biotechnology Ventures: the account of this meeting comes from interviews with Perry, Mendelson, Walton, and West. According to a summary of West's presentation in the meeting program, "Geron has discovered molecular and genetic events regulating a process termed 'cellular senescence' that underlies many aspects of human aging and age-related disease." The program does not mention telomere biology; West had already trademarked an inhibitor of cellular aging he called Senstatin, which he claimed "is over 2,500 times more effective than Retin-A in the reversal of skin aging."

4. "MONEY FOR JAM"

INTERVIEWS: Richard C. Allsopp, Stanford University, Dec. 11, 2000; William Andrews, vice president of research, Sierra Sciences, Inc., Reno, Nev., June 8, 2002 (telephone); Elizabeth Blackburn, Nov. 12, 1999; Titia de Lange, Oct. 13, 1999, and Aug. 22, 2000; Carol Greider, Oct. 20, 1999, Nov. 21, 2001 (e-mail), and Nov. 29, 2001 (telephone); Calvin Harley, Nov. 15, 1999; Leonard Hayflick, Dec. 7, 2000; Joachim Lingner, Swiss Institute for Experimental Cancer Research, Epalinges, Nov. 14, 2001 (e-mail); Vicki Lundblad, June 25, 2002 (telephone); John P. Maroney, director, Office of Technology Transfer and Inhouse Council, Cold Spring Harbor Laboratory, Cold Spring Harbor, N.Y., July 9, 2002 (telephone), and July 16, 2002 (telephone); Nancy Robinson, vice president of communications, Geron Corp., Menlo Park, Calif., Nov. 15, 1999; Homayoun Vaziri, research scientist, Whitehead Institute, Cambridge, Mass., Sept. 28, 2000, and Feb. 8, 2002 (telephone); Bryant

Villeponteau, chief scientific officer, HealthSpan Sciences, Carlsbad, Calif., Nov. 15, 1999 (telephone), Jan. 7, 2000 (telephone), and June 13, 2002 (telephone); Robert Weinberg, Whitehead Institute, Cambridge, Mass., Dec. 23, 2002 (telephone).

82 competing theories aspired to explain: Austad, Hayflick, and Kirkwood all offer good surveys of various theories about aging. For Bruce Ames's work on theories of oxidative damage in aging, see M. K. Shigenaga, T. M. Hazen, and B. N. Ames, "Oxidative Damage and Mitochondrial Decay in Aging," *PNAS* 91 (Nov. 1994): 10771–78.

83 "All of our vice presidents": N. Robinson, during a walking tour of the labs.

86 "We were meeting to talk": C. Greider, personal communication, Nov. 21, 2001.

87 "Throughout fiction and mythology": L. M. Fisher, "Enzyme May Offer Target in Tumors," *New York Times*, Apr. 12, 1994, p. C-10.

87 researchers had cloned the smaller, RNA component: see J. Feng et al., "The RNA Component of Human Telomerase," *Science* 269 (Sept. 1, 1995): 1236–41; and M. A. Blasco et al., "Functional Characterization and Developmental Regulation of Mouse Telomerase RNA," ibid., pp. 1267–70. A measure of the intensity of this research effort is that the first paper had sixteen authors; an indication of how the turf battle between the Geron and Cold Spring Harbor Laboratory researchers was resolved is that the last, and therefore senior, author on the human gene paper was Bryant Villeponteau.

87 "immortalizing enzyme" and "established a leadership": "First Cloning of Human Telomerase Reported in *Science*; Gene Regulates Telomerase Resulting in Death of Cancer Cells," Geron's press release, Sept. 1, 1995.

87 "The span of human life is determined": N. Wade, "New Light Shed on How Enzyme May Play Crucial Role in Cancer," *New York Times*, Sept. 3, 1995, p. C-3.

88 when the patent on the discovery was finally issued: a patent application for the RNA component of mammalian telomerase was filed Oct. 27, 1994, ten months before the scientific publication; the patent, number 5,583,016, was issued Dec. 10, 1996.

90 Greider and Collins published their results: K. Collins, R. Kobayashi, and C. W. Greider, "Purification of *Tetrahymena* Telomerase and Cloning of Genes Encoding the Two Protein Components of the Enzyme," *Cell* 81 (June 2, 1995): 677–86.

5. CONTROLLING THE HEADWATERS

INTERVIEWS: R. Alta Charo, professor of law and bioethics, University of Wisconsin, Madison, July 30, 2001 (telephone), July 31, 2001 (telephone),and Jan. 15, 2002 (telephone); John C. Fletcher, professor emeritus of bioethics, University of

Virginia, Charlottesville, June 4, 2002 (telephone); John Gearhart, Dept. of Obstetrics/Gynecology, Johns Hopkins Medical Institute, Baltimore, Oct. 5, 1999 (telephone), and Oct. 20, 1999; Robin Lovell-Badge, National Institute for Medical Research, London, Aug. 28, 2001 (New York); Ron McKay, NINDS, Bethesda, Md., Oct. 19, 1999, Sept. 18, 2000 (telephone), and Jan. 31, 2002 (telephone); Barbara Mishkin, Hogan and Hartson, Washington, D.C., Nov. 21, 2002 (telephone); Jim Murai, staff scientist, Immunetech, Menlo Park, Calif., Sept. 23, 2002 (telephone); Roger Pedersen, Dept. of Reproductive Medicine, University of California, San Francisco, Nov. 16, 1999, and July 26, 2002 (e-mail); Mark Pittenger, scientific director, Osiris Therapeutics, Inc., Baltimore, Sept. 6, 2002 (telephone); Irving Weissman, professor of cancer biology, Stanford University, Aug. 13, 2001 (telephone); Michael West, Oct. 8, 1999 (telephone), and Dec. 7, 1999 (telephone).

92 The tumor is known as a teratoma: on teratomas and teratocarcinomas, see V. T. DeVita, Jr., S. Helman, and S. A. Rosenberg, *Cancer: Principles and Practice of Oncology*, sixth edition (Philadelphia: Lippincott, Williams and Wilkins, 2001), pp. 2200–7. The authors note that these are primarily pediatric malignancies and that the occurrence in adults is "relatively infrequent."

93 "I looked in textbooks": in fact, as John Gearhart explained in an interview, there was a considerable literature on teratomas dating back to the 1970s, including research by François Jacob, G. Barry Pierce, and Boris Ephrussi documenting the behavior of teratocarcinomas — cancers arising from embryonic cells — and how their power to differentiate might explain fundamental aspects of development. Pierce, for example, had demonstrated early on that a single tumor cell from these embryonal cancers could generate virtually all cell tissue types.

93 West began to familiarize himself with the history: one of the most thoroughly documented recent surveys of stem cell research, both adult and embryonic, is in National Institutes of Health, *Stem Cells*.

93 Strictly speaking, the origins: the key embryonic stem cell papers are M. J. Evans and M. H. Kaufman, "Establishment in Culture of Pluripotential Cells from Mouse Embryos," *Nature* 292 (July 9, 1981): 154–56; and G. R. Martin, "Isolation of a Pluripotent Cell Line from Early Mouse Embryos Cultured in Medium Conditioned by Teratocarcinoma Stem Cells," *PNAS* 78 (Dec. 1981): 7634–38. As revolutionary as this discovery is now recognized to be, Evans revealed to an interviewer that he was hard-pressed to get funding to continue his research on mouse cells in the mid-1980s (see Albert Lasker Award 2001, Transcript of Conversation with Martin Evans, Sept. 20, 2001, at http://www.laskerfoundation.com/awards/library/2001evans_int.shtml).

94 hints . . . in the purely observational annals: on nineteenth- and early-twentieth-century clues about cells related to regeneration, see J. Cohnheim, *Lectures on General Pathology* (London: New Sydenham Society, 1889); S. G. Harvey, "The Healing of the Wound as a Biologic Phenomenon," *Surgery* 25

(May 1949): 655; and L. B. Arey, "Wound Healing," *Physiological Reviews* 16 (July 1936): 327. Cohnheim wrote of a "regenerative process" that "is effected by the activity of special tissue-cells" (p. 368) and said that "wandering cells" associated with the inflammatory response to a wound achieved this regeneration.

95 Every now and then, he would drop in: West kept visiting UCSF, according to a former Geron scientist, because he was pushing the company to launch a program on benign prostate hypertrophy, a condition suffered by many older men.

96 these were called knockout mice: the revolutionary use of embryonic stem cell technology to create knockout mice was described in A. Bradley, M. Evans, M. H. Kaufman, and E. Robertson, "Formation of Germ-Line Chimaeras from Embryo-Derived Teratocarcinoma Cell Lines," *Nature* 309 (May 17, 1984): 255–56. For an indication of the impact of knockout mice on research, see J. Knight and A. Abbott, "Full House," *Nature* 417 (June 20, 2002): 785–86. According to this account, some three thousand strains of knockout mice have been created, and animal facilities are filling to capacity with mutant mice designed to advance the study of human genetics.

98 "it's not appropriate for private investors": Alta Charo notes that "it is wrong to assume that privately funded work is going to control the headwaters and publicly funded work will not. If private money is used, the issue is over exclusive licensing" (A. Charo, personal communication, Oct. 21, 2002). It is true, as Charo notes, that even if the federal government funded human embryonic stem cell research, the patents would go the universities, not the government. However, the early involvement of companies sponsoring research has two important ramifications. One, early basic research may be discouraged or curtailed by the "reach-through" rights of companies that have commercial licenses to the university patents because of sponsored-research agreements. Two, the sponsoring company often enjoys privileged and expedited access to university research materials, which allows the creation of secondary patents that build up the company's intellectual property estate around a particular technology.

99 From that moment forward: for the intersection of stem cell research and reproductive medicine, see R. G. Edwards, "IVF and the History of Stem Cells," *Nature* 413 (Sept. 27, 2001): 349–51. "Fondly believed to be a recent development," Edwards writes of embryonic stem cells, "they have in fact been part and parcel of human in vitro fertilization (IVF) from as long ago as 1962."

99 an estimated 100,000 children: for figures on IVF births, see R. M. Schultz and C. J. Williams, "The Science of ART," *Science* 296 (June 21, 2002): 2188–90. The authors estimate that from 35 to 70 million couples worldwide "are infertile and have turned to ART [assisted reproductive technology] to overcome their infertility."

99 There is no better exhibit: on Soupart, see Andrews, *Clone Age*, pp. 32–33;

and John C. Fletcher, "Deliberating Incrementally on Human Pluripotential Stem Cell Research," paper commissioned by the National Bioethics Advisory Commission, in *Ethical Issues in Human Stem Cell Research*, vol. 2 (Jan. 2000), p. E11, available at www.georgetown.edu/research/nrcbl/nbac/stemcell2.pdf. A slightly different version appears in Kass, *Toward a More Natural Science*, pp. 100–101.

99   neurological researchers in Finland: for details of early fetal research, including immersion in salt solution, see Fletcher, "Deliberating Incrementally," p. E40, n. 77. The immersion study, Fletcher writes, "contributed to the design of artificial life-support systems for premature infants."

100   HEW established the Ethics Advisory Board: for the history of the board in the 1970s, see Fletcher, "Deliberating Incrementally," and Jonsen, *Birth of Bioethics*, pp. 106–7. In a different version of the EAB's demise, its former chairman, Barbara Mishkin, told a meeting of the NBAC in 1999 that an ill-informed deputy secretary of HEW told congressmen that he "couldn't think of any reason why" Congress should continue to support the ethics board when a separate presidential bioethics panel was about to be formed (B. Mishkin, testimony to NBAC, Feb. 3, 1999, pp. 23–24).

101   "it was summarily disbanded": Jonsen, *Birth of Bioethics*, p. 107. For Harris's role in the demise of the EAB, see Fletcher, "Deliberating Incrementally," pp. E11, E43.

101   parliamentary committee headed by . . . Warnock: "Report of the Committee of Inquiry into Human Fertilisation and Embryology" (London: Her Majesty's Stationery Office, 1984).

102   20,000 attempts at in vitro: Andrews, *Clone Age*, pp. 218–19.

102   400 percent increase: on the increased numbers of triplets and other recent trends in assisted reproduction, see P. Braude, "Measuring Success in Assisted Reproductive Technology," *Science* 296 (June 21, 2002): 2101.

6. "The White House Was Nervous . . ."

Interviews: R. Alta Charo, July 30, 2001 (telephone), and July 31, 2001 (telephone); William Galston, University of Maryland, College Park, June 13, 2002 (telephone); Ronald Green, director, Ethics Institute, Dartmouth College, Hanover, N.H., Oct. 10, 1999 (telephone), and Mar. 8, 2002 (Worcester, Mass.); Kathi Hanna, consultant, National Academy of Sciences, Washington, July 31, 2001 (telephone), and June 5, 2002 (telephone); Brigid Hogan, professor of molecular oncology, Vanderbilt University, Nashville, Tenn., Nov. 11, 1999 (telephone); Mark Hughes, director, Center for Molecular Medicine and Genetics, Wayne State University, Detroit, Mich., Dec. 19, 2002 (telephone); Thomas Murray, president, Hastings Center, Garrison, N.Y., Nov. 26, 2001 (telephone), and Nov. 30, 2001 (telephone); Harold Varmus, president, Memorial Sloan-Kettering Cancer Center, New York, Aug. 25, 2001; Michael West, Dec. 7, 1999 (telephone); Dan Wikler, professor of ethics and population health, Harvard School of Public Health, Boston, June 6, 2002 (telephone).

107 Varmus brought real scientific stature to the job: on Varmus's tenure at NIH, see T. Beardsley, "Setting the Course for the Nation's Health," *Scientific American*, Jan. 2000, pp. 30–32.

109 legislators . . . have never had the stomach: Alta Charo notes that unlike the situation in England, the federal government has very little authority over the practice of medicine or the conduct of private research.

110 "If human ES cells could be obtained": in National Institutes of Health, *Report of the Human Embryo Research Panel* (Bethesda, Md.: NIH, 1994), p. 27.

110 Ronald Green . . . prepared: see Green, *Human Embryo Research Debates*, pp. 6–13, 25–53.

111 "Reproductive embryologists report": ibid., p. 37.

111 "Humans are not particularly good": J. Cohen, "Sorting Out Chromosome Errors," *Science* 296 (June 21, 2002): 2164–66. Chromosomal abnormalities "explain half of all miscarriages," according to this survey of recent research.

112 commission as "a special interest group": Doerflinger is quoted in Green, *Human Embryo Research Debates*, p. 18. "If this narrative has a villain," Green writes, "it is Richard Doerflinger" (p. 17).

112 "Pat King was convinced": Charo says she "discounted" some of King's political warnings "because she appeared to be speaking from personal views, i.e., without any analytical basis," and that her position was "a post hoc rationalization for her feelings. In retrospect, I think my reading was correct — she *was* speaking from personal views, not analysis, but I was wrong to discount the political fallout nonetheless." King did not respond to several requests for an interview.

113 "The fertilization of human oocytes": "Statement of Patricia A. King," Human Embryo Research Panel report, Appendix A, p. A-3.

114 a front-page headline: Richard A. Knox, "U.S. Panel May OK Human Embryo Study; Bid to Allow Funding Seen Drawing Fire," *Boston Globe*, Aug. 19, 1994, p. A-1. The political fallout from this news leak followed a predictable pattern: less than two weeks after the *Globe* story, a Virginia-based antiabortion group called on Congress to withhold all funds from the NIH pending an investigation of the embryo panel, Harold Varmus, and the entire NIH, accusing NIH officials of "unblushing greed" and a "voracious lust to create and manipulate human beings" (quoted in R. A. Knox, "Embryo Research Funding Assailed," *Boston Globe*, Aug. 27, 1994, p. A-4). See also E. Marshall, "Rules on Embryo Research Due Out," *Science* 265 (Aug. 19, 1994): 1024–26.

115 creation of research embryos "unconscionable": "Embryos: Drawing the Line," *Washington Post*, Oct. 2, 1994, p. C-6. Interestingly, the *Post* editorially reversed itself eight years later and supported the creation of cloned human embryos solely for research (see "The Promise of Cloning," *Washington Post*, Apr. 19, 2002, p. A-24).

116   the White House formed an ad hoc working group: the embryo panel controversy was by no means a predominant White House concern during the fall of 1994. In his memoir, George Stephanopoulos never mentions the embryo report; he describes the Haiti invasion and health care reform as the dominant issues leading up to the 1994 congressional elections (Stephanopoulos, *All Too Human*, pp. 303–27).

116   "By a huge majority": Muller quoted in William A. Galston, "Ethics and Public Policy in a Democracy: The Case of Human Embryo Research," in *New Dimensions in Bioethics: Science, Ethics and the Formulation of Public Policy*, ed. W. A. Galston and E. G. Shurr (Boston: Kluwer Academic, 2001), pp. 193–207.

117   When Galston spoke about the embryo report: Dan Wikler, a bioethicist at the Harvard School of Public Health, attended Galston's talk and recalled: "It was just political. I wondered, while listening to Galston talk, 'Do you believe a word of what you're saying?' He stood up with a straight face and said the reason that the Clinton White House knocked down the embryo panel recommendation was this 'yuck factor' crap."

117   "We had every reason to believe": Green, *Human Embryo Research Debates*, p. 101; "profiles in courage": ibid., p. 103.

118   copy of a White House statement: the executive order stated, "I do not believe that Federal funds should be used to support the creation of human embryos for research purposes, and I have directed that the NIH not allocate any resources for such purposes" (statement by the president, Dec. 2, 1994).

119   work "in which a human embryo": for the language of the Dickey-Wicker amendment, see Fletcher, "Deliberating Incrementally," p. E-39. Beyond the strict legal constraints of the Dickey-Wicker amendment, congressional legislation on this issue also draws attention to an absurd ethical paradox: embryos that might be destroyed by publicly funded researchers apparently have a higher moral status than embryos that are routinely destroyed by the private sector. Since the early 1980s, countless thousands of embryos created at private in vitro fertilization clinics have been routinely frozen, abandoned, and discarded, without a hint of legislative activity. When asked once about this apparent paradox, a spokesman for Jay Dickey replied, "Our concern is federal dollars." When asked about private companies that discard embryos or create cow-human hybrids, he said that the congressman was "not going to jump in and dictate what these private companies are doing" (Rob Johnson, spokesman for former congressman Jay Dickey, Washington, Jan. 7, 2000, by telephone).

121   "On NBAC, there were three members": Hanna may be overstating; Thomas Murray, for example, said that he did not become a proponent of research cloning until after his time on NBAC.

121   legal opinion . . . from Harriet Rabb: see E. Marshall, "Ruling May Free NIH to Fund Stem Cell Studies," *Science* 283 (Jan. 22, 1999): 465–67.

125 Geron's initial investment in Thomson's research: a former Geron researcher said, "When it started working, it was closer to half a million a year to Thomson's lab."

7. Cloning in Silico

INTERVIEWS: William Andrews, June 8, 2002 (telephone); Thomas R. Cech, president, Howard Hughes Medical Institute, Chevy Chase, Md., Jan. 7, 2002 (telephone); Kathleen Collins, associate professor of biochemistry and molecular biology, University of California, Berkeley, Dec. 12, 2002 (telephone); Christopher Counter, Dept. of Pharmacology and Cancer Biology, Duke University Medical Center, Durham, N.C., July 9, 2002 (telephone); Titia de Lange, Aug. 22, 2000; Carol Greider, June 7, 2002 (telephone); Joachim Lingner, Nov. 14, 2001 (e-mail); Victoria Lundblad, June 25, 2002; Gregg Morin, vice president of biology, MDS Proteomics, Toronto, Oct. 3, 2002 (telephone); Bryant Villeponteau, June 13, 2002 (telephone); Robert Weinberg, Sept. 28, 2000, and Dec. 6, 2002 (e-mail); Michael West, Nov. 5, 1999.

127 The issue contained a paper: K. Collins, R. Kobayashi, and C. W. Greider, "Purification of *Tetrahymena* Telomerase and Cloning of Genes Encoding the Two Protein Components of the Enzyme," *Cell* 81 (June 2, 1995): 677–86. Collins pointed out in an interview that telomerase has turned out to be much more complex than was understood in 1995, and that the 1995 paper was not as wrong as it may have seemed at the time. "We were *so* ignorant of the complexity that lay ahead," she said. "A lot of people in the field had a financial motivation, and ego, to make a big deal out of it. People were fighting for prominence and overstating the data . . . I never had the feeling that if I interpreted the data the way I saw it, I would be blamed for being wrong." Was she? "Absolutely."

129 It was a crushing amount of work: for details of the process of isolating the telomerase protein, see J. Lingner and T. R. Cech, "Purification of Telomerase from Euplotes aediculatus: Requirement of a Primer 3′ Overhang," *PNAS* 93 (Oct. 1, 1996): 10712–17.

130 Geron made preparations to take the company public: in describing the potential market for its drugs in development, the Geron Company Prospectus (July 30, 1996) noted that "with the progressive 'graying' of the population, the incidence of cancer and other age-related diseases and conditions is expected to increase and to place a steadily growing financial burden on the health care system. By the year 2010, the over-65 population in the United States is expected to double to approximately 64 million people and worldwide this population will increase to over one billion" (p. 24).

132 [Lundblad] had a paper in press: T. S. Lendvay, D. K. Morris, J. Sah, B.

Balasubramanian, and V. Lundblad, "Senescence Mutants of *Saccharomyces cerevisiae* with a Defect in Telomere Replication Identify Three Additional EST Genes," *Genetics* 144 (Dec. 1996): 1399–412.

132 "We cloned two components": I am grateful to Carol Greider for sharing a detailed time line of the sequence of events leading to the cloning of the human telomerase gene.

134 net losses totaled more than $150 million: according to annual reports filed with the Securities and Exchange Commission, Geron had net losses of $42.1 million in 2001, $45.8 million in 2000, $46.4 million in 1999, $10.8 million in 1998, $9.6 million in 1997, $10.7 million in 1996, $8.2 million in 1995, and $10.2 million in 1994.

134 By sharing their data: the fruits of the "secret sharing" of telomerase data resulted in J. Lingner, T. R. Hughes, A. Shevchenko, M. Mann, V. Lundblad, and T. R. Cech, "Reverse Transcriptase Motifs in the Catalytic Subunit of Telomerase," *Science* 276 (Apr. 25, 1997): 561–67.

137 they were close on the heels of Cech and Lundblad: Counter admitted that the Whitehead group's work on the yeast telomerase gene was "preliminary" and that the later publication had much more convincing data. It appeared as C. Counter, M. Meyerson, E. N. Wheaton, and R. A. Weinberg, "The Catalytic Subunit of Yeast Telomerase," *PNAS* 94 (Aug. 19, 1997): 9202–7; that date was exactly three days before the same group published its results on the cloning of the human gene.

139 Geron and the Cech group submitted a paper: T. M. Nakamura, G. B. Morin, K. B. Chapman, S. L. Weinrich, W. H. Andrews, J. Lingner, C. B. Harley, and T. R. Cech, "Telomerase Catalytic Subunit Homologs from Fission Yeast and Human," *Science* 277 (Aug. 15, 1997): 911–12. The Whitehead team published their result in M. Meyerson et al., "hEST2, the Putative Human Telomerase Subunit Gene, Is Up-regulated in Tumor Cells and During Immortalization," *Cell* 90 (Aug. 22, 1997): 785–95. Both Cech and Weinberg gave behind-the-scenes accounts of the work in "Hot Papers in Telomerase," *Scientist* 13, no. 20 (Oct. 11, 1999): 14. Cech acknowledged that "some luck was involved" in finding the human DNA sequence in the public databases. "We knew that it would eventually show up," he said. "We had no idea whether we would have to wait months or years." For his part, Weinberg has consistently credited Lingner's work with allowing the Whitehead group to isolate the human gene for telomerase's catalytic protein.

139 public reaction to the news: in a front-page story, the *San Francisco Chronicle* reported in the very first sentence that scientists "have cloned an ancient 'immortality gene' that gives human cells the ability to reproduce seemingly without limit" (C. Petit, "Regenerating Gene Cloned by Scientists; New Light Shed on Cancer Cells," *San Francisco Chronicle*, Aug. 15, 1997, p. 1). Within a day of the announcement, Geron stock rose from $6.50 to $14, with 12.5 million shares traded; see L. M. Fisher, "Biotechnology Company Says It Has Cloned a Cancer Gene," *New York Times*, Aug. 18, 1997, p. D-9.

141 researchers pursued a more nuanced view: on telomere shortening as a tumor suppressor mechanism, see C. W. Greider, "Cellular Responses to Telomere Shortening: Cellular Senescence as a Tumor Suppressor Mechanism," *Harvey Lectures*, series 96, 2002); on a possible role for telomerase in cirrhosis, see the work of Ron de Pinho's group at Dana-Farber Cancer Institute, in K. L. Rudolph et al., "Inhibition of Experimental Liver Cirrhosis in Mice by Telomerase Gene Delivery," *Science* 287 (Feb. 18, 2000): 1253–58; for the more complex mechanism of replicative senescence involving telomere-associated proteins, see J. Karlseder et al., "Senescence Induced by Altered Telomere State, Not Telomere Loss," *Science* 295 (Mar. 29, 2002): 2446–49.

142 turning on the telomerase gene predisposed a cell to become cancerous: W. C. Hahn et al., "Creation of Human Tumour Cells with Defined Genetic Elements," *Nature* 400 (July 29, 1999): 464–68. On cancer vaccine trials, see S. K. Hair et al., "Induction of Cytotoxic T Cell Responses and Tumor Immunity Against Unrelated Tumors Using Telomerase Reverse Transcriptase RNA Transfected Dendritic Cells," *Nature Medicine* 6 (Sept. 2000): 1011–17. For an excellent summary of the "good" and "bad" sides of telomerase, see C. J. Prescott and E. H. Blackburn, "Telomerase: Dr. Jekyll or Mr. Hyde?" *Current Opinion in Genetics and Development* 9 (1999): 368–73.

143 she published a follow-up paper: D. X. Mason, C. Autexier, and C. W. Greider, "*Tetrahymena* Proteins p80 and p95 Are Not Core Telomerase Components," *PNAS* 98 (Oct. 23, 2001) 12,368–73. The article that triggered concern at Geron was C. W. Greider, "Telomere Length Regulation," *Annual Review of Biochemistry* 65 (1996): 337–65. The offending sentence, on p. 357, was "Although the role of telomere length in signaling cellular senescence is not fully understood, telomere length is clearly not directly correlated with organismal aging." The relationship between telomere shortening and aging remains a controversial topic. In 1998, Vincent J. Cristofalo and colleagues challenged the notion that the age of a cell donor did not necessarily correspond to the replicative life span of the cells in culture (see V. J. Cristofalo et al., "Relationship Between Donor Age and the Replicative Lifespan of Human Cells in Culture: A Reevaluation," *PNAS* 95 (Sept. 1998): 10614–19. For two recent sides of the argument, see W. E. Wright and J. W. Shay, "Historical Claims and Current Interpretations of Replicative Aging," *Nature Biotechnology* 20 (July 2002): 682–88; and H. Rubin, "The Disparity Between Human Cell Senescence in Vitro and Lifelong Replication in Vivo," *Nature Biotechnology* 20 (July 2002): 675–81. Wright and Shay concede that "thus far there is only very limited direct evidence for actual physiological effects of replicative aging" (p. 682). Rubin asserts that "there is increasing skepticism that the Hayflick limit of cells in culture accurately reflects their replicative potential in the organism" (p. 675) and warns that a telomerase inhibitor might be especially dangerous at creating secondary skin cancers in people who have had frequent exposure to the sun, which is known to create mutations in a key tumor-suppressing gene.

143   one other researcher received scolding phone calls: Titia de Lange recalled several instances in which Geron officials objected to language in scientific articles or commentaries she had written as being too negative; Cal Harley, in an interview, defended these phone calls, stating, "Someone who writes — and I think it may be from the point of view of getting attention for things that they're seeing — sort of taking a really extreme, negative perspective on this science is harming the science and harming the potential for the advancement of medicine. So I'll point that out to people."

8. HAYFLICK UNLIMITED

INTERVIEWS: William Andrews, June 8, 2002 (telephone); Günter Blobel, Laboratory of Cell Biology, Rockefeller University, New York, Mar. 4, 2002; Leonard Hayflick, Dec. 1, 1999, and June 10, 2002 (telephone); Gregg Morin, Oct. 3, 2002 (telephone); Jim Murai, senior scientist, Immunetech, Menlo Park, Sept. 23, 2002 (telephone); Thomas Okarma, Nov. 15, 1999; Bryant Villeponteau, Nov. 15, 1999 (telephone), Jan. 7, 2000 (telephone), and June 13, 2002 (telephone); Michael West, Dec. 7, 1999 (telephone), and Dec. 4, 2001 (Washington, D.C.); Woodring Wright, July 5, 2001 (telephone).

146   a Canadian film crew: *Old Before Their Time,* a forty-eight-minute documentary on aging research, was produced and directed by David Way for TV Ontario in 1998. In a voice-over one scientist is quoted as saying, "I think the long-term prospects still are that we *will* come to a point where we should be able to slow down the process [of aging] and become immortal."

146   the company . . . that was working on a "Zeus juice": D. Stipp, "The Hunt for the Youth Pill," *Fortune,* Oct. 11, 1999, p. 199.

147   A mere four years later: see Hall, *Invisible Frontiers,* pp. 293–94.

150   The paper that ultimately emerged: A. G. Bodnar, M. Ouellette, M. Frolkis, S. E. Holt, C. P. Chiu, G. B. Morin, C. B. Harley, J. W. Shay, S. Lichtsteiner, and W. E. Wright, "Extension of Life-Span by Introduction of Telomerase into Normal Human Cells," *Science* 279 (Jan. 16, 1998): 349–52. Even an agnostic on aging like Titia de Lange had to admit that the Bodnar experiment put an end to doubts about the connection between telomere shortening and the inability of cells to replicate. "All's well that ends well, and so it goes for the decade-old debate on the role of telomere shortening in the senescence of cells," she wrote in a commentary on the Bodnar work (T. de Lange, "Telomeres and Senescence: Ending the Debate," *Science* 279 [Jan. 16, 1998]: 334–35).

150   Geron's stock . . . nearly quadrupled: for an account of the Bodnar telomerase paper's premature influence on the stock market because of a broken embargo, see E. Marshall, "Trading in Science: A Volatile Mix of Stock Prices and Embargoed Data," *Science* 282 (Oct. 30, 1998): 865.

150   "the highest buzz-to-equity ratio": Stipp, "Hunt for the Youth Pill."
150   "Adding telomerase elongates the telomeres": T. Jacobs, "Geron Reports,"
       MotleyFool.com, Aug. 15, 2000.
154   Geron hired a new vice president: for Okarma's background, see "Geron Corporation Appoints Thomas B. Okarma President and CEO," Geron press release, July 23, 1999.
155   "I didn't think I'd live long enough": Hayflick quoted in C. T. Hall, "Non-Aging Human Cells Created in Lab; Bay Firm's Stock Soars on Hopes of Medical Advances," *San Francisco Chronicle*, Jan. 14, 1998, p. A-1, which also includes an account of Geron's stock performance. For similar front-page press coverage of the "life span" paper, see N. Wade, "Cells' Life Stretched in Lab," *New York Times*, Jan. 14, 1998, p. A-1; and T. Monmaney, "Scientists Give Cell Apparent Immortality," *Los Angeles Times*, Jan. 14, 1998, p. A-1. For follow-up features, see N. Wade, "Cells Unlocked: Longevity's New Lease on Life," *New York Times*, Jan. 18, 1998 (Week in Review, p. 1); N. Wade, "Cell Rejuvenation May Yield Rush of Medical Advances," *New York Times*, Jan. 20, 1998, p. F-1; and T. Monmaney, "A Real Blast: Defusing 'Genetic Time Bomb,'" *Los Angeles Times*, Jan. 18, 1998, p. A-1.

9. "MAMAS, DON'T LET YOUR BABIES GROW UP TO BE COWBOYS . . ."

INTERVIEWS: Alta Charo, July 30, 2001 (telephone); David Cox, Dept. of Genetics, Stanford University, Nov. 23, 1999 (telephone); Neal First, Dept. of Animal Sciences, University of Wisconsin, Madison, Dec. 2, 2002 (telephone); Norman Fost, professor of bioethics, University of Wisconsin, Madison, June 18, 2002 (telephone); John Gearhart, Oct. 20, 1999; Carol Greider, Dec. 2, 2002 (telephone); Joseph Itskovitz-Eldor, Rambam Medical Center, Technion University, Haifa, Israel, Aug. 22, 2002 (telephone); Robin Lovell-Badge, Aug. 28, 2001 (New York); Douglas Melton, chairman, Dept. of Molecular and Cellular Biology, Harvard University, Nov. 26, 2001 (telephone); Thomas Murray, Nov. 3, 1999 (telephone); Roger Pedersen, Nov. 16, 1999; Eric Schon, Columbia University, New York, Nov. 27, 2002 (telephone); Michael Shamblott, Johns Hopkins Medical Institute, Baltimore, Oct. 20 and 21, 1999; James Thomson, University of Wisconsin, Madison, Dec. 13, 1999 (telephone); Shirley Tilghman, Princeton University, Dec. 20, 1999; Alan Trounson, Sept. 10, 2001 (telephone); Michael West, Oct. 8, 1999 (telephone); Ian Wilmut, Dec. 3, 2001.

159   Thomson . . . set up shop: for more details on how Wisconsin set up its stem cell research program, see S. Lueck, "Wisconsin Shows Stem-Cell Quest Can Become a Juggernaut," *Wall Street Journal*, Aug. 23, 2001, p. A-18. On Thomson, see S. G. Stolberg, "Reserved Scientist Creates an Uproar with His Work on Stem Cells," *New York Times*, July 10, 2001, p. A-12.
159   his lab's creation of monkey stem cells: on rhesus monkey embryonic stem

cells, see J. A. Thomson et al., "Isolation of a Primate Embryonic Stem Cell Line," *PNAS* 92 (Aug. 1995): 7844–48.

160   Trounson . . . had been asked to organize a session: the session was actually at a symposium sponsored by the pharmaceutical company Serono in conjunction with the ASRM meeting. The speed with which the research was progressing became apparent at a meeting in Utah in July 1997, when Roger Pedersen chaired a panel on the ethics of research on human embryonic stem cells, and Gearhart reported that the Hopkins group had isolated embryonic germ-line (EG) cells and maintained them in culture for a number of months (see R. Lewis, "Embryonic Stem Cells Debut amid Little Media Attention," *Scientist* 11, no. 19 [Sept. 29, 1997]: 1).

160   Bongso, an expert in fertility research: see R. Frank and V. Brooks, "A Stem-Cell Line in Singapore Lab Nears a Payoff," *Wall Street Journal,* Aug. 31, 2001, p. B-1; and D. Gay, "Asia's Stem Cell Savant," *Asiaweek,* Aug. 24, 2001. For a description of the early work in human embryonic stem cell research, as well as a list of Bongso's early publications, see the National Institutes of Health, *Stem Cells,* pp. 11–21.

162   The Wisconsin group started out with thirty-six: for the attrition rate from the initial thirty-six embryos and other details, see James A. Thomson et al., "Embryonic Stem Cell Lines Derived from Human Blastocysts," *Science* 282 (Nov. 6, 1998): 1145–47.

163   Gearhart and Shamblott . . . appeared to be slightly behind: Michael J. Shamblott et al., "Derivation of Pluripotent Stem Cells from Cultured Human Primordial Germ Cells," *PNAS* 95 (Nov. 1998): 13726–31.

164   The Australians . . . may have been much closer: on the status of the Australian research team, see G. Vogel, "In the Mideast, Pushing Back the Stem Cell Frontier," *Science* 295 (Mar. 8, 2002): 1818–20. In addition to detailing the progress of the Australian-Israeli collaboration on human ES cells by the fall of 1998, the article also describes the considerable achievements of Israeli scientists in the stem cell field, including Itskovitz-Eldor in Haifa and Benjamin Reubinoff and Nissim Benvenisty of Hebrew University in Jerusalem.

165   "Jose worked in my lab independently": e-mail from Itskovitz-Eldor to author, Aug. 26, 2002.

165   "The study of aging is undergoing": N. Wade, "Immortality, of a Sort, Beckons to Biologists," *New York Times,* Nov. 17, 1998, p. F-1. For front-page coverage, see N. Wade, "Scientists Cultivate Cells at Root of Human Life," *New York Times,* Nov. 6, 1998, p. A-1; and T. H. Maugh, "Scientists Move Closer to Ability to Grow Tissue," *Los Angeles Times,* Nov. 6, 1998, p. A-1.

165   describing the work as a "breakthrough": "First Derivation of Human Embryonic Stem Cells Reported in *Science,*" press release, Geron Corp., Nov. 5, 1998.

165   "déjà vu all over again": D. Malakoff, "Reaction to Stem Cells: A Tale of the Ticker," *Science* 282 (Nov. 13, 1998): 12. For the performance of Geron stock,

see also A. Pollack, "Small Company Gains High Profile in the Scientific World," *New York Times*, Nov. 6, 1998, p. A-24. This *Times* story, incidentally, noted that Michael West had quit Geron in February 1998 "to start a company to apply telomere theory to animals, an effort that Geron executives had declined to pursue."

166 Okarma was testifying before Congress: "Hearings Before a Subcommittee of the Committee on Appropriations, United States Senate," Dec. 2, 1998, pp. 51–59. As an example of the identity crisis going on at the time at Geron, it is fascinating to note a kind of double-speak from company officials about the combination of telomerase therapy and embryonic stem cell therapy. In one instance, Okarma told the *Times* that making the body immortal "is indeed fanciful and certainly something we don't even contemplate here. We are much more interested in dealing with the 20 or so degenerative diseases that have no present treatment and could be addressed with immortalized cells" (Wade, "Immortality, of a Sort"). Barely a week earlier, the *Times* had reported, "Geron biologists believe they can manipulate the telomeres of the human embryonic stem cells so that the cells stay immortal even as they turn into specialized tissues. Can the mortal body therefore be repaired with new tissues that remain youthful indefinitely? 'Exactly,' Dr. Okarma said" (Wade, "Scientists Cultivate Cells").

166 "That would make me": D. Melton, personal communication.

168 ACT and Jose had filed patents: "Embryonic or Stem-like Cell Lines Produced by Cross Species Nuclear Transplantation," international patent application filed by the University of Massachusetts, July 28, 1997.

170 "It's hard to say this is a total sham": Pedersen quoted in N. Wade, "Researchers Claim Embryonic Cell Mix of Human and Cow," *New York Times*, Nov. 12, 1998, p. A-1.

171 Bill Clinton denounced the research: the relevant portion of President Clinton's letter to the National Bioethics Advisory Commission is in the transcript of the NBAC meeting, Nov. 18, 1998, p. 99.

171 Scientists dismissed the research as doomed: see E. Marshall, "Claim of Human-Cow Embryo Greeted with Skepticism," *Science* 282 (Nov. 20, 1998): 1390–91. Even colleagues of West's expressed doubts about the timing of the announcement; an executive of a public relations firm that once worked with ACT later said that the pattern of me-too announcements by West seemed "too fortuitously coincidental to be accidental."

171 a cow-human clone was unlikely to yield viable embryos: scientists familiar with cell biology suggest two reasons the interspecies work might be expected not to succeed. A cow egg, even after its nucleus had been removed, would still contain dozens of the little organelles known as mitochondria, which extract energy biochemically to power the cell's activities. Mitochondria contain their own DNA, and in the ACT experiments, the mitochondrial DNA from cows would probably be incompatible with the human

DNA of the adult donor cell, leading to dysfunction and death. Neal First, an expert in animal cloning at the University of Wisconsin, has done experiments demonstrating the difficulty of doing interspecies cloning, probably because of mitochondrial incompatibility. However, Eric Schon of Columbia University, an expert on mitochondria, said in an interview that he believed nuclear mitochondrial DNA — that is, from the donor (in this case, human) cells — would in effect trump the mitochondrial DNA of the egg cell, thus supporting the view that interspecies nuclear transfer would not be impossible because of mitochondrial incompatibility.

Another problem, in both reproductive and therapeutic cloning, involves the phenomenon known as "imprinting." When sperm meets egg in normal sexual reproduction, the DNA of the two cells is usually labeled, or enhanced, in certain areas, depending on whether the DNA comes from the male or the female. These imprinted genes confer an advantage to either the male or female contributor of a given gene. Cloning is believed to strip away the physical apparatus that allows genes to bear imprinting, and thus impairs the normal process of gene integration and expression in a developing fertilized egg. According to George Daley, a stem cell researcher at the Whitehead Institute, flaws in imprinting would likely have more severe ramifications for reproductive cloning, in which the timing of gene activation throughout gestation is crucial, than for therapeutic cloning, in which the period of crucial gene activation is about one week.

172 rumors that the company was sponsoring: see R. Weiss, "Embryo Work Raises Specter of Human Harvesting; Medical Research Teams Draw Closer to Cloning," *Washington Post,* June 14, 1999, p. A-1. This story quoted Calvin Harley of Geron as saying the company was working on cloned human embryos. "'The work is not going on at Geron's headquarters in California but in the laboratory of a scientist who is funded by Geron,' Harley said," according to the story.

172 "I do want to tell the commissioners": transcript of National Bioethics Advisory Commission hearing, Nov. 17, 1998, pp. 121–34, available at www.georgetown.edu/research/nrcbl/nbac/transcripts/index.html#nov98. Of West's appearance before the commission, Thomas Murray said, "He had not been asked to come and speak to us. He just showed up."

173 "venturing deep into uncharted realms": Wade, "Researchers Claim Embryonic Cell Mix."

10. Dead in the Water

Interviews: Ali Hemmati Brivanlou, Laboratory of Molecular Vertebrate Embryology, Rockefeller University, New York, Nov. 10, 2000, June 8, 2001 (telephone), and June 26, 2001 (telephone); Thomas Cech, Jan. 7, 2002 (telephone); Andrew Cohn, Wisconsin Alumni Research Foundation, Madison, Dec. 4, 2002 (telephone); George Daley, Whitehead Institute, Cambridge, Mass., Nov. 27, 2000 (telephone), and Dec. 4, 2002 (telephone); Carl E. Gulbrandsen, managing director, Wisconsin Alumni Research Foundation, Jan. 10, 2000 (telephone), and Dec. 20, 2002 (telephone); Ron McKay, Sept. 18, 2000; Douglas Melton, Nov. 26, 2001 (telephone), and Feb. 27, 2002 (telephone).

176 candidate interviews conducted by *Science*: see "Gore and Bush Offer Their Views on Science," *Science* 290 (Oct. 13, 2000): 262–69.

176 [Gore's and Bush's] views of embryonic stem cell research: following the publication of final NIH guidelines for stem cell research in the *Federal Register* on Aug. 23, 2000, both presidential campaigns issued comments. A spokesman for the George W. Bush campaign said, "The governor opposes federal funding for stem cell research that involves destroying a living human embryo." A spokesperson for Albert Gore said he "supports the recommendations," and the Democratic Party platform included the following statement: "We should allow stem cell research to make important new discoveries" (R. Weiss, "Clinton Hails Embryo Cell Test Rules," *Washington Post*, Aug. 24, 2000, p. A-11).

176 "Taxpayer funds should not underwrite": R. Lacayo, "How Bush Got There," *Time*, Aug. 20, 2001, p. 18.

176 an op-ed piece by actor Michael J. Fox: M. J. Fox, "A Crucial Election for Medical Research," *New York Times*, Nov. 1, 2000, p. A-34.

177 described in 1924 by Hans Spemann: see T. J. Horder and P. J. Weindling, in Horder, Witkowski, and Wylie, *History of Embryology*, pp. 183–242.

178 "nobody can stop scientists": although Brivanlou in fact refrained from any experiments, there is specific precedent for this sentiment from the last major scientific controversy. During the voluntary moratorium against recombinant DNA experimentation in the United States between 1975 and 1978, several groups of biologists exploited loopholes in the research guidelines to circumvent the moratorium, either traveling overseas to conduct research or adopting technical approaches not specifically proscribed by the moratorium. As a result of their actions, genetically engineered insulin and many other life-sustaining products are today used by millions of patients throughout the world. See Hall, *Invisible Frontiers*, pp. 81–82, 249–65, 270–71.

178 tension between "molding and beholding": see William May, transcript of the President's Council on Bioethics, morning session, Feb. 14, 2002, session 6, available at www.bioethics.gov.

178 conservatives have characterized the research: see D. Johnson, National Right to Life Committee, in R. Weiss, "House Votes Broad Ban on Cloning," *Washington Post*, Aug. 1, 2001, p. A-1, and C. Krauthammer, "Crossing Lines: A Secular Argument Against Research Cloning," *New Republic*, Apr. 29, 2002

("embryo farming"); C. Connolly, "3 GOP Leaders Warn Bush on Stem Cell Studies," *Washington Post*, July 3, 2001, p. A-2 ("industry of death"); C. Krauthammer, transcript, President's Council on Bioethics, Jan. 17, 2002, session 3, and Feb. 14, 2002, session 6 ("harvested" and "strip-mined"); and L. R. Kass, "Preventing a Brave New World," *New Republic*, May 21, 2001 (the *Brave New World* scenario).

182   Knocking out the action: see A. Hemmati-Brivanlou and D. A. Melton, "A Truncated Activin Receptor Dominantly Inhibits the Induction of Mesoderm in Xenopus Embryos," *Nature* 359 (1992): 609–14; and A. Hemmati-Brivanlou and D. A. Melton, "Inhibition of Activin Receptor Signaling Promotes Neuralization in *Xenopus laevis*," *Cell* 77 (1994): 273–81.

182   Melton underwent a dramatic . . . conversion: a good account of Melton's background is A. Allen, "God and Science," *Washington Post Magazine*, Oct. 15, 2000, p. 8.

183   experiments . . . "will change human embryology": A. Hemmati-Brivanlou, "A Proposal for Molecular Human Embryology," unpublished research proposal, 2002, p. 19 (courtesy of Brivanlou).

184   "breakthrough" of the year: see G. Vogel, "Capturing the Promise of Youth," *Science* 28 (Dec. 17, 1999): 2238–39.

184   the National Bioethics Advisory Commission recommended: see E. Marshall, "Ethicists Back Stem Cell Research, White House Treads Cautiously," *Science* 285 (July 23, 1999): 502; for the full report, see the NBAC Web site, archived at Georgetown University: www.georgetown.edu/research/nrcbl/nbac/.

184   "President Clinton thanked the commission": R. Weiss, "Embryonic Breakthroughs? Stem Cell Studies Race Ahead as U.S. Policy Languishes," *Washington Post*, Apr. 19, 2000, p. A-1.

184   "knock-down, drag-out battle": see "Ready to Rumble," *Science* 288 (Apr. 7, 2000): 27; on Specter's failure to bring a stem cell bill to the Senate floor, see "No-Show Showdown," *Science* 290 (Oct. 13, 2000): 261.

184   the NIH issued final guidelines: see G. Vogel, "Researchers Get Green Light for Work on Stem Cells," *Science* 289 (Sept. 1, 2000): 1442–43.

184   "potentially staggering benefits": Clinton quoted in N. Wade, "New Rules on Use of Human Embryos in Cell Research," *New York Times*, Aug. 24, 2000, p. A-1.

184   only a handful [of researchers]: see Vogel, "Researchers Get Green Light," p. 1443.

184   Geron controlled all uses of the cell line: on Johns Hopkins's and WARF's distribution policies, see G. Vogel, "Wisconsin to Distribute Embryonic Cell Lines," *Science* 287 (Feb. 11, 2000): 948–49; and E. Marshall, "The Business of Stem Cells," *Science* 287 (Feb. 25, 2000): 1419–21. For the history of WARF, see M. Penn, "Working Beyond Eureka!" from *On Wisconsin* and "History of WARF" at the organization's Web site, www.warf.org.

185   "I was surprised . . . by the . . . restrictions": according to the terms of an early

version of the Wisconsin MTA, distributed in February 2000, any researcher using the cells "agrees to cease any use of the materials for any purpose upon ninety (90) days notice from WARF, at which time all of the Materials and Derivative Materials shall be entirely destroyed."

186 "This is where the center of the universe": Okarma quoted in A. Pollock, "The Promise in Selling Stem Cells," *New York Times,* Aug. 26, 2001, Business p. 1. Okarma was also quoted as saying of Geron, "This company is going to dominate regenerative medicine."

188 "It is probably no accident": Kass, *Toward a More Natural Science,* p. 316.

188 The New York tabloids reported: on the East River homicide, see G. Gittrich, "Slain Couple Mourned," *New York Daily News,* Nov. 26, 2000, p. 8.

11. ELIXIR

INTERVIEWS: William Andrews, June 8, 2002 (telephone); Leonard Guarente, Dept. of Biology, Massachusetts Institute of Technology, Cambridge, Sept. 27, 2000, and Jan. 14, 2002 (telephone); Cynthia Kenyon, San Francisco, Dec. 12, 2000; David Sinclair, Harvard Medical School, Boston, Dec. 13, 2002 (telephone); Alan Walton, June 18, 2002 (telephone).

190 "The process of aging influences our poetry": C. Kenyon, "Ponce d'elegans: Genetic Quest for the Fountain of Youth," *Cell* 84 (1996): 501–4. In addition to its whimsical title, this mini-review also has an uncharacteristically tongue-in-cheek conclusion. Kenyon notes that "we have no idea how many life span genes will be found," but adds, "We should get going before it is too late."

191 "The king of the state": from the Chinese text *Zhan Guo Ce,* quoted in Teresi, *Lost Discoveries,* p. 311; see also pp. 285, 310–11, on Chinese alchemy. Teresi writes, "Another goal of alchemists was to find an elixir of eternal youth" (p. 285), and "Chinese alchemy had three objectives: 1) the search for the elixir of life using semichemical methods; 2) production of artificial gold and silver, but for therapy, not for wealth; and 3) pharmacology and botanical research" (p. 310).

191 "Interest in the fountain of youth": see G. J. Gruman, "A History of Ideas About the Prolongation of Life: The Evolution of Prolongevity Hypotheses to 1800," *Transactions of the American Philosophical Society,* new ser. (1966): vol. 56, pt. 9, pp. 1–102, an excellent history of mankind's interest in the idea of life extension.

192 Ponce de León gave this myth: on the "fountain of youth" myth, see Fuson, *Ponce de León,* pp. 118–21. Peter Martyr is quoted in Stimpson, *A Book about American History,* pp. 182–83. Fuson argues that it was those old reliables of conquistadorial motivation, glory and gold, rather than the fountain of

youth, that led Ponce de León on his 1513 mission; he suggests that the truly interested party was King Ferdinand of Spain, then sixty years old, who had taken a bride thirty-five years younger and intensely wanted to produce a male heir. "At best, the quest of the magic fountain was a secondary motive for the exploration," Fuson writes. "Clearly, erasing old age — and in those days sixty years was indeed old age — was possibly an ever-present thought in the king's mind which led him to make inquiries about the 'secrets' of all the newly discovered lands in and around the West Indies. If the search for a magic fountain was motivation for the voyage to Bimini, then it was Ferdinand's idea — not Ponce de León's" (p. 118).

193 Brown-Sequard claimed he could rejuvenate: see International Longevity Center, *Is There an "Anti-Aging" Medicine?* (New York: International Longevity Center, 2002), p. 1.

195 Kenyon wasn't wandering through virgin scientific territory: for other research exploring the genetics of aging in the 1980s and 1990s, see T. E. Johnson and W. B. Wood, "Genetic Analysis of Life Span in *Caenorhabditis elegans*," *PNAS* 79 (1982): 6603–7; M. R. Rose and T. J. Nusbaum, "Prospects for Postponing Human Aging," *FASEB Journal* 8 (Sept. 1994): 925–28; and Y. Lin, L. Seroude, and S. Benzer, "Extended Life-Span and Stress Resistance in the *Drosophila* Mutant *Methuselah*," *Science* 282 (Oct. 30, 1998): 943–46. Taking another avenue into the same territory, geneticist Gary Ruvkun of Massachusetts General Hospital and Harvard Medical School has elegantly laid out the role of hormonal signaling in the aging of model organisms.

195 The longevity gene they discovered: on the *daf-2* discovery, see C. Kenyon, J. Chang, E. Gensch, A. Rudner, and R. Tabtiang, "A *C. elegans* Mutant That Lives Twice as Long as Wild Type," *Nature* 366 (1993): 461–64. On the *daf-16* discovery, see K. Lin, J. B. Dorman, A. Rodan, and C. Kenyon, "Daf-16: An HNF-3/forkhead Family Member That Can Function to Double the Life-Span of Caenorhabditis elegans," *Science* 278 (1997): 1319–22.

196 As the UCSF group teased apart the biology: concerning the biology of these long-lived mutants, the term "fountain of youth gene" has turned out to be a bit of a misnomer for more than the obvious reason. In continuing experimentation, Kenyon's group has learned, in fact, that a number of genes seem to act in concert, following sundry cues from the reproductive system, to achieve a greater life span. For the outlines of this complicated hormonal signaling system, see H. Hsin and C. Kenyon, "Signals from the Reproductive System Regulate the Lifespan of C. elegans," *Nature* 399 (May 27, 1999): 362–66.

198 Guarente's lab identified and cloned a gene: see B. K. Kennedy, N. R. Austriaco, Jr., J. Zhang, and L. Guarente, "Mutation in the Silencing Gene SIR4 Can Delay Aging in Saccharomyces cerevisiae," *Cell* 80 (1995): 485–96; and B. Kennedy et al., "Redistribution of Silencing Proteins from Telomeres to the Nucleolus Is Associated with Extension of Life Span in S. cerevisiae," *Cell* 89 (1997): 381–92.

199 one spectacular wild-goose chase: by the end of 1997, the MIT researchers felt convinced by the yeast data that there was a universal marker for aging inside cells: the presence of circles in something called ribosomal DNA. And so they spent two years looking for circles — in tissue samples from elderly people, in young and old mice, in different organs. "We looked for circles in general, any circles," Guarente recalled. "And just saw nothing convincing."

199 The actual purpose of the gene: on the link between *sir-2* and metabolism, see S.-I. Imai, C. M. Armstrong, M. Kaeberlein, and L. Guarente, "Transcriptional Silencing and Longevity Protein Sir2 Is an NAD-dependent Histone Deacetylase," *Nature* 403 (Feb. 17, 2000): 795–800; and S. J. Lin, P.-A. Defossez, and L. Guarente, "Life Span Extension by Calorie Restriction Requires NAD and SIR2," *Science* 289 (Sept. 22, 2000): 2126–28. For a more popular, but nonetheless dense, account of this biochemistry, see N. Wade, "Searching for Genes to Slow the Hands of Biological Time," *New York Times,* Sept. 26, 2000, p. D-1.

200 experiments conducted by Clive McCay: for a good account of this work, see Austad, *Why We Age,* pp. 80–83. A more modern, and perhaps more reliable, iteration of this line of experimentation is R. Weindruch and R. L. Walford, "Dietary Restriction in Mice Beginning at 1 Year of Age: Effect on Life-Span and Spontaneous Cancer Incidence," *Science* 215 (Mar. 12, 1982): 1415–18. For a popular account of Walford's own starvation diet, see G. Taubes, "Staying Alive," *Discover,* Feb. 2000. For the cultural ramifications, see L. Johannes, "The Surprising Rise of a Radical Diet: 'Calorie Restriction,'" *Wall Street Journal,* June 3, 2002, p. A-1.

200 Guarente's group showed that by inserting: on *sir-2* in worms, see H. A. Tissenbaum and L. Guarente, "Increased Dosage of a *sir-2* Gene Extends Lifespan in *Caenorhabditis elegans,*" *Nature* 410 (Mar. 8, 2001): 227–30. On cloning the *daf-2* gene, see W. Roush, "Worm Longevity Gene Cloned," *Science* 277 (Aug. 15, 1997): 897–98; and K. D. Kimura, H. A. Tissenbaum, Y. Liu, and G. Ruvkun, "*Daf-2,* an Insulin Receptor–like Gene That Regulates Longevity and Diapause in *Caenorhabditis elegans,*" ibid., pp. 942–46.

200 That evolutionary view has gained: for other genetic studies on aging, see Lin, Seroude, and Benzer, "Extended Life-Span and Stress Resistance"; and D. J. Clancy et al., "Extension of Life-Span by Loss of CHICO, a Drosophila Insulin Receptor Substrate Protein," *Science* 292 (Apr. 6, 2001): 104–6. In this latter study, Linda Partridge's group at University College London showed that when researchers knocked out a gene called *chico* in fruit flies, the insects lived up to 48 percent longer; this gene, like *daf-2,* is involved in insulin/insulin-like growth factor signaling. In fruit flies, at least, manipulating the insulin-related pathway not only seems to extend longevity but in certain cases results in an interesting side effect: some mutations produce not only notable extensions of life span (up to 85 percent) but also dwarfism (see M.

Tatar et al., *Science* 292 [Apr. 6, 2001]: 107–10). For the French study extending the IGF-1 regulation of life span to mice, see M. Holzenberger et al., "IGF-1 Receptor Regulates Lifespan and Resistance to Oxidative Stress in Mice," *Nature* advanced on-line publication, Dec. 4, 2002 (doi:10.1038/nature01298).

202 "Apparently there was a drug": on the scandal associated with the original Elixir, see L. Bren, "Frances Oldham Kelsey: FDA Medical Reviewer Leaves Her Mark on History," *FDA Consumer*, Mar.–Apr. 2001; and Fried, *Bitter Pills*, pp. 130–33.

203 A group headed by Thomas Perls: on the centenarian studies, see A. A. Puca et al., "A Genome-wide Scan for Linkage to Human Exceptional Longevity Identifies a Locus on Chromosome 4," *PNAS* 98 (Aug. 28, 2001): 10505–8.

203 "When single genes are changed": L. Guarente and C. Kenyon, "Genetic Pathways That Regulate Aging in Model Organisms," *Nature* 408 (Nov. 9, 2000): 255–62.

203 "equivalent to extending": Elixir press release, June 3, 2002.

203 "There are no death or aging genes": S. Jay Olshansky, respondent's remarks at the meeting "Living to 100 and Beyond: Survival at Advanced Ages," held in Florida, Jan. 2002.

203 "God gave us this wonderful response": J. Campisi, remarks at Society for Regenerative Medicine meeting, Washington, D.C., Dec. 4, 2001.

12. Unk!

INTERVIEWS: Günter Blobel, Mar. 4, 2002; Arthur D. Caplan, director, Center for Bioethics, University of Pennsylvania, Philadelphia, Feb. 18, 2002; Jose Cibelli, vice president of research, Advanced Cell Technology (ACT), Worcester, Mass., Nov. 5, 1999, and Dec. 15, 1999; Philippe Collas, Institute of Molecular Biochemistry, University of Oslo, Norway, June 19, 2002 (telephone); K. C. Cunniff, research technician, ACT, Dec. 15, 1999; John Gearhart, Oct. 20, 1999; Leonard Hayflick, Dec. 1, 1999; Robert P. Lanza, vice president of medical and scientific development, ACT, Nov. 5, 1999; F. Abel Ponce de León, Dept. of Animal Science, University of Minnesota, Minneapolis, Sept. 11, 2002 (telephone); Miller Quarles, June 12, 2002 (telephone); James Robl, president, Hematech, Sioux Falls, S.D., Dec. 9, 1999 (telephone), and Dec. 4, 2001 (telephone); Nancy Sawyer, research technician, ACT, Dec. 15, 1999; Steven Stice, Animal and Dairy Science Dept., University of Georgia, Athens, Dec. 27, 1999 (telephone); James Thomson, Dec. 13, 1999 (telephone); Michael West, Nov. 5, 1999, and Dec. 15, 1999.

208 As I watched a magnified view: as members of the National Bioethics Advisory Commission made clear when they discussed the cow-human cloning experiments at their meeting on Nov. 18, 1998, there is nothing unusual

about fusing cells of two different species or of mingling DNA from two different species; it has formed the bedrock of thousands of experiments over the last few decades, and, in the case of viral transfer of genes between species, has probably been a feature of life on earth for millions of years. The difference here, as NBAC member David Cox put it, is that this particular form of fusion might result in an embryo.

209 an example of . . . the "yuck factor": see Caplan, *Am I My Brother's Keeper?*, 211–18; "wisdom of repugnance," see L. R. Kass, "The Wisdom of Repugnance," *New Republic*, June 2, 1997.

210 "Does a blastocyst warrant the same rights": R. P. Lanza, J. B. Cibelli, and M. D. West, "Human Therapeutic Cloning," *Nature Medicine* 5 (Sept. 1999): 975–77.

211 "Even if the cow nuclear transfer": as noted earlier, there is no scientific agreement on whether mitochondrial incompatibility would be a hurdle to cow-human nuclear transfer experiments, although incomplete reprogramming could be.

212 a team of Scottish scientists: I. Wilmut, A. E. Schnieke, J. McWhir, A. J. Kind, and K. H. Campbell, "Viable Offspring Derived from Fetal and Adult Mammalian Cells," *Nature* 385 (1997): 810–13.

212 Clinton urgently requested "a moratorium on the cloning": for this and other Clinton comments, see Kolata, *Clone*, p. 35. For the NBAC report on cloning, see National Bioethics Advisory Commission, "Cloning Human Beings," Rockville, Md., June 1997 (available at www.georgetown.edu/research/nrcbl/ nbac/pubs/cloning1/cloning.pdf).

214 a paper sat on [Cibelli's] desk: D. Chen et al., "The Giant Panda (*Ailuropoda melanoleuca*) Somatic Nucleus Can Dedifferentiate in Rabbit Ooplasm and Support Early Development of the Reconstructed Egg," *Science in China*, ser. C, vol. 42 (Aug. 1999), no. 4, pp. 346–53. The Chinese scientists claimed a success rate of 11.7 percent in creating blastocysts in cloning experiments using 219 mammary gland cells. In the spring of 2002, an American biologist told me he had seen the abstract of a paper submitted to *Nature* in which Chinese scientists claimed to have created embryos by transferring human cells into rabbit oocytes and then harvesting the stem cells. For a later news report on the Chinese work on human cloning to obtain stem cells, see K. Leggett and A. Regalado, "Fertile Ground: As West Mulls Ethics, China Forges Ahead in Stem-Cell Research," *Wall Street Journal*, Mar. 6, 2002, p. A-1.

215 the creation and harvest of dopaminergic neurons: on the creation of embryonic stem cells for Parkinson's patients, see W. M. Zawada et al., "Somatic Cell Cloned Transgenic Bovine Neurons for Transplantation in Parkinsonian Rats," *Nature Medicine* 4 (May 1998): 569–74.

219 There is a photograph of . . . this *thing*: while the scientific implications of the blastocyst photograph may be unclear, ACT's claims for it are not. According to a caption attached to the back of a copy the company gave to me, the

photo depicted "human embryonic stem cells produced using nuclear transfer." The caption continued: "Advanced Cell Technology, Inc. of Worcester, Ma. produced the primitive cells, which are believed to have the ability to be grown into virtually any tissue, by fusing a human skin cell with a cow's egg from which the nucleus has been removed."

219 the international patent application: J. Robl, J. Cibelli, and S. Stice, "Embryonic or Stem-like Cell Lines Produced by Cross Species Nuclear Transplantation," international patent application submitted by the University of Massachusetts, filed on July 28, 1997, and claiming priority as of Aug. 19, 1996. For a recent news report on this technology, see A. Regalado and M. Song, "Furor over Cross-Species Cloning," *Wall Street Journal*, Mar. 19, 2002.

219 a daring project to create a cloned cow: for ACT's ongoing program on both cow cloning and the insertion of foreign genes into these clones, see J. B. Cibelli et al., "Cloned Transgenic Calves Produced from Nonquiescent Fetal Fibroblasts," *Science* 280 (May 22, 1998): 1256–58; and J. B. Cibelli et al., "Transgenic Bovine Chimeric Offspring Produced from Somatic Cell–Derived Stem-like Cells," *Nature Biotechnology* 16 (July 1998): 642–46.

220 The company has created human antibodies: on reprogramming skin cells to act like immune cells, see A.-M. Hakelien, H. B. Landsverk, J. M. Robl, B. S. Skalhegg, and P. Collas, "Reprogramming Fibroblasts to Express T-Cell Functions Using Cell Extracts," *Nature Biotechnology* 20 (May 2002): 460–66.

221 scientific missionary work among the lay public: in addition to the *Nature Medicine* commentary, see the roundup of Nobel laureates reported in R. Weiss, "Nobel Laureates Back Stem Cell Research," *Washington Post*, Feb. 22, 2001, p. A-2.

222 This phenomenon is called imprinting: see A. Efstratiadis, "Parental Imprinting of Autosomal Mammalian Genes," *Current Opinion in Genetics and Development* 4 (1994): 265–80. To be fair, other biologists now believe that bits of human mitochondrial DNA that are transferred with somatic cell DNA during cloning may be able to trump, or overcome, mitochondria contributed by the egg cell, thus supporting West's position.

223 "they have developed at the 'predictable rate'": J. Cibelli, e-mail to author, Dec. 26, 1999. "Not at all": J. Cibelli, presentation to National Academy of Sciences workshop on human cloning, Aug. 7, 2001, Washington, D.C.

13. STREET-FIGHTIN' MAN

INTERVIEWS: Piero Anversa, New York Medical College, Valhalla, Aug. 24, 2001 (telephone); Frank Barry, research scientist, Osiris Therapeutics, Baltimore, May 8, 2001, and Nov. 11, 2002 (telephone); Arnold Caplan, Dept. of Biology, Case Western Reserve University, Cleveland, Apr. 16, 2001; Robert Deans, former vice president of research, Osiris Therapeutics, Mar. 12, 2001 (telephone); Jeff Gimble, vice president of tissue engineering, Artecel Sciences, Inc., Durham, N.C., June 13, 2001 (telephone); Edwin M. Horwitz, St. Jude Children's Research Hospital, Memphis, July 2, 2001 (telephone), and Nov. 4, 2002 (telephone); Diane Krause, Yale University School of Medicine, June 13, 2001 (telephone); Daniel Marshak, former vice president of science, Osiris Therapeutics, Oct. 11, 1999 (telephone), Oct. 21, 1999, and Mar. 24, 2001 (Cold Spring Harbor, N.Y.); Bradley J. Martin, research scientist, Osiris Therapeutics, May 7, 2001, May 8, 2001, July 2, 2001 (telephone), and June 11, 2002 (telephone); Ron McKay, NINDS, June 13, 2001 (telephone); Annemarie Mosely, former president, Osiris Therapeutics, Aug. 14, 2001 (telephone); David Prentice, professor of life sciences, Indiana State University, Terre Haute, Dec. 16, 2002 (telephone); Darwin J. Prockop, Center for Gene Therapy, Tulane University Health Sciences Center, New Orleans, Apr. 24, 2001 (telephone); Evan Snyder, Dept. of Neurology, Children's Hospital, Boston, Sept. 28, 2000, and June 13, 2001 (telephone); Neil Theise, Dept. of Pathology, New York University School of Medicine, July 2, 2001; Ann Tsukamoto, chief scientific officer, StemCells, Inc., Sunnyvale, Calif., Dec. 11, 2000; John E. Wagner, Jr., University of Minnesota Cancer Center, Minneapolis, June 21, 2001 (telephone); Irving Weissman, Dept. of Pathology, Stanford University School of Medicine, Aug. 13, 2001 (telephone).

226   "It seems like there is a really good alternative": Prentice quoted in R. Weiss, "In Cell 'Alchemy,' an Alternative to Embryo Studies," *Washington Post*, Apr. 24, 2000, p. A-11.

227   Prentice was almost alone: bioethicist Tom Murray has argued that the strategy of using one or two scientists to contradict mainstream thinking and create the illusion of scientific disagreement was first used to counter allegations that smoking was dangerous to health. Murray writes, "Taking a cue from the tobacco industry, pro-life operatives learned that you do not need masses of scientists on your side" (T. H. Murray, "Hard Cell," *American Prospect*, Sept. 24, 2001).

227   Osiris researchers had accumulated . . . data: see J. G. Shake et al., "Mesenchymal Stem Cell Implantation in a Swine Myocardial Infarct Model: Engraftment and Functional Effects," *Annals of Thoracic Surgery* 73 (2002): 1919–26.

228   the cardiac stem cell story is far from settled: the stem cells that take up residence in the scar tissue have all the markers of cardiomyocytes, the muscle cells unique to heart muscle, Osiris scientists say, but do not appear to be organized in the same way and do not exhibit the typical contractile properties of heart muscle. "We're not saying they're cardiomyocytes," said Deans, "but they maintain the thickness of the heart wall." "In the pig model," said Martin, "we've seen such good results in terms of function that we didn't care if they were myocytes or not."

228   Adult stem cell research has made impressive strides: for overviews of the research as it has unfolded, see G. Vogel, "Can Old Cells Learn New Tricks?"

Science 287 (Feb. 25, 2000): 1418–19; E. Marshall, "The Business of Stem Cells," ibid., pp. 1419–21; G. Vogel, "Stem Cells: New Excitement, Persistent Questions," *Science* 290 (Dec. 1, 2000): 1672–74; G. Vogel, "Can Adult Stem Cells Suffice?" *Science* 292 (June 8, 2001): 1820–22; and S. S. Hall, "Adult Stem Cells," *Technology Review,* Nov. 2001, pp. 42–49.

229 Bone marrow transplants . . . achieved routine success: see "History of Stem Cell Transplants," National Marrow Donor Program Web site (www.marrow.org/NMDP/history_stem_ cell_transplants.html); see also Mizel and Jaret, *In Self-Defense,* pp. 129–30; the procedure is performed about 18,000 times a year worldwide, according to a recent review, with success rates as high as 85 percent against certain forms of cancer (F. R. Appelbaum, "Haematopoietic Cell Transplantation as Immunotherapy," *Nature* 411 [May 17, 2001]: 385–89).

230 The existence of that marvelously prolific cell: on the original stem cell research, see J. E. Till and E. A. McCulloch, "A Direct Measurement of the Radiation Sensitivity of Normal Mouse Bone Marrow Cells," *Radiation Research* 14 (1961): 213; for the isolation of the human hematopoietic stem cell, see C. M. Baum, I. L. Weissman, A. S. Tsukamoto, A. M. Buckle, and B. Peault, "Isolation of a Candidate Human Hematopoietic Stem-Cell Population," *PNAS* 89 (Apr. 1992): 2804–8.

232 They called it a mesenchymal stem cell: for an early review of the MSC, see H. M. Lazarus, S. E. Haynesworth, S. L. Gerson, N. Rosenthal, and A. I. Caplan, "Ex-Vivo Expansion and Subsequent Infusion of Human Bone Marrow–Derived Stromal Progenitor Cells: Implications for Therapeutic Use," *Bone Marrow Transplantation* 16 (1995): 557–64.

234 The harvesting process, detailed in a *Science* article: M. F. Pittenger et al., "Multilineage Potential of Adult Human Mesenchymal Stem Cells," *Science* 284 (1999): 143–46. In 1995, Osiris began the first of several clinical trials involving adult stem cells, having doctors inject the cells into patients with advanced breast cancer who were receiving high-dose chemotherapy. The idea, according to the company, was that these particular stem cells upholster the inner lining (or "stroma") of bones, thereby facilitating the activity of hematopoietic stem cells in rebuilding the blood after chemotherapy. Initial testing suggested that large infusions of stem cells were safe, and phase II data suggested that the treatment modestly accelerated the recovery of normal blood clotting values and immune function. But this first corporate study also illustrates the perils of these expensive clinical trials for small biotech companies. Since Osiris's trial began, doctors have moved away from high-dose chemotherapy for breast cancer. The company has since decided to test the cells as a way of accelerating blood recovery in patients with leukemia and lymphoma following bone marrow transplants.

234 a factor that actively inhibits a normal immune response: on immunology issues, see R. Deans, "Clinical Promise of Mesenchymal Stem Cell Transplantation," paper read at Cold Spring Harbor Laboratory meeting, Mar. 22, 2001.

In the abstract Deans stated, "Not only do MSCs fail to elicit an allogeneic T cell response in vitro. The addition of MSCs to T cells and allogeneic PBMC blocks the induction of a primary or ongoing MLR [mixed lymphocyte response]." In 2002, by contrast, Nissim Benvenisty of Hebrew University in Jerusalem and colleagues published a report demonstrating that embryonic stem cells increasingly appear visible to the immune system as they become more mature, differentiated cells; see M. Drukker et al., "Characterization of the Expression of MHC Proteins in Human Embryonic Stem Cells," *PNAS* 99 (July 23, 2002): 9864–69.

235 the cells "seem to be totally immunoprivileged": Shake et al., "Mesenchymal Stem Cell Implantation," p. 1926.

235 damage caused by heart attacks . . . could be repaired: for Anversa's work, see D. Orlic et al., "Bone Marrow Cells Regenerate Infarcted Myocardium," *Nature* 410 (Apr. 5, 2001): 701–5; Itescu's work is A. A. Kocher et al., "Neovascularization of Ischemic Myocardium by Human Bone–Derived Angioblasts Prevents Cardiomyocyte Apoptosis, Reduces Remodeling and Improves Cardiac Function," *Nature Medicine* 7 (Apr. 2001): 430–36. For scientific commentaries, see M. Sussman, "Hearts and Bones," *Nature* 410 (Apr. 5, 2001): 640–41; and N. Rosenthal and L. Tsao, "Helping the Heart to Heal with Stem Cells," *Nature Medicine* 7 (Apr. 2001): 412–13. These reports received extensive coverage in the lay press, including stories in the *New York Times, Washington Post, Wall Street Journal,* and network news.

236 Osiris began a cardiac program: Osiris researchers, in collaboration with scientists at Johns Hopkins, have published several papers, including Shake et al., "Mesenchymal Stem Cell Implantation"; B. J. Martin et al., "Mesenchymal Stem Cell (MSC) Implantation Improves Regional Function in Infarcted Swine Myocardium," *Circulation* 201, no. 18 (2000): I-54; and D. J. Caparrelli et al., "Cellular Myoplasty with Mesenchymal Stem Cell Results in Improved Cardiac Performance in a Swine Model of Myocardial Infarction," *Circulation* 104, no. 17 (2001): II-599. Additional Osiris publications on the stem cells include R. J. Deans and A. B. Moseley, "Mesenchymal Stem Cells: Biology and Potential Clinical Uses," *Experimental Hematology* 28 (2000): 875–84; and K. W. Liechty et al., "Human Mesenchymal Stem Cells Engraft and Demonstrate Site-Specific Differentiation after *in Utero* Transplantation in Sheep," *Nature Medicine* 6 (Nov. 2000): 1282–86.

237 goats that have sustained severe knee damage: an abstract describing the initial treatment in goats is M. Murphy et al., "Injected Mesenchymal Stem Cells Stimulate Meniscal Repair and Protection of Articular Cartilage," presented at annual meeting, Orthopaedic Research Society, San Francisco, 2001.

238 the cells migrating in such orderly fashion: see N. Uchida et al., "Direct Isolation of Human Central Nervous System Stem Cells," *PNAS* 97 (Dec. 19, 2000): 14720–25.

238 work at the Salk Institute: on the isolation of adult brain stem cells, see T. D.

Palmer et al., "Progenitor Cells from Human Brain after Death," *Nature* 411 (May 3, 2001): 42–43.

239 Kondziolka has treated a dozen or so stroke victims: for the clinical trial using embryonic cancer cells, see D. Kondziolka, "Transplantation of Cultured Human Neuronal Cells for Patients with Stroke," *Neurology* 55 (Aug. 22, 2000): 565–69; and "University of Pittsburgh to Begin Second Phase of Cell Implant Study for Stroke Patients," Univ. of Pittsburgh press release, Mar. 15, 2001.

239 Snyder's group at Harvard Medical School: on Snyder's work using a stem cell line isolated from a human fetus, see J. D. Flax et al., "Engraftable Human Neural Stem Cells Respond to Developmental Cues, Replace Neurons, and Express Foreign Genes," *Nature Biotechnology* 16 (Nov. 1998): 1033–39.

239 use of adult stem cells to regenerate the liver: see E. Lagasse et al., "Purified Hematopoietic Stem Cells Can Differentiate into Hepatocytes in Vivo," *Nature Medicine* 6 (Nov. 2000): 1229–34.

240 Horwitz described preliminary results: see E. Horwitz et al., "Isolated Allogeneic Bone Marrow–Derived Mesenchymal Cells Engraft and Stimulate Growth in Children with Osteogenesis Imperfecta: Implications for Cell Therapy of Bone," *PNAS* 99 (June 25, 2002): 8932–37.

241 In another early trial at the University of Minnesota: see J. Wagner's work presented as abstracts at the American Society of Hematology annual meeting, Philadelphia, Dec. 2002.

241 Neil Theise . . . and Diane Krause . . . published a report: D. S. Krause et al., "Multi-Organ, Multi-Lineage Engraftment by a Single Bone Marrow–Derived Stem Cell," *Cell* 105 (2001): 369–77. Papers from leading laboratories argued that the versatility of adult cells was much higher than previously believed; such papers included reports on bone marrow cells turning into neurons (D. Woodbury, E. J. Schwarz, D. J. Prockop, and I. B. Black et al., "Adult Rat and Human Bone Marrow Stromal Cells Differentiate into Neurons," *Journal of Neuroscience Research* 61 [Aug. 15, 2000]: 364–67); nerve cells into blood (C. R. R. Bjornson et al., "Turning Brain into Blood: A Hematopoietic Fate Adopted by Adult Neural Stem Cells in Vivo," *Science* 283 [Jan. 22, 1999]: 534–37); and neural stem cells into almost anything (D. L. Clarke et al., "Generalized Potential of Adult Neural Stem Cells," *Science* 288 [June 2, 2001]: 1660–63). The report on stem cells in fat was P. A. Zuk et al., "Multilineage Cells from Human Adipose Tissue: Implications for Cell-Based Therapies," *Tissue Engineering* 7 (Apr. 2001): 211–28. In the spring of 2002, two papers published in *Nature* cast doubt on the earlier claims; they came from the labs of Bryon E. Petersen of the University of Florida and Austin Smith of the University of Edinburgh. For a review of the controversy, see A. E. Wurmser and F. H. Gage, "Cell Fusion Causes Confusion," *Nature* 416 (Apr. 4, 2002): 485–87; and C. Holden and G. Vogel, "Plasticity: Time for a Reappraisal?" *Science* 296 (June 21, 2002): 2126–29. Moreover, Irving Weissman said his lab

had been unable to reproduce the Krause-Theise work after nearly a year of trying (I. Weissman, interview, Jan. 24, 2002).

242 a remarkably potent adult stem cell isolated: on the Catherine Verfaillie flap, see S. P. Westphal, "Is This the One?" *New Scientist,* Jan. 26, 2002, pp. 4–5. The Do No Harm press release stated that Verfaillie's research could "render obsolete" the need for human cloning to obtain embryonic stem cells, cited in J. Marcotty and M. Lerner, "U Researcher Drawn into Embryo Debate," *Minneapolis Star Tribune,* Jan. 25, 2002. Brownback was quoted in the *New York Times* as saying, "Science continues to prove that destructive embryonic stem cell research is unnecessary"; see N. Wade with S. G. Stolberg, "Scientists Herald a Versatile Adult Cell," *New York Times,* Jan. 25, 2002. Verfaillie's work, ultimately published as Y. Jiang et al., "Pluripotency of Mesenchymal Stem Cells Derived from Adult Marrow," *Nature* 418 (July 4, 2002): 41–56, was widely noted in the press.

242 Verfaillie later sent a letter to U.S. senators: the letter, a copy of which was provided by the University of Minnesota, was dated Feb. 4, 2002. The letter states that "it is far too early to say whether they [adult stem cells] will stack up when compared to embryonic stem cells in longevity and function. Further, we will not know which stem cells, adult or embryonic, are most useful in treating a particular disease without side by side comparison of adult and embryonic stem cells."

243 "an internationally recognized expert": David A. Prentice's Web site at Indiana State University is http://mama.indstate.edu/dls/facstaff/prentice.html. Prentice said in an interview that although he has to date published no research involving stem cells, he has a paper in preparation.

14. THE SNOWFLAKE INTERVENTION

INTERVIEWS: Ali Brivanlou, June 8, 2001 (telephone); Daniel Callahan, founding fellow, Hastings Center, Garrison, N.Y., Feb. 21, 2002; Thomas Cech, Jan. 7, 2002 (telephone); Julie Furer, attorney, Schiff, Hardin & Waite, Chicago, Mar. 15, 2002 (telephone); Thomas Hungar, partner, Gibson, Dunn & Crutcher, Washington, D.C., July 31, 2002 (telephone); Leon R. Kass, chairman, President's Council on Bioethics, Washington, Feb. 19, 2002; Dan Kaufman, University of Wisconsin, Mar. 21, 2002 (e-mail); Clarence T. Kipps, Jr., attorney, Miller & Chevalier, Washington, Mar. 18, 2002 (telephone); Michael Manganiello, chairman, Coalition for the Advancement of Medical Research (CAMR), Washington, Nov. 5, 2002 (telephone); Jeffrey C. Martin, partner, Shea & Gardner, Washington, Dec. 4, 2001, Feb. 19, 2002, and Mar. 26, 2002 (telephone); Anthony Mazzaschi, associate vice president for research, Association of American Medical Colleges, Washington, Aug. 21, 2002 (telephone); Ronald McKay, Jan. 31, 2002 (telephone), and Aug. 21, 2002 (telephone); Douglas Melton, Nov. 26, 2001 (telephone), and Feb. 27, 2002 (telephone); Roger

Pedersen, Cambridge University, July 25, 2002 (e-mail); Harriet Rabb, general counsel, Rockefeller University, New York, Aug. 1, 2002 (telephone); Lawrence Soler, ex-chairman, CAMR, Washington, Jan. 25, 2002 (telephone), and Mar. 22, 2002 (telephone); Chris Tackett, president, Wisconsin Merchants Association, Madison, July 24, 2002 (telephone); Tommy Thompson, Secretary, Dept. of Health and Human Services, Washington, Nov. 21, 2002 (telephone); Alan Trounson, Sept. 10, 2001 (telephone); Frankie Trull, president, National Association for Biomedical Research, Washington, Nov. 8, 2002 (telephone); LeRoy Walters, Kennedy Center for Bioethics, Georgetown University, Washington, Dec. 6, 2002 (telephone), and Dec. 19, 2002 (telephone).

245 "I believe we can find stem cells": for Bush's statement and Fleischer's clarification, see R. Fournier, "Bush Opposes Stem Cell Research Funding," Associated Press, Jan. 26, 2001. His exact comment to reporters was: "I believe there's some wonderful opportunities for adult stem cell research. I believe we can find stem cells from fetuses that died a natural death, but I do not support research from aborted fetuses." The White House did not respond to repeated calls for additional information.

245 president who had been "agonizing": C. Connolly, "Bush 'Agonizing' over Funding of Embryo Research," *Washington Post*, July 15, 2001, p. A-1.

246 flashed its right-to-life credentials: see R. Weiss, "Fetal Cell Research Funds Are at Risk," *Washington Post*, Jan. 26, 2001, p. A-3.

248 Martin launched his PowerPoint presentation: on the AAAS meeting, see P. Recer, "Parkinson's Cure May Be Near," Associated Press, Feb. 16, 2001. In the same session at which Jeff Martin spoke, Ole Isaacson of Harvard Medical School showed unpublished data demonstrating, for example, that his laboratory could restore function in mice suffering a disorder analogous to Parkinson's disease with a therapy derived from embryonic stem cells.

249 Rabb . . . had written a legal memorandum: see E. Marshall, "Ruling May Free NIH to Fund Stem Cell Studies," *Science* 283 (Jan. 22, 1999): 465–67.

249 "No matter how much one hates disease": E. Parens, personal communication.

250 "illegal, immoral and unnecessary": see N. Wade, "New Rules on Use of Human Embryos in Cell Research," *New York Times*, Aug. 24, 2000, p. A-1.

250 a group . . . filed a little-noticed lawsuit: *Nightlight Christian Adoptions et al. v. Thompson et al.*, Complaint for Declaratory and Injunctive Relief, United States District Court for the District of Columbia, filed Mar. 8, 2001.

250 "derives its name from the idea": JoAnn L. Davidson, testimony before the U.S. House of Representatives Committee on Government Reform, Subcommittee on Criminal Justice, Drug Policy, and Human Resources, July 17, 2001. See also the Nightlight Christian Adoptions Web site: www.nightlight. org. On "embryo adoption" as a contract, see K. Kerr, "Battle over Stem Cell Research," *Newsday*, June 1, 2001.

251 it gave bioethicists like Leon Kass pause: see Kass, *Toward a More Natural Science*, pp. 110–15. Kass writes, "Clarity about your origins is crucial for self-identity, itself important for self-respect. It would be, in my view, deplorable public policy to erode further such fundamental beliefs, values, institutions and practices. This means, concretely, no encouragement of embryo adoption or especially of surrogate pregnancy" (p. 113).

251 devoted to "family values, religious freedom": see Alliance Defense Fund Web site: www.alliancedefensefund.org.

251 the press conference . . . was "hosted": on Brownback's role, see "ProLife Groups Sue U.S. over Stem Cell Research," Life Lines News Archives, Mar. 10, 2001. Casey's statement "Destroying living human beings" is quoted in "Christian Medical Association on Stem Cell Lawsuit," Pro-Life Infonet, Mar. 8, 2001: www.prolifeinfo.org.

252 disappointing results of a fetal-cell transplant technique: on the "failed" Parkinson's treatment, see G. Kolata, "Parkinson's Research Is Set Back by Failure of Fetal Cell Implants," *New York Times*, Mar. 8, 2001, p. A-1.

252 mixed messages coming out of the administration: see L. Meckler, "Thompson Troubled by Stem Cell Ban," Associated Press, Mar. 6, 2001. Thompson was quoted as saying to Sen. Gordon Smith, "I am troubled, as you probably are, by the law, and I know Congress is trying to change the law. But there is a law on the books. It is troublesome." A spokesman for Thompson later claimed that the secretary did not find Dickey-Wicker troublesome. Thompson's role had already become controversial in the stem cell debate, as detailed in L. McGinley, "HHS Choice Upsets Antiabortion Allies over Stem Cells," *Wall Street Journal*, Jan. 16, 2001, p. A-28.

256 the meeting had been abruptly canceled: on HHS's cancellation of NIH's Human Pluripotent Stem Cell Review Group meeting, see L. McGinley, "NIH's Move to Cancel Meeting Worries Backers of Embryo Stem Cell Research," *Wall Street Journal*, Apr. 13, 2001, p. B-6; R. Weiss, "Bush Administration Order Halts Stem Cell Meeting," *Washington Post*, Apr. 21, 2001, p. A-2; and G. Vogel, "NIH Pulls Plug on Ethics Review," *Science* 292 (Apr. 20, 2001): 415.

258 a White House on record against any research: prior to the election Bush had opposed "federal funding for stem cell research that involves destroying living human embryos," letter from George W. Bush to the Culture of Life Foundation, reported in C. Connolly, "Conservative Pressure for Stem Cell Funds Builds," *Washington Post*, July 2, 2001, p. A-1. Throughout the spring and summer of 2001, Bush responded to patient advocates with a similar letter, stating, "I oppose Federal funding for stem-cell research that involves destroying living human embryos. I support innovative medical research on life-threatening and debilitating diseases, including promising research on stem cells from adult tissue" (copy of letter from George W. Bush to constituent, dated May 18, 2001, courtesy of J. Martin).

259 an animal rights group sued: it was the Alternative Research and Develop-

ment Foundation that sued the Department of Agriculture (USDA), and in September 2000, USDA agreed to a settlement and initiated the process to change the rules, to the alarm of the biomedical research community, which had not been consulted on the terms of the proposed settlement. Organizations like the National Association for Biomedical Research mounted a two-year lobbying effort, which resulted in a Senate amendment to the farm bill in 2002 that permanently excluded mice, rats, and birds from federal protection (F. Trull, personal communication).

259 Martin . . . filed papers on May 2: the formal title of the intervention papers was Cross-Claim for Declaratory and Injunctive Relief, *Nightlight Christian Adoptions et al. v. Thompson et al.*, in the United States District Court for the District of Columbia, civil action no. 01 CV502 (RCL), May 2, 2001; the formal lawsuit was *Thomson et al. v. Thompson et al.*, Complaint for Declaratory and Injunctive Relief, civil action no. 01 CV0973, U.S. District Court for the District of Columbia, filed May 8, 2001. For coverage, see G. Vogel, "Court Asked to Declare NIH Guidelines Legal," *Science* 292 (May 25, 2001): 1463.

260 "The fact is, the government": although Rabb thought it unlikely that the litigation would result in any change of government policy, she echoed other lawyers who suggested that the judge in the case, Royce Lamberth, was independent-minded and unpredictable. In 2002, Lamberth found a cabinet member, Interior Secretary Gale A. Norton, in contempt of court, the third time in modern memory that a cabinet-level official had been held in contempt — in each case by Lamberth (see H. Rumbelow and N. Tucker, "Interior's Norton Cited for Contempt in Trust Suit," *Washington Post*, Sept. 18, 2002, p. A-1).

260 The euphemistic inventiveness of the right-to-life movement: Samuel Casey used the term "preborn children," quoted in Kerr, "Battle over Stem Cell Research,"; "embryonic babies" appeared in *RNC for Life Report*, July/Aug. 2001; and "microscopic Americans" in D. Murdock, "The Adoption Option," National Review Online, posted Aug. 27, 2001.

260 Lefkowitz . . . played a key role in shaping: on Lefkowitz's background, see D. Milbank, "A Hard-Nosed Litigator Becomes Bush's Policy Point Man," *Washington Post*, Apr. 30, 2002, p. A-17.

261 uncertainty surrounding the Bush administration's intentions: see C. Connolly, "Stem Cell Research Divides Administration," *Washington Post*, June 12, 2001, p. A-8, which also discusses the role of Karl Rove; R. Pear, "Bush Administration Is Split over Stem Cell Research," *New York Times*, June 13, 2001, p. A-29; and E. Thomas and E. Clift, "Battle for Bush's Soul," *Newsweek*, July 9, 2001, pp. 28–30.

261 a meal Thompson had at the White House with Bush: see R. Lacayo, "How Bush Got There," *Time*, Aug. 20, 2001, pp. 17–23.

261 Soler opened the weekly conference call: on the June "scare" at the CAMR

meeting, see also L. McGinley, "Stem Cell Research Stirs Passionate Debate and Changing Politics," *Wall Street Journal*, July 9, 2001, p. A-30.

263 "There are serious problems that would be generated": J. C. Martin to T. G. Thompson, June 15, 2001.

265 If Bush sneaked a peek at the poll results: for Juvenile Diabetes Research Foundation results, see S. G. Stolberg, "Stem Cell Research Advocates in Limbo," *New York Times*, Jan. 20, 2001, p. A-17; and C. Connolly, "Stem Cell Research," *Washington Post*, June 12, 2001. The CAMR poll showed 77 percent of Americans in favor of harvesting ES cells from frozen embryos, including 69 percent of adults who identified themselves as "pro-life." For the ABC News/Beliefnet data, see L. Meckler, "Americans Support Stem Cell Research," Associated Press, June 26, 2001.

265 no newspaper was too small to editorialize: on the media coverage, see H. Kurtz, "Politics, Science Collide," *Washington Post*, July 16, 2001.

266 The most influential . . . was Hatch: In his letter to the president, Hatch framed Bush's stem cell decision not as an ethical issue but as a test of personal and global leadership, invoking the fall of the Berlin Wall during "your father's Presidency" and adding, "It seems to me that leading the way in finding new cures for disease is precisely the type of activity that accrues to our benefit both at home and abroad" (O. Hatch, letter to president, June 13, 2001).

266 Kass had decried the eating of ice cream in public: C. Adams, "Bioethics Appointee Says He Is No Indoctrinator," *Wall Street Journal*, Aug. 17, 2001, p. B-1.

267 "science essentially endangers" and following quote: Kass, *Toward a More Natural Science*, p. 4.

267 "runaway train": L. R. Kass, "Preventing a Brave New World," *New Republic*, May 21, 2001. "Now may be as good a chance as we will ever have," Kass writes, "to get our hands on the wheel of the runaway train now headed for a post-human world and steer it toward a more dignified human future."

267 Bush had asked Kass to bring along: See K. Q. Seelye with F. Bruni, "A Long Process That Led Bush to His Decision," *New York Times*, Aug. 11, 2001, p. A-1. Based on White House sources, the *Times* reported, "Mr. Bush had asked Dr. Kass to bring along someone who had a different viewpoint from his own. He thought Dr. Callahan would have a different view, but as it turned out he too disliked the idea of destroying embryos and had a position similar to Dr. Kass. This apparently made a big impression on Mr. Bush."

268 first reports of a potential . . . "compromise": see S. G. Stolberg and D. E. Sanger, "Bush Aides Seek Compromise on Embryonic Cell Research," *New York Times*, July 4, 2001, p. A-1; and Thomas and Clift, "Battle for Bush's Soul."

271 the NIH stem cell report . . . let the air out: the report's executive summary concludes, "Predicting the future of stem cell applications is impossible, par-

ticularly given the very early stage of the science of stem cell biology. To date, it is impossible to predict which stem cells — those derived from the embryo, the fetus, or the adult — or which methods of manipulating the cells, will best meet the needs of basic research and clinical applications. The answers clearly lie in conducting more research" (p. ES-10).

15. THE BREATH OF LIFE

INTERVIEWS: Arthur Caplan, Feb. 18, 2002; Alta Charo, Jan. 15, 2002 (telephone); Rudolf Jaenisch, Whitehead Institute for Biomedical Research, Cambridge, Mass., Oct. 7, 2002 (telephone); Peter Mombaerts, Laboratory of Developmental Biology and Neurogenetics, Rockefeller University, Aug. 14, 2001 (e-mail), and Mar. 11, 2002; Daniel Perry, Aug. 21, 2001 (telephone); William Safire, chairman, Dana Foundation, New York, Dec. 4, 2000 (telephone); Maxine Singer, president, Carnegie Institution of Washington, Aug. 7, 2001; Irving Weissman, Feb. 16, 2002 (telephone); Ian Wilmut, director, Dept. of Gene Expression and Development, Roslin Institute, Edinburgh, at Renaissance Hotel, Washington, Dec. 3, 2001.

272 A few days earlier, two private biomedical companies: on the experiments in the summer of 2001 to create human embryos, either by IVF techniques or "therapeutic cloning," for stem cell research, see R. Weiss, "Scientists Use Embryos Made Only for Research," *Washington Post*, July 11, 2001, p. A-1; S. G. Stolberg, "Scientists Create Scores of Embryos to Harvest Cells," *New York Times*, July 11, 2001, p. A-1; R. Weiss, "Firm Aims to Clone Embryos for Stem Cells," *Washington Post*, July 12, 2001, p. A-1; and A. Regalado, "Experiments in Controversy," *Wall Street Journal*, July 13, 2001, p. B-1.

273 "you can't equate embryos": testimony of Orrin G. Hatch, Subcommittee on Labor, Health and Human Services, Education, and Related Agencies, Senate Appropriations Committee (hereinafter referred to as the Harkin-Specter subcommittee), July 18, 2001. On Frist's position, see L. McGinley, "Influential GOP Sen. Frist Supports Stem-Cell Research," *Wall Street Journal*, July 18, 2001, p. A-20; "We all agree that this embryo": testimony of Sam Brownback, July 18, 2001.

273 "I thought, in the spirit of trying": testimony of Gordon Smith, July 18, 2001, text provided by J. Sheffo. For earlier discussions of Smith's views, see B. Davis, "GOP Avoids Abortion for Now, but Science Is Stirring the Debate," *Wall Street Journal*, Aug. 1, 2000, p. A-1. For Catholic criticism of Smith's remarks, see D. Clark, "The Mormon Stem-Cell Choir," Slate.com, posted Aug. 2, 2001; and A. Zitner, "Bible Guides Senate on Stem Cell Studies," *Los Angeles Times*, July 19, 2001, p. A-1, in which Richard Doerflinger is quoted as calling Smith's biblical interpretations "amateur theology."

276 "I agree . . . about the ad nauseam part": A. Caplan, "Is It Ethical to Try to

Significantly Extend the Human Life Span?" paper given at annual meeting, Society for Regenerative Medicine, Washington, D.C., Dec. 3, 2001. For similar views, see A. Caplan, "Attack of the Anti-Cloners," *Nation*, June 17, 2002, pp. 5–6; Caplan writes, "The debate over human cloning and stem cell research has not been one of this nation's finest moral hours. Pseudoscience, ideology, and plain fearmongering have been much in evidence."

277   "No Recombination Without Representation": see Hall, *Invisible Frontiers*, pp. 125–28.

277   "human-cloning coalition": A. Regalado, "Scientists Convene to Discuss Future of Human Cloning," *Wall Street Journal*, Aug. 7, 2001, p. B-7.

277   the purpose of the meeting was to hear: in the report that resulted from the workshop, "Scientific and Medical Aspects of Human Reproductive Cloning," the academy panel concluded that reproductive cloning is "dangerous and likely to fail. The panel therefore unanimously supports the proposal that there should be a legally enforceable ban on the practice of human reproductive cloning." The panel added, however, that "the scientific and medical considerations that justify a ban on human reproductive cloning at this time are not applicable to nuclear transplantation to produce stem cells" (Executive Summary, pp. 1–2).

277   conversations . . . at the . . . meeting in Davos: Ian Wilmut recalled that the idea was discussed by Bruce Alberts, president of the National Academy of Sciences, Irving Weissman, Fred Gage, and Wilmut, with the encouragement of *New York Times* columnist William Safire, who has taken an active interest in bioethics as chairman of the Dana Foundation.

278   "the loony cloning element": Caplan, Society of Regenerative Medicine talk; "puzzling collapse of standards": D. Magnus and A. Caplan, "NAS Cloning Hearing Disappoints Participants," letter to the editor, *Science* 294 (Nov. 23, 2001): 1651.

279   he lashed out at . . . Jaenisch: the article that aroused Zavos's ire was N. Gibbs, "Baby, It's You! And You, and You," *Time*, Feb. 19, 2001, pp. 46–58. For more scientific discussions of the dangers of cloning, see R. Jaenisch and I. Wilmut, "Don't Clone Humans!" *Science* 291 (Mar. 30, 2001): 2552; and D. Humpherys et al., "Epigenetic Instability in ES Cells and Cloned Mice," *Science* 293 (July 6, 2001): 95–97, in which Jaenisch's lab demonstrated "extremely unstable" epigenetic, or inherited, variations in cloned mice; this paper, which was accepted only two months after its submission, prompted Zavos to complain that *Science* published it rapidly to influence congressional discussions and potential legislation on the issue. For a more recent paper expanding on the same themes, see D. Humpherys et al., "Abnormal Gene Expression in Cloned Mice Derived from ES Cell and Cumulus Cell Nuclei," *PNAS* Early Edition, www.pnas.org/cgi/doi/10.1073/pnas192433399. For a popular account of cloning difficulties, see G. Kolata, "In Cloning, Failure Far Exceeds Success," *New York Times*, Dec. 11, 2001, p. F-1.

280 voted to expel Antinori for "disreputable conduct": quoted in A. Coghlan, "Kicked Out," *New Scientist,* Sept. 22, 2001, p. 15.

281 to the average citizen . . . reading the paper: for coverage of the cloning workshop, see A. Zitner, "Researchers Defend Human Cloning Plans," *Los Angeles Times,* Aug. 8, 2001, p. A-1; S. G. Stolberg, "Despite Opposition, Three Vow to Pursue Cloning of Humans," *New York Times,* Aug. 8, 2001, p. A-1; and R. Weiss, "Scientists Declare Progress on Human Cloning," *Washington Post,* Aug. 8, 2001, p. A-2. In addition to providing an excellent analysis of the meeting, an op-ed piece by Ronald Bailey was among the few to refer to the would-be cloners as "a few quacks" (R. Bailey, "There'll Never Be Another You," *Wall Street Journal,* Aug. 10, 2001, p. A-8).

282 embryo research is "the kind of subject": Dan Rather quoted in T. Shales, "Bush's Rare Appearance in the National Pulpit," *Washington Post,* Aug. 10, 2001, p. C-1.

282 "a nuclear bomb to kill a fly": R. Alta Charo, presentation at NAS meeting, Aug. 7, 2001.

282 House of Representatives held a floor debate: remarks from the debate on cloning are in *Congressional Record — House,* July 31, 2001, pp. H4933–40. On Deutsch's remark that the debate "may be the lowest level of knowledge," see "Clone Homework Skimped," *Science* 293 (Aug. 10, 2001): 1043.

284 scientists . . . had indeed used so-called therapeutic cloning: see T. Wakayama, V. Tabar, I. Rodriguez, A. C. F. Perry, L. Studer, and P. Mombaerts, "Differentiation of Embryonic Stem Cell Lines Generated from Adult Somatic Cells by Nuclear Transfer," *Science* 292 (Apr. 27, 2001): 740–43; and W. M. Rideout III, K. Hochedlinger, M. Kyba, G. Q. Daley, and R. Jaenisch, "Correction of a Genetic Defect by Nuclear Transplantation and Combined Cell and Gene Therapy," *Cell* 109 (Apr. 5, 2002): 1–20.

284 Jaenisch . . . said there are "major differences": in taking exception to Weldon's remarks, Jaenisch said, "Clearly there is data that epigenetic methylation status is not normal in cloned embryos, and clearly there is data that reprogramming is not normal; if allowed to develop, there is major gene dysregulation. We have looked at only 10,000 genes, and in 4 percent of them we see abnormal expression. *All* of the embryos are abnormal." The larger point, he said, is that with cloned embryos, there is not a new combination of genetic material (as occurs with sexual fertilization) and not any potential for life. "The potential to make a normal baby is nil," he said.

285 Parkinson's disease could be cured: testimony by Prockop at Harkin-Specter subcommittee, Sept. 14, 2000.

285 "some diseases are better": testimony by Russell E. Saltzman, pastor, Ruskin Heights Lutheran Church, Kansas City, Mo., Harkin-Specter subcommittee, Sept. 14, 2000.

286 "a six-month delay in scientific research": testimony by Michael D. West, Harkin-Specter subcommittee, Dec. 4, 2001.

286 "We need to make it very clear": testimony by Frist, Harkin-Specter sub-committe, July 18, 2001.

287 "From the maw of this 'morality'": A. Bartlett Giamatti, "A Liberal Education and the New Coercion," *Yale Alumni Magazine*, Oct. 1981, from Giamatti's address to the class of 1985. *Newsweek* columnist Fareed Zakaria mentioned the speech in a recent column, for which I am grateful.

290 Perhaps that's what motivated a plaintive letter: on the reluctance of the scientific community to speak out, see C. A. Morella, "Stem Cell Research Needs United Support," letter to the editor, *Science* 293 (July 6, 2001): 47; and A. Specter, remarks at Harkin-Specter subcommittee, Dec. 4, 2001. David Magnus and Arthur Caplan made much the same point in their Nov. 23, 2001, letter ("NAS Cloning Hearing Disappoints") to *Science*: "What is happening in the discussion of cloning in American public policy, as the NAS panel made sadly evident, is that the scientific community has become too lax about making sure that the public and policy-makers can hear them clearly."

292 Specter had apparently misspoken: according to Specter's spokesman William Reynolds, Specter did not mean to suggest that he opposed cloning for research.

292 "As the apostle Paul said": testimony of Michael D. West, Harkin-Specter subcommittee, July 18, 2001.

16. FREE THE BUSH 64!

INTERVIEWS: Joyce Brinton, Office for Technology and Trademark Licensing, Harvard University, Cambridge, Mass., Dec. 2, 2002 (telephone); Ali Brivanlou, Aug. 21, 2001 (telephone), Aug. 27, 2001 (telephone), Sept. 10, 2001 (telephone), and Oct. 11, 2001 (telephone); Andrew Cohn, Dec. 4, 2002 (telephone); George Daley, Dec. 4, 2002 (telephone); Carl Gulbrandsen, Dec. 19, 2002 (telephone); Leonard Hayflick, Apr. 29, 2002 (telephone); Jeff Martin, Feb. 19, 2002; Anthony Mazzaschi, Aug. 21, 2002 (telephone); Douglas Melton, Feb. 27, 2002 (telephone), and Aug. 19, 2002 (telephone); Lawrence Soler, Mar. 22, 2002 (telephone); Harold Varmus, Aug. 25, 2001; LeRoy Walters, Dec. 6, 2002 (telephone); and several sources who requested anonymity.

294 Bush . . . planned to announce: M. Allen, "Bush Suggests an August 21 Decision on Stem Cell Research," *Washington Post*, Aug. 9, 2001, p. A-5. Bush's meeting with Hughes and Lefkowitz was described in a White House press briefing by Scott McClellan, Aug. 9, 2001. McClellan told reporters, "The President reached a final decision yesterday and he decided to move forward on that decision yesterday," but it is likely that a tentative decision had been made earlier.

295   the stocks of Geron, StemCells, and Aastrom: "Stem Cell Stocks Up Ahead of Decision," CBS.MarketWatch.com, Aug. 9, 2001.

295   Pope John Paul II . . . remarked: on the pope's statement and the president's trip to Italy, see A. Stanley, "Bush Hears Pope Condemn Research in Human Embryos," *New York Times*, July 24, 2001, p. A-1; and M. Allen, "Pope Tells Bush Views on Embryos," *Washington Post*, July 24, 2001, p. A-1.

296   meeting "crystallized" the president's thinking: K. Q. Seelye with F. Bruni, "A Long Process That Led Bush to His Decision," *New York Times*, Aug. 11, 2001, p. A-1. For good background accounts on the decision, see also H. Fineman, D. Rosenberg, and M. Brant, "Bush Draws a Stem Cell Line," *Newsweek*, Aug. 20, 2001, pp. 16–21; and R. Lacayo, "How Bush Got There," *Time*, Aug. 20, 2001, pp. 17–23.

297   bioethicist Thomas Murray later likened: T. H. Murray, "Hard Cell," *American Prospect*, Sept. 24–Oct. 8, 2001. "It was fascinating to watch this slow, lurching journey down the lane and to wonder where the ball would actually strike," Murray wrote.

298   "there is deep in his eyes": T. Shales, "Bush's Rare Appearance in the National Pulpit," *Washington Post*, Aug. 10, 2001, p. C-1.

298   "snowflake" and "each of these embryos": for the text of the speech, see "Bush's Address on Federal Financing for Research with Embryonic Stem Cells," *New York Times*, Aug. 10, 2001, p. A-16.

298   "The president has introduced": Connor quoted in D. Milbank, "Clear Break from the Right May Be Brief," *Washington Post*, Aug. 10, 2001, p. A-1 (in retrospect it is clear that this article overstated the degree to which Bush "broke" with the Republican party's right wing on the stem cell issue).

299   "essentially the most restrictive use": A. Goldstein and M. Allen, "Bush Backs Partial Stem Cell Funding," *Washington Post*, Aug. 10, 2001, p. A-1; "By limiting research": A. Caplan, "President's Stem-Cell Plan Adds Up to a Ban," *Philadelphia Inquirer*, Aug. 12, 2001.

299   Will wrote a glowing review: G. F. Will, "Bush Draws the Line on Stem Cells," *Washington Post*, Aug. 14, 2001, p. A-15. Will described Bush's position as "so measured and principled that his critics are in danger of embracing extremism." Eight months later, in "Cranky Conservatives," *Washington Post*, Apr. 25, 2002, p. A-29, Will wrote that "conservatives cannot fault either the substance of Bush's decision on biomedical matters (cloning, stem cells) or the seriousness with which he has arrived at that substance."

299   Another telling reaction: Okarma's statement in Geron press release, Aug. 10, 2001.

300   the president had . . . reinforced "an oligopoly": Murray, "Hard Cell."

300   these concerns made their public debut: Gearhart and Melton quoted in S. G. Stolberg, "Disappointed by Limits, Scientists Doubt Estimate of Available Cell Lines," *New York Times*, Aug. 10, 2001, p. A-17; Pedersen quoted in J. Knight, "Bush Compromise Raises Doubts over Stem-Cell Resilience," *Nature* 412 (Aug. 16, 2001): 665.

301 "They're diverse, they're robust": Thompson quoted in S. G. Stolberg, "U.S. Acts Quickly to Put Stem Cell Policy in Effect," *New York Times*, Aug. 11, 2001, p. A-1.

302 "If it is the case that the cells": Sen. Edward Kennedy, Senate Health, Education, Labor, and Pensions Committee hearing, Sept. 5, 2001.

302 "That is the real distinguishing line": Thompson quoted in R. Brownstein, "Bush Won't Budge on Stem Cell Position, Health Secretary Says," *Los Angeles Times*, Aug. 13, 2001, p. A-1.

302 "I've never seen such a thing": unnamed official quoted in Seelye with Bruni, "A Long Process."

303 "while it is unethical to end": G. W. Bush, "Stem Cell Science and the Preservation of Life," *New York Times*, Aug. 12, 2001, Week in Review, p. 13.

303 as many as 600,000 embryos: see Murray, "Hard Cell"; and N. Wade, "An Old Question Becomes New Again," *New York Times*, Aug. 15, 2001, p. A-20.

303 Skepticism about the Bush policy mounted: see H. Varmus and D. Melton, "The Stem Cell Compromise," *Wall Street Journal*, Aug. 14, 2001, p. A-14; on the AAAS letter, see P. Recer, "Science Group Demands White House Identify Stem-Cell Lines," *Philadelphia Inquirer*, Aug. 18, 2001; for Chatterbox, see T. Noah, "Show Me the Stem Cells!" Slate.com, posted Aug. 19, 2001; for Fleischer's "burden of proof" remark, see C. Connolly, J. Gillis, and R. Weiss, "Viability of Stem Cell Plan Doubted," *Washington Post*, Aug. 20, 2001, p. A-1.

304 NIH finally revealed the identity: on the "Bush 64" announcement, see C. Connolly and R. Weiss, "Stem Cell Colonies' Viability Unproven," *Washington Post*, Aug. 28, 2001, p. A-1.

305 "Wisconsin's cells are the gold standard": quoted in M. Marchione, "64 Cell Lines Listed for Study," *Milwaukee Journal Sentinel*, Aug. 27, 2001.

306 "probably hundreds have got as far": Martin Evans, personal communication, Sept. 23, 2002.

308 Wisconsin "desires to license": quoted in P. Elias, "Biotech Firm Is Sued over Rights to Stem-Cell Research," Associated Press, Aug. 15, 2001. For other accounts of the Geron-WARF patent dispute, which was resolved in Jan. 2002, see S. G. Stolberg, "Suit Seeks to Expand Access to Stem Cells," *New York Times*, Aug. 14, 2001, p. C-2; and D. P. Hamilton and A. Regalado, "Geron Gives Up Some Stem-Cell Rights," *Wall Street Journal*, Jan. 10, 2002, p. A-3.

308 "Patents are what patents are": Okarma quoted in A. Shadid, "Some Scientists See Limitations from Rules," *Boston Globe*, Aug. 10, 2001, p. A-17.

310 "Our challenge now is to move out": testimony of Tommy G. Thompson on embryonic stem cell research, Senate Health, Education, Labor, and Pensions Committee hearing, Sept. 5, 2001.

310 Memorandum of Understanding and Agreement: the interest in the number of cell lines at the Senate hearing on September 5, 2001, overshadowed a significant development concerning intellectual property claims on human embryonic stem cells: the relaxation of several important conditions in the University of Wisconsin's original material transfer agreement. After Jamie

Thomson's discovery of human ES cells in 1998, Wisconsin patented the cells and the method of their derivation and subsequently developed an MTA that set the conditions for the use of ES cells by other researchers. This initial MTA, although not unlike other MTAs formulated by universities at the time, struck both researchers and government officials as overreaching. Wisconsin demanded commercial rights, for example, to any discoveries made by academic researchers using the Wisconsin cells. In addition, any academic research with clinical potential in the six therapeutic areas to which Geron held an exclusive license granted by Wisconsin meant the universities might ultimately have to negotiate with Geron, too. Douglas Melton of Harvard led the academic revolt against these conditions.

After President Bush announced his stem cell policy on August 9, 2001, government negotiators and the University of Wisconsin reached agreement on a less restrictive "memorandum of understanding and agreement," which not only clarified the rights of NIH researchers using the Wisconsin cells, but was intended to serve as a model for Wisconsin's interactions with other researchers. The main change was that Wisconsin agreed to relinquish its claim to commercial rights to discoveries made by other researchers using the Wisconsin cells. "Early on, we asked for commercial reach-through rights, and we gave up on that," explained Carl Gulbrandsen, managing director of WARF. "And that was a pretty good idea. It was a nightmare to administrate, and I've slept better since we gave it up." What gave the government leverage in this instance, according to a person familiar with the negotiations, was that the initial Wisconsin patent on Thomson's work involved the 1995 isolation of primate embryonic stem cells from rhesus monkeys. Although this is the research that initially attracted Michael West's interest (and led to Geron's corporate sponsorship), the work had been funded by the NIH and, according to current law, the government has the right to patented work if it funded the conception of the discovery or its reduction to practice. Wisconsin's later patent application for human embryonic stem cells, based on the research funded by Geron, was essentially the same as the earlier patent on primate ES cells, according to this same source, which allowed government negotiators to assert that NIH researchers should have unfettered access to the Wisconsin cells for research. This led to the agreement that Tommy Thompson announced with great fanfare at the Senate hearing in September.

311 the next day's headlines: see L. McGinley, J. Carroll, and A. Regalado, "White House Cuts Estimate of Available Stem Cells," *Wall Street Journal*, Sept. 6, 2001, p. A-2; S. G. Stolberg, "U.S. Concedes Some Cell Lines Are Not Ready," *New York Times*, Sept. 6, 2001, p. A-1; C. Connolly and J. Gillis, "Thompson: Stem Cell Work Viable," *Washington Post*, Sept. 6, 2001, p. A-1; and D. Vergano, "Thompson: Stem Cells Not Ready," *USA Today*, Sept. 6, 2001, p. A-5.

312   "There are plenty of cells": Thompson quoted in S. G. Stolberg, "Trying to Get Past Numbers on Stem Cells," *New York Times*, Sept. 7, 2001, p. A-15.

313   a year after the policy was announced: for one-year assessments of Bush's stem cell policy, see C. Holden and G. Vogel, "'Show Us the Cells,' U.S. Researchers Say," *Science* 297 (Aug. 9, 2002): 923–25. In testimony to the Harkin-Specter subcommittee in Sept. 2002, Curt Civin of the Johns Hopkins School of Medicine said embryonic stem cell research in the U.S. was "crawling like a caterpillar" (quoted in S. G. Stolberg, "Stem Cell Research Is Slowed by Restrictions, Scientists Say," *New York Times*, Sept. 26, 2002, p. A-27).

313   "President Bush Cloned": *Weekly World News*, Sept. 11, 2001.

313   "You're going to see a legislative free-for-all": Specter quoted in M. Hall, "Key Lawmakers Are Rethinking Stem-Cell Plan," *USA Today*, Aug. 29, 2001, p. A-11.

17. BEATITUDE

INTERVIEWS: Arthur Caplan, Feb. 17, 2002 (Philadelphia); Ronald Green, Mar. 8, 2002 (Worcester, Mass.); Ann Kiessling, Harvard Medical School, Boston, Jan. 8, 2003 (telephone); Karen Lebacqz, Pacific School of Religion, Berkeley, Nov. 19, 1999 (telephone); Douglas Melton, Nov. 26, 2001 (telephone); Peter Mombaerts, Nov. 2, 2001, and Mar. 11, 2002; Thomas Murray, Feb. 21, 2002; Thomas Okarma, Nov. 15, 1999; Lorenz Studer, research scientist, Memorial Sloan-Kettering Cancer Center, New York, July 9, 2001; Irving Weissman, Feb. 16, 2002 (telephone); Michael West, Washington, Dec. 2, 2001, Feb. 1, 2002 (telephone), and Worcester, Mar. 8, 2002; Ian Wilmut, Dec. 4, 2001.

314   fountain of youth "is likely to be found": Haseltine's speech and all other quotes are from second annual meeting, Society for Regenerative Medicine, Renaissance Hotel, Washington, D.C., Dec. 2–4, 2001.

318   the kindred technology known as therapeutic cloning: see J. B. Gurdon and A. Colman, "The Future of Cloning," *Nature* 402 (Dec. 16, 1999): 743–46.

319   Kiessling . . . "agreed within five minutes": A. Regalado, "Experiments in Controversy," *Wall Street Journal*, July 13, 2001, p. B-1. The language from the egg donor ads is also quoted in this story.

320   reports suggested that Geron was creating human embryos: see R. Weiss, "Embryo Work Raises Specter of Human Harvesting; Medical Research Teams Draw Closer to Cloning," *Washington Post*, June 14, 1999, p. A-1. Weiss quoted Calvin Harley, Geron's vice president for science, as saying that the company was supporting the creation of cloned human embryos at another laboratory. Geron officials denied the report in a story that appeared the following day (N. Wade, "No Research on Cloning of Embryos, Geron Says," *New York Times*, June 15, 1999, p. A-21). For a later version of events, see D. P.

Hamilton and A. Regalado, "California School Attempted to Clone Human Embryos," *Wall Street Journal*, May 24, 2002, p. A-3. Hamilton and Regalado obtained California state records that documented a substantial effort by Roger Pedersen, with funding from Geron, to insert adult human cells into human oocytes that had been rejected for use in in vitro fertilization. The *Journal* story stated, "Geron's chief executive, Thomas Okarma, has also denied funding embryo-cloning research — largely for semantic reasons, he says now."

322 "Because it is not sentient": Green, *Human Embryo Research Debates*, p. 45.

322 "Advanced Cell Technology . . . assembled a board": R. M. Green, introduction to "The Ethical Considerations," *Scientific American*, Jan. 2002, pp. 48–49.

323 ethical protections for an egg donor program: since the potential to exploit egg donors has become a serious issue to feminists, bioethicists, and even President Bush, it is worth hearing from Ron Green on this point. "One of the lamentable features of the lack of federal support for reproductive research is that nobody has figured out successfully yet how to really freeze an egg," he said. "If eggs could be frozen and matured in vitro — if you could take immature eggs, put them aside, and freeze them — it would provide a wholly new reproductive option for women who are currently facing the biological clock . . . It would also eliminate completely the need for this donor program. We could go to one woman, harvest ten thousand eggs, and do all the research that anybody wanted with a single, or even a cadavered, donation of eggs. But none of that can be done, because there's no federal money for this research."

324 A major architect of the policy: in an interview Kiessling acknowledged her dual role as researcher and ethics board member, but said she did not view it as a conflict. The issue was not mentioned in a scholarly article written by Advanced Cell's EAB; see R. M. Green et al., "Overseeing Research on Therapeutic Cloning: A Private Ethics Board Responds to Its Critics," *Hastings Center Report*, May–June, 2002, pp. 2–7.

325 a group headed by Peter Mombaerts . . . published: see T. Wakayama et al., "Differentiation of Embryonic Stem Cell Lines Generated from Adult Somatic Cells by Nuclear Transfer," *Science* 292 (Apr. 27, 2001): 740–43.

326 "He manipulates a microinjector": P. Mombaerts, talk presented to NAS workshop on human cloning, Aug. 7, 2001. Studer's work with parthenogenetic cells appeared as J. B. Cibelli et al., "Parthenogenetic Stem Cells in Nonhuman Primates," *Science* 295 (Feb. 1, 2002): 819.

327 the Jones Institute . . . had published: see S. E. Lazendorf et al., "Use of Human Gametes Obtained from Anonymous Donors for the Production of Human Embryonic Stem Cell Lines," *Fertility and Sterility* 76 (July 2001): 132–37; see also R. Weiss, "Scientists Use Embryos Made Only for Research," *Washington Post*, July 11, 2001, p A-1. For press reports on ACT's therapeutic cloning project, see R. Weiss, "Firm Aims to Clone Embryos for Stem Cells,"

Washington Post, July 12, 2001, p. A-1; and S. G. Stolberg, "Company Using Cloning to Yield Stem Cells," *New York Times,* July 13, 2001, p. A-14.

328 Jose Cibelli had begun human cloning: for the two principal insider accounts of the ACT cloning experiments, see J. Fischer, "Scientists Have Finally Cloned a Human Embryo," *U.S. News & World Report,* Dec. 3, 2001, pp. 50–63; and J. B. Cibelli, R. P. Lanza, and M. D. West, with C. Ezzell, "The First Human Cloned Embryo," *Scientific American,* posted on-line Nov. 25, 2001, and then in the Jan. 2002 print edition, pp. 44–51. See also K. Dunn, "Cloning Trevor," *Atlantic Monthly,* June 2002, pp. 31–48.

329 West and his colleagues ran to publish: J. B. Cibelli et al., "Somatic Cell Nuclear Transfer in Humans: Pronuclear and Early Embryonic Development," *e-biomed: The Journal of Regenerative Medicine* 2 (Nov. 26, 2001): 25–31.

330 gossip mavens Rush and Molloy: see G. Rush and E. Molloy with S. Botton, "Gossip," *New York Daily News,* Nov. 28, 2001, p. 34.

332 reports about the "first human clones": see E. Marshall and G. Vogel, "Cloning Announcement Sparks Debate and Scientific Skepticism," *Science* 294 (Nov. 30, 2001): 1802–3; D. Adam, "First Human Clones Get a Cool Response," *Nature* 414 (Nov. 29, 2001): 477. See also R. A. Weinberg, "Of Clones and Clowns," *Atlantic Monthly,* June 2002.

333 "Although its executives claimed": H. Varmus, "The Weaknesses of Science for Profit," *New York Times,* Dec. 4, 2001, p. A-23.

334 equated West . . . with Osama bin Laden: E. Cohen and W. Kristol, "Dr. West and Mr. Bin Laden," *Weekly Standard,* Dec. 17, 2001.

334 "My fear . . . is that this tiny step": in "Don't Expect Any Miracles," *New Scientist,* Dec. 1, 2001, pp. 4–6.

335 The company ultimately agreed to pay: see S. G. Stolberg, "Cloning Company Misspent U. S. Grants, Auditors Say; It Will Repay," *New York Times,* May 15, 2002, p. A-16.

EPILOGUE: FINITUDE

341 "It may be too big for Geron": G. Blobel, Mar. 4, 2002, personal interview.

341 Geron laid off 30 percent: on the economic woes of Geron and Advanced Cell Technology, see D. Firn and V. Griffith, "Stem Cell Science on a Shoestring," *Financial Times,* July 18, 2002; for specifics of the Geron layoffs, see "Geron to Focus on Product Pipeline," company press release, June 24, 2002. For the economic situation at both ACT and Geron, a venture capitalist who asked not to be named said, "The model for making money with a stem cell business is unknown. I don't think Geron has it. I don't think ACT has it. Where's the business?"

341 The NIH had disbursed: presentation by Wendy Baldwin, Office of Extramural Research, NIH, to President's Council on Bioethics, session 3, July 11, 2002, and postsession remarks to reporters.

341 Research on telomerase: see "Geron and Memorial Sloan-Kettering Cancer Center Report on GRN163, an Inhibitor of Telomerase for the Treatment of Cancer," company press release, June 13, 2002. Geron reported "compelling efficacy" in preclinical studies and hoped to launch safety tests in human patients by early 2003.

342 the yeast cells indeed achieved: for the "panting" update on Leonard Guarente's research on extension of life span in yeast, see S.-J. Lin et al., "Calorie Restriction Extends Saccharomyces cerevisiae Lifespan by Increasing Respiration," *Nature* 41 (July 18, 2002): 344–48. For an amusing and informed commentary, see S. S. Lee and G. Ruvkun, "Don't Hold Your Breath," *Nature* 418 (July 18, 2002): 287–88. The authors write, "Prompted by the continuing stream of press reports about the rodent experiments, a variety of optimists, hucksters and fanatics have started to promote [calorie restriction] and to practise it with the aim of living longer themselves — or at least transferring wealth from the rest of us."

344 "To praise mortality must seem": Kass, *Toward a More Natural Science*, p. 308; "work opportunities, retirement plans," p. 303; "prolongation of healthy and vigorous life," p. 300; "For most of us," p. 306.

345 "immortalizing therapies will be available": B. Alexander, "Immortality Reality Check," *Wired*, Jan. 2000, p. 184.

345 "large increases in life expectancy": International Longevity Center, "Is There an 'Anti-Aging' Medicine?" workshop report, 2001, p. 3.

346 "The prospect of immortality": "Position Statement on Human Aging," courtesy of L. Hayflick.

347 "not only all the Follies": Swift, *Gulliver's Travels and Other Writings* (New York: Modern Library, 1958), p. 170.

347 these assumptions are often an amalgam: In *Toward a More Natural Science*, Kass writes, "I shall here assume what is held to be the most attractive prospect, an increase in life span with parallel increases in vigor, for ten to twenty years, but perhaps longer. I shall also assume an antiaging technology that is easy to administer, inexpensive, and not burdensome or distasteful to the users — a technology that will be widely demanded and used" (p. 301).

348 Even stem cells and other forms: stem cell treatments for certain diseases may take longer than for others. Type I diabetes, for example, is recognized to be an autoimmune disease, in which the patient's immune cells attack, for unknown reasons, the insulin-producing islet cells in the pancreas; mere replenishment of the destroyed islet cells would merely feed the disease mechanism, so stem cells would probably need to be modified, a task likely to add more complexity and require more time. Similarly, because Alzheimer's disease is widely disseminated through the brain and the disease process is not thoroughly understood, both the delivery and mechanism of action of stem cell therapy remain hurdles to swift treatment.

348 "suggests that reductions in mortality": J. Oeppen and J. W. Vaupel, "Broken Limits to Life Expectancy," *Science* 296 (May 10, 2002): 1029–31.

349 there will be a million centenarians: according to a U.S. Census report published in 2001; see G. C. Armas, "Number of Americans 100 or Older Rises," Associated Press, Oct. 3, 2001.

349 the Society of Actuaries held a meeting: see G. Held, "Plastic Omega," paper presented at "Living to 100 and Beyond: Survival at Advanced Ages," meeting sponsored by the Society of Actuaries, Lake Buena Vista, Florida, January 17–18, 2002. The text of this and other papers presented at the conference can be found at http://www.soa.org.research/living.html.

350 "a demographic effect": on AIDS and life expectancy curves, see D. Brown, "Study: AIDS Shortening Life in 51 Nations," Washington Post, July 8, 2002, p. A-2.

351 Women . . . "are less likely than men": Fukuyama, Our Posthuman Future, p. 63; for his thoughts on life extension, see pp. 57–71.

352 shrinking . . . "will cause an even greater upheaval": P. Drucker, "The Next Society," Economist, Nov. 1, 2001.

353 "richer and deeper public bioethics": Kass, opening remarks, President's Council on Bioethics, Jan. 17, 2002.

353 "Now may be as good a chance": L. Kass, "Preventing a Brave New World," New Republic, May 21, 2001 (available at www.tnr.com/052101/kass052101.html).

353 "didn't want to embarrass the president": Kass quoted in S. S. Hall, "President's Bioethics Council Delivers," Science 297 (July 19, 2002): 322–24. Four members of the council later signed an editorial lamenting the majority decision of the panel (see J. D. Rowley, E. Blackburn, M. S. Gazzaniga, and D. W. Foster, "Harmful Moratorium on Stem Cell Research," Science 297 [Sept. 20, 2002]: 1957).

354 One of the most insidious words: on the use of "inevitable," see, for example, C. Krauthammer, "Crossing Lines: A Secular Argument Against Research Cloning," New Republic, Apr. 29, 2002.

356 "would surely effect fundamental . . . changes": Kass, "Preventing a Brave New World."

357 "What are you going to say": for the Kass-Meilaender exchange, see transcript of the President's Council on Bioethics, afternoon session, Feb. 14, 2002 (transcript available at www.bioethics.gov).

358 "each in his narrow cell": Thomas Gray, "Elegy Written in a Country Church-Yard," in F. T. Palgrave, The Golden Treasury, p. 121.

359 "We're never going to agree": interview with Laurie Zoloth, Berkeley, Nov. 13, 1999.

BIBLIOGRAPHY

Anderson, David, Richard Gardner, and Daniel Marshak, eds. *Stem and Progenitor Cells: Biology and Applications.* Abstracts of papers presented at a meeting, Mar. 22–25, 2001, Cold Spring Harbor Laboratory, Cold Spring Harbor, N.Y., 2001.

Andrews, Lori B. *The Clone Age: Adventures in the New World of Reproductive Technology.* New York: Henry Holt, 1999.

Andrews, Lori, and Dorothy Nelkin. *Body Bazaar: The Market for Human Tissue in the Biotechnology Age.* New York: Crown, 2001.

Austad, Steven N. *Why We Age: What Science Is Discovering About the Body's Journey Through Life.* New York: John Wiley & Sons, 1997.

Bova, Ben. *Immortality: How Science Is Extending Your Life Span — and Changing the World.* New York: Avon, 1998.

Caplan, Arthur L. *If I Were a Rich Man, Could I Buy a Pancreas? and Other Essays on the Ethics of Health Care.* Bloomington: Indiana University Press, 1992.

———. *Am I My Brother's Keeper? The Ethical Frontiers of Biomedicine.* Bloomington: Indiana University Press, 1997.

Clark, William A. *A Means to an End: The Biological Basis of Aging and Death.* New York: Oxford University Press, 1999.

Comfort, Nathaniel C. *The Tangled Field: Barbara McClintock's Search for the Patterns of Genetic Control.* Cambridge, Mass.: Harvard University Press, 2001.

Connolly, Francis X. *Wisdom of the Saints.* New York: Pocket Books, 1963.

Darnton, John. *The Experiment.* New York: Dutton, 1999.

Farris, Edmond J. *Human Fertility and Problems of the Male.* White Plains, N.Y.: Author's Press, 1950.

———. *Human Ovulation and Fertility.* Philadelphia: J. B. Lippincott, 1956.

Finch, Caleb, and Thomas Kirkwood. *Chance, Development and Aging.* New York: Oxford University Press, 2000.

Fox, Michael J. *Lucky Man: A Memoir.* New York: Hyperion, 2002.

Fried, Stephen. *Bitter Pills: Inside the Hazardous World of Legal Drugs.* New York: Bantam, 1998.

Fukuyama, Francis. *Our Posthuman Future: Consequences of the Biotechnology Revolution.* New York: Farrar, Straus and Giroux, 2002.

Fuson, Robert H. *Juan Ponce de León and the Spanish Discovery of Puerto Rico and Florida.* Blacksburg, Va.: McDonald & Woodward, 2000.

George, Robert P. *The Clash of Orthodoxies: Law, Religion, and Morality in Crisis.* Wilmington, Del.: ISI Books, 2001.

Goethe. *The Collected Works.* Vol. II: *Faust I and II.* Ed. and trans. Stuart Atkins. Princeton, N.J.: Princeton University Press, 1984.

Green, Ronald M. *The Human Embryo Research Debates: Bioethics in the Vortex of Controversy.* New York: Oxford University Press, 2001.

Guarente, Lenny. *Ageless Quest: One Scientist's Search for the Genes That Prolong Youth.* Woodbury, N.Y.: Cold Spring Harbor Laboratory Press, 2002.

Hall, Stephen S. *Invisible Frontiers: The Race to Synthesize a Human Gene.* New York: Atlantic Monthly Press, 1987.

———. *A Commotion in the Blood: Life, Death, and the Immune System.* New York: Henry Holt, 1997.

Hamburger, Victor. *The Heritage of Experimental Embryology: Hans Spemann and the Organizer.* New York: Oxford University Press, 1988.

Hayflick, Leonard. *How and Why We Age.* New York: Ballantine, 1994.

Hooper, Edward. *The River: A Journey to the Source of HIV and AIDS.* Boston: Little, Brown, 1999.

Horder, T. J., J. A. Witkowski, and C. L. Wylie, eds. *A History of Embryology: The Eighth Symposium of the British Society for Developmental Biology.* Cambridge: Cambridge University Press, 1986.

Institute of Medicine. *Society's Choices.* Ed. R. E. Bulger, E. M. Bobby, and H. V. Fineberg. Washington, D.C.: National Academy Press, 1995.

Jonsen, Albert R. *The Birth of Bioethics.* New York: Oxford University Press, 1998.

Kass, Leon R. *Toward a More Natural Science: Biology and Human Affairs.* New York: Free Press, 1985.

Kass, Leon R., and James Q. Wilson. *The Ethics of Human Cloning.* Washington, D.C.: AEI Press, 1998.

Kirkwood, Tom. *Time of Our Lives: The Science of Human Aging.* New York: Oxford University Press, 1999.

Kolata, Gina. *Clone: The Road to Dolly and the Path Ahead.* New York: William Morrow, 1998.

Kondracke, Morton. *Saving Milly: Love, Politics, and Parkinson's Disease.* New York: Public Affairs, 2001.

Kundera, Milan. *Immortality.* New York: Grove Press, 1991.

Lear, John. *Recombinant DNA: The Untold Story.* New York: Crown, 1978.

Mizel, Steven B., and Peter Jaret. *In Self-Defense: The Human Immune System — the New Frontier in Medicine.* San Diego: Harcourt Brace Jovanovich, 1985.

National Academy of Sciences. *Scientific and Medical Aspects of Human Reproductive Cloning.* Washington, D.C.: National Academy Press, 2002.

National Institutes of Health. *Stem Cells: Scientific Progress and Future Research Directions*. Bethesda, Md.: NIH, June 2001.

National Research Council. *Between Zeus and the Salmon: The Biodemography of Longevity*. Ed. K. W. Wachter and C. E. Finch. Washington, D.C.: National Academy Press, 1997.

Olshansky, S. Jay, and Bruce A. Carnes. *The Quest for Immortality: Science at the Frontiers of Aging*. New York: Norton, 2001.

Pfeffer, Naomi. *The Stork and the Syringe: A Political History of Reproductive Medicine*. Cambridge, Eng.: Polity, 1993.

President's Council on Bioethics. *Human Cloning and Human Dignity: An Ethical Inquiry*. Washington, D.C.: Government Printing Office, 2002 (also available at www.bioethics.gov).

Ridley, Matt. *Genome: The Autobiography of a Species in 23 Chapters*. New York: HarperCollins, 2000.

Silver, Lee M. *Remaking Eden: How Genetic Engineering and Cloning Will Transform the American Family*. New York: Avon, 1997.

Smith, Jane S. *Patenting the Sun: Polio and the Salk Vaccine*. New York: William Morrow, 1990.

Stephanopoulos, George. *All Too Human: A Political Education*. Boston: Little, Brown, 1999.

Stimpson, George W. *A Book About American History*. New York: Harper, 1950.

Stock, Gregory. *Redesigning Humans: Our Inevitable Genetic Future*. Boston: Houghton Mifflin, 2002.

Teresi, Dick. *Lost Discoveries: The Ancient Roots of Modern Science — from the Babylonians to the Mayans*. New York: Simon & Schuster, 2002.

Wade, Nicholas. *The Ultimate Experiment: Man-made Evolution*. New York: Walker, 1977.

———. *Life Script: How the Human Genome Discoveries Will Transform Medicine and Enhance Your Health*. New York: Simon & Schuster, 2001.

Walton, Alan G. *Beneath This Gruff Exterior There Beats a Heart of Plastic: My Life in Genetic Engineering . . . and Beyond*. Naperville, Ill.: Oak Hill, 2000.

Warnock, Mary. *A Question of Life: The Warnock Report on Fertilisation and Embryology*. Oxford: Basil Blackford, 1985.

Weiner, Jonathan. *Time, Love, Memory: A Great Biologist and His Quest for the Origins of Behavior*. New York: Knopf, 1999.

Wilmut, Ian, Keith Campbell, and Colin Tudge. *The Second Creation: Dolly and the Age of Biological Control*. New York: Farrar, Straus and Giroux, 2000.

ACKNOWLEDGMENTS

When Shakespeare spoke in Sonnet 73 of the "bare ruined choir" and of a day that "by and by black night doth take away," he invoked the image of a soundless twilight into which all of us eventually walk. The bard could not have imagined stem cells or immortalizing enzymes, of course, but our enthusiasm for these recent technologies addresses our most ancient human concerns. The inspiration for this book originated with an article I wrote for the *New York Times Magazine* on embryonic stem cells, which appeared in January 2000, and that is as good a place as any to begin the always satisfying task of dispensing thanks. I am deeply grateful to my colleagues at the magazine who recognized that this was an important, timely topic worthy of being a cover story and pushed me to make the article as good as it could be — Adam Moss, Katharine Bouton, Eric Nash, and Jeff Z. Klein. That's what propelled me into the topic.

As in previous books, I marvel at my good luck, because I often find myself stumbling into a story that's much richer, more entertaining, and more profound than anything I had the right to imagine at the outset. The richness of this story derives almost entirely from the scientists and business people who, in granting repeated interviews and speaking with candor, allowed me to bring a fascinating, complicated, ongoing scientific saga to life. I conducted hundreds of interviews for this book, and I am grateful to every interview subject, including the ones who didn't spare me more than a minute or two. Several people, however, responded to multiple pesterings and deserve more than perfunctory and anonymous thanks. I wish in particular to acknowledge the patient cooperation and help of Leonard Hayflick, Michael West, Carol Greider, Harold Varmus, Douglas Melton, Jeff Martin, and Ali Brivanlou.

For logistical help along the way, I'd like to thank Ellen Pure, Frank Hoke, and Nina Long of the Wistar Institute; Nancy Robinson of the Geron Corporation; Sharon Karlsberg of Feinstein Kean Health Partners; Terry Devitt of the University of Wisconsin; Jennifer O'Brien of the University of California at San Francisco; Catherine Yarbrough of Rockefeller University; Erik Parens of the Hastings Center; Christopher Corey of the International Longevity Center; Lisa Onaga and David Malakoff of *Science;* Linda Magyar, formerly of the *New York Times Magazine;* Adam Liptak of the *New York Times;* Sean Tipton of the American Society for Reproductive Medicine; Lawrence Soler of the Juvenile Diabetes Research Foundation; Anthony Mazzaschi of the Association of American Medical Colleges; Michael Manganiello of the Coalition for the Advancement of Medical Research; Frankie Trull of the National Association for Biomedical Research; Diane Gianelli of the President's Council on Bioethics; Tony Jewell and Matt Bloom of Tommy Thompson's office; Carl Hall of the *San Francisco Chronicle;* Greg Smith of the *New York Daily News;* Robin Marantz Henig; Don Polhman of the *Washington Post;* and the folks at the Kaisernet Reproductive Health Newsletter. In addition, I wish to thank the staff of the library at the New York Academy of Medicine, which remains a wonderful resource for biomedical research.

I wish to thank my colleagues at *Technology Review* for their editorial help and guidance on (among other things) our story on adult stem cells, including John Benditt, Rebecca Zacks, David Rotman, Erika Joinetz, and Bob Buderi; similarly, I am grateful to Colin Norman at *Science* for our journey into bioethics. It is a special pleasure to acknowledge the work of colleagues who have done a marvelous job of chasing down the multiplicitous threads of this story, often under severe deadline constraints. They include Antonio Regalado of the *Wall Street Journal;* Sheryl Stolberg and Nicholas Wade of the *New York Times;* Aaron Zitner of the *Los Angeles Times;* Rick Weiss and Ceci Connolly of the *Washington Post;* and Gretchen Vogel of *Science.* My work has materially benefited from their reporting and, in many instances, from their collegiality.

I have made every effort to assure that the material contained in this book is not only factually accurate but true to the larger sense of the people involved and their work; if I have succeeded, it is in no small part because of a corps of diligent readers that would be the envy of any science writer: Ali Hemmati Brivanlou, Alta Charo, Vincent Cristofalo, Titia de Lange, Steven Dickman, John C. Fletcher, Kathi Hanna, Rudolf Jaenisch, Joachim Lingner, Daniel Marshak, Douglas Melton, Thomas Murray, Stanley Plotkin, David Sinclair, Bryant Villeponteau, Robert Weinberg, and several others who preferred to remain anonymous. Special thanks to Harold Varmus not only for scientific counsel but for astute literary analysis worthy of a veteran of Publisher's Row. As is customary, I take full responsibility for any errors that persist, while acknowledging that there would have been many more without the sage input and advice of these critical readers.

For inspiration of a more particular and personal sort, I want to thank Steve and Clarissa Burnett for the use of that cabin with an unusually long view; Dr. Michael Faust, for focusing my attention on the more personal parameters of the aging process; Argiris Efstratiadis for a wonderful tutorial on imprinting; and Steve Tager for thoughtful discussions about the book business. For informative, or merely consoling, conversations that eased my passage over the normal hilly terrain of a book project, it's always a treat to thank old reliables like Nelson Smith, Tom O'Neill, Joe McElroy, Lisa Shea, Mary Hawthorne, Stephen Dubner, Gary Taubes, Michael Pollan, Russ Rymer, Victor McElheny, Charles Mann, Ellen Ruppel Shell, Jerry Roberts, Steve Rubenstein, Sean Elder and Peggy Northrop, David White and Betsy Gardella, Chris Smith, Jeff Goldberg, Phil Greenberg, Bill Grueskin, Lloyd Old, and Robert Cook-Deegan.

This book would have been years longer in the making were it not for the generous support of the Alfred P. Sloan Foundation, and I am immensely grateful to Arthur Singer, Doron Weber, and everyone else at Sloan for their much appreciated support of science writers, which has fostered greater mastery of the craft and materially advanced the cause of scientific literacy in our society.

When *Merchants of Immortality* was first being conceptualized, Pat Strachan at Houghton Mifflin immediately grasped its import and value; although she was unable to serve as midwife at its completion, her initial encouragement, including gently sagacious advice on early drafts, was fundamental to the ultimate shape and success of the book. In picking up the baton, Eamon Dolan assured me that I would not be an orphan in the storm. He has been good to his word, and then some; I have never received more attentive and diligent editing, whether big-picture or line by line. Anne Nolan did a marvelous job as manuscript editor, Susan Innes was a meticulous proofreader, and Liz Duvall and Peg Anderson worked tirelessly to guide the ship home. Thanks, too, to Michaela Sullivan for the jacket design, and to Brigitte Marmion, Lori Glazer, and Whitney Peeling for their help in spreading the word. Emily Little was consistently gracious while pestering the tardy author at crucial junctures.

As ever, in an industry increasingly convulsed by editorial transfers, consolidations, and ownership changes, my agent, Melanie Jackson, continues to be an island of stability, wisdom, and, not least, loyalty in this sea of change. She has been a great friend for two decades now, but also an exemplary spirit to anyone who cares about books, integrity, and taste.

Finally, family. I am blessed to enjoy the continuing company of two loving parents, who never cease to awe me with their vibrant, ambitious, rich lives; as the epilogue attests, their experiences have something to say to us all. It's a pleasure to thank my brother, Eric, and his family for their support, as well as Susan and Jedd Levine, and Gerry and Hindley Mendelsohn.

As the spouses and children of writers know only too well, it is doubly tormenting to have a family member in such proximity while at the same time so inaccessible when a book is in progress. I can't make up for all the lost hours and skipped meals, the forsworn outings to playgrounds and puppet shows, to my beloved Micaela and Sandro, but I can publicly thank them for the astonishing inspiration they bring to my life and my work. As for my wife, Mindy, at the risk of repeating past gratitudes, it is difficult in a line or two to thank her adequately for all the sacrifices she endured and all the psychic and logistical support she managed to muster over the course of this project, so I'll turn the mike back over to Shakespeare (with thanks to Cynthia Kenyon for sending me back to the sonnets) to at least hint at my boundless thanks:

> O know, sweet love, I always write of you,
> And you and love are still my argument.
> So all my best is dressing old words new,
> Spending again what is already spent:
> For as the sun is daily new and old,
> So is my love still telling what is told.

INDEX